JOHN TANAKA
UCONN PS 251

The Structures of the Elements

THE STRUCTURES
OF THE ELEMENTS

JERRY DONOHUE

Department of Chemistry
and
Laboratory for Research on the Structure of Matter
University of Pennsylvania

A WILEY-INTERSCIENCE PUBLICATION

JOHN WILEY & SONS

New York · London · Sydney · Toronto

Library of Congress Cataloging in Publication Data:

Donohue, Jerry, 1920–
    The structures of the elements.

    "A Wiley-Interscience publication."
    Bibliography: p.
    1. Chemical elements. 2. Crystallography.
I. Title.
QD466.D6   546   73-13788
ISBN 0-471-21788-3

Printed in the United States of America

10 9 8 7 6 5 4 3 2 1

FOR PAT

Preface

For a long time I have been planning to write a book on the structures of the elements, and over the years this book has evolved to its present form. It is intended for advanced undergraduates and graduate students in chemistry, metallurgy, physics, and, of course, crystallography, and includes not only descriptions of the elemental structures as accepted today, but also details of all previously proposed structures which subsequently proved to be erroneous. I hope that this emphasis on history will be useful as well as entertaining. Mistakes in structure determinations are not confined to the older literature, but continue, unfortunately, right up to the present.

A section on any given element is designed to stand on its own, without reference to any other sections. This procedure has led to some inevitable duplication of some descriptive data. The major emphasis is on pictorial representations, but numerical data are also given for each structure. These include lattice constants, for which *all* reported values are given in most cases; space group; positional (but not thermal) parameters for the atoms; bond distances and angles; torsion angles where appropriate; atomic volumes; and, for molecular crystals, nonbonded distances. In the many cases where some of the foregoing were not given by the original authors these quantities were calculated from the primary data given by them; no attempt has been made to distinguish such values from literature values. I have also included brief mention of numerous preparations presumed to be new allotropes by their preparators, but for which no structural information is available.

It may seem to some readers that several very remarkable structures are reported here without comment. Subsequent work in which the published data are reinterpreted may well result in entirely different structures. On the other hand, if it becomes possible to gather additional data (and this may often prove to be an insurmountable problem), it may turn out that certain rather bizarre structures are indeed correct.

Preliminary work on the preparation of this book was aided by the existence of earlier compilations such as Wyckoff's *The Structure of Crystals,* Pearson's *The Structure of Metals and Alloys, Strukturbericht,* and *Structure Reports.* Nevertheless, I decided early not to rely solely on such secondary information, but to refer to the original literature whenever possible. This decision turned out to be a wise one. The literature was covered through December 1972.

It is not possible for me to give individual thanks to the dozens of people with whom I corresponded in order to clarify various matters and to obtain unpublished details, as well as those who made comments and suggestions on individual chapters; I am indebted to them all. I would never have started this project had it not been for the tremendous talent of Maryellin Reinecke who, alas, will never see the results of her efforts in print; the majority of the figures are the product of her ability. Finally, special thanks for advice and encouragement go to Drs. Linus Pauling and Stephen Schwartz.

JERRY DONOHUE

Philadelphia, Pennsylvania
June 1973

Contents

Chapter 1

Hydrogen*

HYDROGEN (PROTIUM)

The first work on the crystal structure of solid H_2 was that of Keesom, de Smedt, and Mooy (1930 a,b,c). They observed a total of seven powder lines which could be explained by H_2 molecules in a hexagonal close packed arrangement with a unit cell having a = 3.758 Å, c = 6.136 Å (from kX), c/a = 1.633 at 4.2°K.

This result was later questioned by Kogan, Lazarev, and Bulatova (1956), who stated that the diffraction pattern of solid H_2, including both the previous data and their new data, could be accounted for as arising from a tetragonal lattice, and that the data on the structure of hydrogen previously obtained had been incorrectly interpreted. No lattice constants were given.

Shortly thereafter, Kogan, Lazarev, and Bulatova (1958) reported that they had made improvements in the method of taking photographs and confirmed their previous conclusion that the earlier structure had been incorrectly determined. Again no lattice constants were given; they reported only the values of 3.15 and 2.79 Å for the spacings of the two lines they observed.

Kogan, Lazarev, and Bulatova (1960) then gave the results of some additional work on solid H_2, D_2, and T_2 at 4.2°K. With regard to H_2, they now proposed two possible structures, both with two molecules per unit cell:

tetragonal, $I4-C_4^5$, a = 4.49 Å, c = 3.68 Å, c/a = 0.82, or
hexagonal, $P6_3/mmc-D_{6h}^4$, a = 3.71 Å, c = 6.42 Å, c/a = 1.73

---

*Readers not interested in the rather lengthy history of the crystallography of the isotopes of hydrogen may turn directly to the summary, p. 10.

(The values given above were calculated from the published axial ratios and values of c which had been corrected by extrapolation of data taken with several different wave lengths.) The choice of these two space groups is very curious, as there are numerous others which are consistent with the published data; moreover, because those data consist of only two diffraction lines, at d = 3.25 and 2.89 Å (corrected) or 3.18 and 2.85 Å (uncorrected), the selection of these two space groups can scarcely be regarded as definitive.

Dukhin (1960) apparently preferred the tetragonal over the hexagonal structure on the basis of some calculations, the details of which were not given, concerning the anomalous specific heat and anisotropy of the nuclear resonance of solid H_2.

The results of a neutron diffraction study on solid H_2 and D_2 were reported by Kogan, Lazarev, Ozerov, and Zhdanov (1961b). These results were said to confirm the tetragonal structure for H_2. The lattice constants at $12°K$ were given as a = 4.5 Å, c = 3.6 Å, c/a = 0.81, although some diffraction maxima at small angles required doubling these to $a \approx 9$ Å, c ≈ 7 Å. The authors further pointed out that occurrence of reflections having $h + k + \ell$ odd, a result at variance with the x-ray results, could be explained by different scattering amplitudes for neutrons for ortho and para hydrogen, and by an ordered arrangement of ortho and para molecules in the lattice. The ways in which such an ordering might be achieved were discussed shortly thereafter by Ozerov, Kogan, Zhdanov, and Kukhto (1961), and a unit cell having slightly different lattice constants, but still body centered tetragonal, was later reported by Ozerov, Zhdanov, and Kogan (1962).

It was pointed out by van Kranendonk and Gush (1962) that the infrared spectrum of solid para H_2 was incompatible with the body centered tetragonal structure proposed by Kogan et al. Furthermore, van Kranendonk and Gush observed no changes in the spectrum at $13.6°K$, from the vapor pressure up to 50 atm. This result further argued against the tetragonal structure, because they expected a transition to a close packed structure at a sufficiently high pressure, at most a few atmospheres.

The conclusion of Dukhin (1960) was questioned by Brith and White (1964), who found that it was not possible to differentiate between the hexagonal and tetragonal structures on the basis of heat capacity measurements.

New x-ray diffraction data were reported by Kogan, Bulatov, and Yakimenko (1964). These data were stated to require reevaluation of the structures proposed previously; the old photographs differed from the new because of preferred orientation effects. It was now reported that at liquid helium temperature, H_2 is hexagonal, with a = 3.78 Å, c= 6.16 Å, c/a = 1.63, obtained from three observed lines.

In an electron diffraction study Bostanjoglo (1965) found that between 2.8 and $4.5°K$ thin films of normal H_2, para H_2, and normal D_2 deposited on

Formvar gave diffraction patterns consisting of seven lines, five of which corresponded to a face centered cubic unit cell, having, in the case of both normal and para H_2, $a = 5.35 \pm 0.02$ Å. The two extra lines, it was suggested, arose from diffraction by cross gratings and double diffraction by twinned crystals; they were probably not due to a hexagonal lattice because their relative intensities proved incorrect on that assumption, and because they responded differently to changes in substrate temperature. (If one takes simple averages of the lattice constants corresponding to the spacings listed by Bostanjoglo, a value of a of 5.351 ± 0.018 Å is obtained for normal H_2 and 5.377 ± 0.032 Å for para H_2; the small difference between the two values is not significant.)

In another electron diffraction study, Curzon and Mascall (1965) found that thin films of solid H_2 at $5°K$ gave four sharp rings, which could be indexed on the basis of a face centered cubic unit cell with $a = 5.29 \pm 0.05$ Å, and broad rings consistent with the hexagonal unit cell of Keesom et al. (1930c) and Kogan et al (1964). They did not feel that their results were necessarily inconsistent with the previous work because the materials they examined were thin films deposited on amorphous germanium, whereas in the other studies bulk material was used, and a change in structure in such a case was not without precedent.

Clouter and Gush (1965) reported that they had observed an effect in the infrared spectrum of solid normal H_2 which provided evidence that the crystal structure changes in the neighborhood of $1.5°K$. They concluded that at $1.9°K$ the crystal does not have a center of inversion, but changes on cooling to a structure which does. (If the high temperature form consists of randomly oriented or freely rotating H_2 molecules in a hexagonal close packed array, this conclusion is not strictly true: such a crystal would in fact possess centers of inversion, but the hydrogen molecules would not lie on them.) The transition temperature was found to be strongly dependent on the ortho-para ratio; decrease in the ratio lowered the transition temperature.

X-ray diffraction data were obtained above and below the transition temperature by Mills and Schuch (1965). They found that normal H_2 is hexagonal close packed from 4 down to about $1.3°$ K, with $a = 3.761 \pm 0.007$, $c = 6.105 \pm 0.011$ Å, $c/a = 1.623$; between 1.25 and $1.30°K$ it is face centered cubic, with $a = 5.312 \pm 0.010$ Å. For a sample having composition about 50% ortho, the transition temperature decreased to below $1.25°K$. No molecular arrangement was suggested for the low temperature form, the unit cell of which contains four molecules, but in a discussion of orientation order in solid ortho H_2 Raich and James (1966) assumed that this substance was isostructural with α-N_2, that is, space group $Pa3$-T_h^6, with the molecular centers in the face centered positions and the molecular axes directed in four different directions along the threefold axes of the crystal.

Somewhat different results were then reported by Barrett, Meyer, and Wasserman (1966). They found that normal H_2 usually froze as the hexagonal close packed form, but if gold foil with a cube texture was present both normal and para H_2 froze as the cubic form. Furthermore, plastic deformation of normal H_2 between 3 and $4.2°K$ caused cubic lines to appear in the diffraction pattern of hexagonal H_2. They concluded that the cubic form was the stable one for normal H_2, but that spontaneous transformation from the hexagonal form on cooling may often be incomplete. They did not present any lattice constants for the hexagonal form but merely stated that the peak positions were in at least rough agreement with the values of Mills and Schuch (1965). They did, however, give the observed spacings for three lines at $4.2°K$, and these are found to correspond to $a = 3.778 \pm 0.006$, $c = 6.191 \pm 0.008$ Å, $c/a = 1.639$. For the cubic form at $4.2°K$, Barrett et al. found $a = 5.21$ Å, and because this value was quite different from that of Mills and Schuch, the authors believed that this was not an equilibrium value but was due to some matching of spacings between the gold substrate and the condensate.

The results of another electron diffraction were reported by Bostanjoglo and Kleinschmidt (1967). They found that thin films of H_2 deposited on Formvar at 2.8 to $4.5°K$ are face centered cubic, with $a = 5.338 \pm 0.008$ Å, as obtained from five observed lines. Heating the films up to a temperature at which rapid sublimation set in brought on the appearance of 11 sharp lines corresponding to a hexagonal structure with $a = 3.776 \pm 0.008$, $c = 6.162 \pm 0.01$ Å, $c/a = 1.632$. This hexagonal structure persisted down to $2.8°K$.

The results reported for the cubic and hexagonal lattice constants are collected in Table 1-1. For the present, the best values are taken as the simple averages of the individual values, and the standard errors are those obtained from the individual deviations. More precise values must await data based on diffraction patterns containing a larger number of lines. It is interesting that in the first study of hexagonal H_2 Keesom et al. (1930c) observed 7 lines, whereas only 3 lines were observed in each of the more recent investigations, except the last, in which 11 lines were observed.

HYDROGEN DEUTERIDE

Thin films of HD deposited on Formvar at 2.8 to $4.5°K$ were found to be face centered cubic by Bostanjoglo and Kleinschmidt (1967), with $a = 5.182 \pm 0.009$ Å, on the basis of four observed lines. When these films were heated, five sharp reflections corresponding to a hexagonal structure appeared, and these persisted down to $2.8°K$. The lattice constants found were $a = 3.642 \pm 0.015$, $c = 5.951 \pm 0.01$ Å, $c/a = 1.634$.

Table 1-1. Lattice Constants of Solid Hydrogen, H_2

a, Å	c, Å	c/a	T, °K	Cell Volume, Å³	Reference
		Hexagonal Form			
3.758	6.136	1.633	4.2	75.05	a
3.78	6.16	1.63	liquid He	76.22	b
3.761	6.105	1.623	1.3–4	74.79	c
3.778	6.199	1.639	4.2	76.62	d
3.776	6.162	1.632	2.8–4.5	76.09	e
av. 3.771	6.152	1.631		75.76	
±0.009	±0.031	±0.009		±0.53	
		Cubic Form			
5.351			2.8–4.5	153.22	f
5.377			2.8–4.5	155.46	f
5.290			5	148.04	g
5.312			1.25–1.30	149.89	c
5.338			2.8–4.5	152.10	e
av. 5.334				151.76	
±0.030				±2.55	

[a]Keesom, de Smedt, and Mooy, 1930c.
[b]Kogan, Bulatova, and Yakimenko, 1964.
[c]Mills and Schuch, 1965.
[d]Barrett, Meyer, and Wasserman, 1966.
[e]Bostanjoglo and Kleinschmidt, 1967.
[f]Bostanjoglo, 1965.
[g]Curzon and Mascall, 1965.

DEUTERIUM

The first work on solid D_2 was that of Kogan, Lazarev, and Bulatova (1957). They reported that between 1.5 and 4.1°K, D_2 is tetragonal, with $a = 5.4$, $c = 5.1$ Å, $c/a = 0.94$, on the basis of an x-ray diffraction pattern containing an unspecified number of lines. In the next report from these same authors (1958) it was stated that the question of the accurate determination of the structure of deuterium must remain open because of the difficulty of assigning definite indices to x-ray patterns which contain so few lines. This would appear to be somewhat of an understatement, for the pattern they observed consisted of only one line, at $d = 2.84$ Å.

In their third paper of this series, Kogan, Lazarev, and Bulatova (1960) reported new diffraction data on solid H_2, D_2, and T_2. The diffraction pattern of D_2 at $4.2°K$ still consisted of only one line, now at $d = 2.93$ Å. (This is the value they obtained using FeK_α radiation; a corrected value of 2.895 Å was obtained on the basis of an extrapolation of values obtained with other wave lengths.) This line was considered to be the superposition of (101) and (002) of a tetragonal cell having $a = 3.35$, $c = 5.79$ Å (corrected), $c/a = 1.73$. Lines corresponding to the (001) and (110) of this cell were absent, and the space group was accordingly stated to be $I4-C_4^5$. Surely this proposal must rank among the boldest in the history of x-ray crystallography.

Next, the results of a neutron diffraction study of Kogan, Lazarev, Ozerov, and Zhdanov (1961b) were alleged to confirm the tetragonal structure of solid D_2 as obtained by x-ray diffraction by Kogan et al. (1960). Data obtained at $12°K$ were said to correspond to a tetragonal unit cell with $a = 3.38$, $c = 5.60$ Å, $c/a = 1.66$. They attributed the difference in the axial ratio from the x-ray value of 1.73 to the difference in the temperatures at which the two investigations were carried out. (It was not pointed out that the unit cell from the neutron study, at $12°K$, was *smaller* than the unit cell from the x-ray study, at $4.2°K$.) As in the case of H_2 (see above) some maxima in the neutron diffraction pattern were indexed in violation of the $h + k + \ell = 2n$ restriction required by the body centered space group proposed from the x-ray study, and the same explanation for these discrepancies was offered. As in the case of H_2, ways in which ortho and para molecules might be ordered were suggested somewhat later by Ozerov, Kogan, Zhdanov, and Kukhto (1961); the tetragonal structure, with $a = 3.4$, $c = 5.6$ Å, was mentioned later by Ozerov, Zhdanov, and Kogan (1962).

Brith and White (1964) reported that the anomalous heat capacity of solid D_2 at high temperature could not be accounted for by the tetragonal structure, and, moreover, that the best agreement of theory with experiment was obtained with an ideal hexagonal close packed lattice having $c/a = 1.63$.

Curzon and Pawlowicz (1964a) reported that the electron diffraction pattern of D_2 at $12°K$ differed from those given in the x-ray and neutron diffraction studies of Kogan and co-workers (1960, 1961b), and gave a list of their observed spacings. These results were shortly withdrawn by Curzon and Pawlowicz (1964b), because it had been discovered that their D_2 had been contaminated with air, and the patterns they observed were actually due to α-N_2. The new data, on pure D_2 at about $7°K$, were indexed on the basis of a face centered cubic unit cell, with $a = 5.07 \pm 0.02$ Å. This result was also different from those of the previous x-ray and neutron studies.

New data were reported by Kogan, Bulatova, and Yakimenko (1964), which, as in the case of H_2, required reevaluation of their previous work on D_2. The unit cell of solid D_2, at liquid helium temperature, was reported to be hexagonal, with a = 3.54, c =5.91 Å, c/a = 1.67, as determined from an x-ray diffraction pattern containing three lines.

In another electron diffraction study of D_2, at 5 and 7°K, Curzon and Mascall (1965) observed four sharp lines, which they indexed as being from a face centered cubic unit cell with a = 5.07 ± 0.05 Å at 5°K. As in the case of their work on solid H_2, Curzon and Mascall again examined thin films and not bulk material; thus they did not feel that their results were inconsistent with those of Kogan et al. (1964). A slightly different value for a, 5.069 ± 0.009 Å, results from the average value calculated from the published spacings.

Mucker, Talhouk, Harris, White, and Erickson (1965), after briefly review-ing the earlier work on the crystal structure of both H_2 and D_2, commented that a major problem here was the preparation of randomly oriented crystallite samples. They stated that they had succeeded in preparing such samples of both ortho D_2 (97.8% ortho) and normal D_2 (66.66% ortho) and had ob-tained reproducible neutron diffraction data. At 13°K both preparations gave identical patterns, which Mucker et al. indexed on a hexagonal unit cell with a = 3.63, c = 5.84 Å, c/a = 1.61, based on eight lines. In what appears to be the first discussion of details of the crystal structure of this element, they noted that the intensity distribution was very similar to that of other close packed structures, and that there were only two structures which satisfied this condition. Both have space group $P6_3/mmc-D_{6h}^4$; in the first the individual atoms have specifiable sites, that is, positions $4f$, $(\frac{1}{3}\frac{2}{3}z)$, $(\frac{2}{3}\frac{1}{3}\bar{z})$, $(\frac{2}{3}\frac{1}{3}\frac{1}{2}+z)$, $(\frac{1}{3}\frac{2}{3}\frac{1}{2}-z)$, and the second, the uncorrelated case, compatible with random molecular orientation or free rotation, with the molecular centers in positions $2c$,$(\frac{1}{3}\frac{2}{3}\frac{1}{4})$, $(\frac{2}{3}\frac{1}{3}\frac{3}{4})$. The authors stated that until a detailed analysis of the in-tegrated intensities was made it was not possible to distinguish between them, but they favored the second case because the observed axial ratio was close to ideal value of 1.633 for close packed spheres and there was qualitative agree-ment between observed and calculated intensities.

The discovery that solid H_2 can be either hexagonal close packed or face centered cubic near 1.3°K, depending on the ortho-para ratio, prompted Schuch and Mills (1966) to investigate solid D_2 at very low temperatures. They found that 60 to 65% para D_2 changed from hexagonal to cubic below 1.4°K, whereas 52% para D_2 remained hexagonal. Normal D_2 at 17 and 19.4°K was also found to be hexagonal. Between 1.40 and 1.50°K the hexagonal unit cell had a = 3.600 ± 0.004, c = 5.858 ± 0.007 Å, c/a = 1.627, and between 0.95 and 1.40°K the face centered cubic form had a = 5.081 ± 0.006 Å, based on x-ray diffraction patterns containing eight and four lines, respectively.

In a second neutron diffraction study, Mucker, Talhouk, Harris, White, and Erickson (1966) found that solid D_2, irrespective of the para content, was hexagonal close packed above the λ-transition: the pattern of para en-riched (> 80%) D_2 was found to be the same, within experimental error, of those of ortho and normal D_2 previously obtained by Mucker et al. (1965).

At $1.9°K$, however, 83% para D_2 was found to be cubic, with $a = 5.083 \pm 0.005$ Å. The hexagonal-cubic transformation exhibited considerable hysteresis. A total of five lines was observed, the three strongest of which conformed to face centered cubic restrictions and were used to determine the lattice constant. In addition, two weak lines were observed which violated that restriction: the low temperature form of D_2 is not, therefore, face centered cubic, as had been heretofore stated. Mucker et al. stated that their data were consistent with space group Pa3 [the one assumed by Raich and James (1966) for cubic H_2], but it was not possible for them to tell whether the structure was deformed to space group $P2_13$, as had been suggested in the case of α–N_2. (This deformation is discussed below in the section dealing with nitrogen.)

These points were discussed in greater detail by Mucker (1966). For cubic D_2 at $1.9°K$ of 79% para content he found $a = 5.074 \pm 0.009$ Å. Since two reflections violating the face centered restriction were observed, the probable space group was also given as Pa3, but deformation to $P2_13$ was not mentioned.

Shortly thereafter, Barrett, Meyer, and Wasserman (1966) reported that D_2 usually froze to the hexagonal close packed form, which, unlike that form of H_2, was stable to plastic deformation. At $4.2°K$, the unit cell of hexagonal D_2 had $a = 3.65$, $c = 5.83$ Å, $c/a = 1.60$, based on six observed lines. When, on the other hand, the D_2 was condensed on cubic textured gold foil at $4.2°K$, it was found to be cubic, with $a = 5.21$ Å, the same value as observed for solid H_2 on gold foil; the authors did not believe this to be an equilibrium value, but due to some matching of spacings between substrate and condensate.

Bostanjoglo and Kleinschmidt (1967) reported that thin films of D_2 deposited on Formvar at 2.8 to $4.5°K$ were face centered cubic, with $a = 5.092 \pm 0.007$ Å, as obtained from five observed lines. As in the case of H_2 and HD, a hexagonal structure appeared on warming, and, on the basis of 11 lines lattice constants of $a = 3.601 \pm 0.007$, $c = 5.884 \pm 0.01$ Å, $c/a = 1.634$ were obtained.

Some of the results described by Mucker (1966) were later discussed by Mucker, Harris, White, and Erickson (1968). They gave the same lattice constant of 5.074 Å for the cubic form as given earlier. The two lines which violate the face centered extinction rule were observed, in agreement with space group Pa3. There was, however, no indication of lines forbidden by that space group, but they pointed out that any shift of the deuterium molecules producing the noncentrosymmetric space group $P2_13$ would be extremely difficult to detect in the neutron powder diffraction patterns, and that single crystal work will be necessary to establish the true symmetry. They obtained excellent agreement between the observed and calculated intensities with the eight hydrogen atoms placed in positions 8c of space group Pa3, that is, *xxx*, . . ., with the parameter *x* assigned a value corresponding to a bond distance of 0.7416 Å, the spectroscopic value.

Mucker et al. also reported more extensive work on the high temperature hexagonal form. Excellent agreement between observed and calculated intensities and a structure with the molecules in positions $2c$ of space group $P6_3/$ mmc: it was not possible to distinguish between a model in which the molecules were precessing about the c direction and one in which the molecules were statistically disordered. The ordered model, with the atoms in positions $4f$, was rejected. They also measured the lattice constants of samples having para concentrations of 3, 33, 63, and 80%; no significant differences could be detected.

The results among which there is concordance are collected in Table 1-2. (The values given by Mucker et al., 1965, for the hexagonal form are not included because their work was carried out at a much higher temperature than the others.) Simple averages are chosen as the best values for the lattice constants of the hexagonal form, with the standard errors as obtained from the individual deviations; for the cubic form the given uncertainties were taken into account in calculating the average.

TRITIUM

In the first study of solid T_2, Kogan, Lazarev, and Bulatova (1960) assigned to it a tetragonal unit cell with $a = 3.30$, $c = 5.71$ Å, $c/a = 1.73$, on the basis of a single line at d(corrected) = 2.89 Å, which was indexed as the superposition of (101) and (002). (Cf. the results from this same reference on D_2, above.)

In a somewhat later second study Kogan, Bulatova, and Yakimenko (1964) revised these conclusions and, on the basis of three lines, concluded that solid T_2 at liquid helium temperature is hexagonal, with $a = 3.68$, $c = 6.06$ Å, $c/a = 1.66$. They also stated that different lattice constants of $a = 3.47$, $c = 5.80$ Å resulted if one assumed that the one line observed in the previous work was (002). (It was also apparently assumed that c/a had the value 1.67, perhaps by analogy with the value obtained for D_2 in the same paper; otherwise two lattice constants could not have been derived from a single observation.) The lattice constants of the first set above are suspect because they are *larger* than the corresponding values for D_2 (see Table 1-2), and the second are based on assumptions so tenuous that they should not be accepted either. Additional experimental data are obviously needed to establish the structure of solid T_2.

Table 1-2. Lattice Constants of Solid Deuterium, D_2

a, Å	c, Å	c/a	T, °K	Cell Volume, Å³	Reference
			Hexagonal Form		
3.54	5.91	1.67	liquid He	64.14	a
3.600	5.858	1.627	1.40–1.50	65.75	b
3.65	5.83	1.60	4.2	67.26	c
3.601	5.884	1.634	4.2	66.08	d
3.63*	5.84*	1.61	13	66.64	i
3.604	5.869	1.628	2.0–12.9	66.02	j
av. 3.599	5.870	1.631		65.85	
±0.035	±0.027	±0.016		±1.06	
			Cubic Form		
5.07 ± 0.02			7	130.32	e
5.069 ± 0.009			5	130.25	f
5.074 ± 0.009			1.9	130.63	h,j
5.081 ± 0.006			0.95–1.40	131.17	b
5.083 ± 0.005			1.9	131.33	g
5.092 ± 0.007			4.2	132.03	d
av. 5.080 ± 0.008				131.10 ± 0.62	

[a]Kogan, Bulatova, and Yakimenko, 1964.
[b]Schuch and Mills, 1966.
[c]Barrett, Meyer, and Wasserman, 1966.
[d]Bostanjoglo and Kleinschmidt, 1967.
[e]Curzon and Pawlowicz, 1964b.
[f]Curzon and Mascall, 1965.
[g]Mucker, Talhouk, Harris, White, and Erickson, 1966.
[h]Mucker, 1966.
[i]Mucker, Talhouk, Harris, White, and Erickson, 1965.
[j]Mucker, Harris, White, and Erickson, 1968.
*Omitted in the averaging.

OTHER

At the present time, no results on the remaining two combinations, HT and DT, have been reported.

SUMMARY

1. The low temperature forms of H_2, HD, and D_2 are cubic, and the high temperature forms are hexagonal. Lattice constants and some derived quantities for these are presented in Table 1-3.

2. The nature of the transitions in H_2 of hexagonal $\rightarrow$ cubic on cooling, and of cubic $\rightarrow$ hexagonal on warming have been studied in detail by Mills, Schuch, and Depatie (1966), who found that both transition temperatures could be accurately expressed as linear functions of the fraction of ortho present. They further found that the transition temperatures were not always sharp and well defined, a result which may well explain the apparently discordant results of Bostanjoglo (1965), Barrett, Meyer, and Wasserman (1966), and Bostanjoglo and Kleinschmidt (1967), all of whom observed cubic H_2 above the transition temperature.

3. No detectable change in the lattice constants with change in the ortho-para ratio was observed by Mucker et al. (1965), Schuch and Mills (1966), and Mucker et al. (1966, 1968), in the case of D_2. This direct comparison does not yet appear to have been made in the case of H_2.

4. The high temperature forms are hexagonal close packed and seem likely to consist of freely or nearly freely rotating molecules. If this is true, the intermolecular distances (Table 1-3) are far larger than the value of 3.14 Å predicted by the use of the 0.74 Å bond distance and the customary van der Waals radius of 1.2 Å.

5. There is disagreement on the nature of the lattice of the low temperature cubic forms: H_2 has been stated to be face centered cubic (cubic close packed) by Bostanjoglo (1965), Curzon and Mascall (1965), Mills and Schuch (1965), Barrett, Meyer, and Wasserman (1966), and Bostanjoglo and Kleinschmidt (1967); D_2 has been similarly characterized by Curzon and Pawlowicz (1964), Bostanjoglo (1965), Curzon and Marshall (1965), Schuch and Mills (1966), Barrett, Meyer, and Wasserman (1966), and Bostanjoglo and Kleinschmidt (1967; HD, observed only by Bostanjoglo and Kleinschmidt (1967), was stated by them to be face centered cubic.

In contrast to this, Raich and James (1966) assumed that cubic H_2 had a *primitive* lattice, space group Pa3-T_h^6, by analogy with α-N_2, and Mucker et al. (1966 and 1968) observed two lines in the diffraction pattern of cubic D_2 which violated the face centered extinction rule but were consistent with the Pa3 structure. It thus seems likely that cubic hydrogen is not face centered cubic, as has commonly been stated, but whether the structure has space group Pa3 or is distorted to space group $P2_1 3$, as has been suggested in the case of α-N_2 (see section on Nitrogen in Chapter 8), cannot be decided from the data now available. The observation of Clouter and Gush (1965) that in the low temperature form the H_2 molecules must lie on centers of symmetry (as in space group Pa3) is probably of no help here: if the distortion to the noncentric space group $P2_1 3$ were quite small, the infrared lines forbidden by the centric structure might be too faint for unequivocal detection. Additional data are needed to settle this question.

6. It is generally assumed that in the high temperature hexagonal form rotating molecules of hydrogen are present, whereas in the low temperature cubic

Table 1-3. Lattice Constants of Hydrogen Isotopes

	H_2 (1.3–4.5°K)	HD (2.8–4.5°K)	D_2 (1.40–4.5°K)
a	3.771 ± 0.009 Å	3.642 ± 0.015 Å	3.604 ± 0.027 Å
c	6.152 ± 0.031 Å	5.951 ± 0.010 Å	5.865 ± 0.027 Å
c/a	1.631 ± 0.009	1.634 ± 0.007	1.627 ± 0.020
V[a]	37.88 ± 0.27 Å³	34.18 ± 0.29 Å³	33.02 ± 0.56 Å³
d[b]	3.77 ± 0.01 Å	3.64 ± 0.01 Å	3.60 ± 0.02 Å
ρ[c]	0.0884 ± 0.006	0.1468 ± 0.0012	0.2032 ± 0.0046

Cubic Form

	H_2 (1.25–5°K)	HD (2.8–4.5°K)	D_2 (0.95–7°K)
a	5.334 ± 0.030 Å	5.182 ± 0.009 Å	5.080 ± 0.008 Å
V[a]	37.94 ± 0.64 Å³	34.79 ± 0.18 Å³	32.78 ± 0.15 Å³
d[b]	3.77 ± 0.02 Å	3.66 ± 0.01 Å	3.59 ± 0.01 Å
ρ[c]	0.0882 ± 0.015	0.1442 ± 0.0008	0.2040 ± 0.0010

[a] V, the volume per molecule.
[b] d, a molecular diameter, which, for close packed structures, equals $2^{1/6} V^{1/3}$
[c] ρ, the calculated density, in g cm^{-3}.

form some sort of ordering, or cooperative ordering of molecular rotation, takes place. It is notewirthy, therefore, that the intermolecular distances are identical, within their respective limits of error, in both forms (see Table 1-3).

7. It has been repeatedly observed, in the case of both H_2 and D_2, that the temperature of the transition nexagonal $\rightleftharpoons$ cubic is strongly dependent on the ortho-para ratio (Clouter and Gush, 1965, Mills and Schuch, 1965, Schuch and Mills, 1966, Mills, Schuch and Depatie, 1966). It will therefore be of interest to learn the temperature dependence of the hexagonal $\rightleftharpoons$ cubic transition in HD, a species in which the ortho-para equilibrium does not exist.

Chapter 2

Group 0, The Rare Gases

HELIUM

^{4}He

Solid helium was first studied by x-ray diffraction by Keesom and Taconis (1938). They observed two powder lines and concluded that the material (at 1.45°K and 37 atm) is hexagonal close packed, with a = 3.577, c = 5.842 Å (from kX), c/a = 1.633.

In a study of the melting properties and thermodynamic functions of solid helium, Dugdale and Simon (1953) found a first order transition; the equilibrium line cut the melting curve at 14.9°K and 1092 atm and moved to higher temperatures at higher pressures. They termed the low temperature modification α, the high temperature β, and, because "substances with short-range forces of the van der Waals type are stable only in close-packed lattices", conjectured that β-He was cubic close packed, the α modification already having been shown by Keesom and Taconis to be hexagonal close packed.

The hexagonal close packed structure for the low temperature form was verified by Henshaw (1958a), who observed six lines in a neutron diffraction pattern. The lattice constants, as revised by Donohue (1959a), were found to be a = 3.531 ± 0.006, c = 5.693 ± 0.010 Å, c/a = 1.612 ± 0.004 at 1.15°K and 66 atm.

Schuch (1958) also verified the hexagonal close packed structure, finding c/a = 1.594 at 1.7°K and 175 atm from Laue x-ray data.

Somewhat later Schuch and Mills (1961a) reported lattice constants of a = 3.555, c = 5.798 Å, c/a = 1.631 at 2.51°K and 61.2 atm for α-^{4}He, but gave neither the number of observed lines on which these values were based nor the estimated uncertainties.

An x-ray diffraction investigation by Mills and Schuch (1961) then established that the high temperature β form is in fact cubic close packed. They observed five lines, and at $15.80 \pm 0.35°$K and 1255 atm the lattice constant was found to be $a = 4.240 \pm 0.016$ Å.

At about the same time Vignos and Fairbank (1961) announced that they had discovered a third phase in solid ^{4}He, which they termed γ. This form, which was detected by measurements of the longitudinal velocity of first sound in the solid, was found to be stable between 1.45 and $1.78°$K in a narrow range of pressure adjoining the melting curve. At $1.50°$K the change in volume at the transition $\alpha \to \gamma$ was $+0.3\%$. By analogy with the observed phase transition in ^{3}He at low temperature (see below) Vignos and Fairbank conjectured that their γ phase was body centered cubic.

This conjecture was shortly thereafter verified by Schuch and Mills (1962). At $1.73°$K and 29.01 atm solid γ-^{4}He gave two diffraction maxima, corresponding to a body centered cubic unit cell containing two atoms, with $a = 4.110 \pm 0.005$ Å. They also determined the lattice constants of the α (hexagonal close packed) phase at $1.73°$K and 29.70 atm and obtained, on the basis of an unstated number of observed lines, $a = 3.650 \pm 0.012$, $c = 5.945 \pm 0.014$ Å, $c/a = 1.629 \pm 0.006$.

The phase diagram for ^{4}He is presented in Figs. 2-1 and 2-2. (In the case of the low temperature transitions, Fig. 2-2, the phase boundaries of Grilly and Mills (1962) were used; their results gave these as 0.2 to 0.3 atm lower than those reported by Vignos and Fairbank.) Discussion of the volume changes at the transitions is combined with those in ^{3}He, below .

^{3}He

A transition in solid ^{3}He was first reported by Mills and Grilly (1958). The transition line was found to intersect the melting curve at $3.15°$K and 136 atm. The low pressure form was termed α, and the high pressure form was termed β. These two forms were investigated by x-ray diffraction by Schuch Grilly, and Mills (1958), who found that at $1.9°$K and 97 atm the α form is, as shown by its Laue pattern, body centered cubic, with $a = 4.01 \pm 0.03$ Å as determined from the one observed powder line, while at $3.3°$K and 177 atm the β form, also on the basis of a Laue pattern, is hexagonal close packed, with $a = 3.46 \pm 0.03$, $c = 5.60 \pm 0.03$ Å, $c/a = 1.619 \pm 0.016$, as determined from three observed powder lines.

The slightly different values of $a = 3.963$ Å at $2.88°$K and 125 atm for body centered cubic α-^{3}He, and $a = 3.501$, $c = 5.721$ Å, $c/a = 1.634$ at $3.48°$K and 163 atm for hexagonal close packed β-^{3}He were later reported by Schuch and Mills (1961a).

By analogy with the transition above $14.9°$K and 1100 atm in ^{4}He, $\alpha \to \beta$, that is, hexagonal close packed $\to$ cubic close packed, it was expected that a

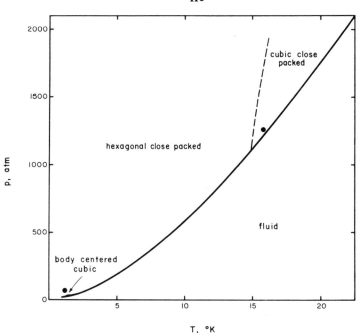

Fig. 2-1.   Phase diagram for ⁴He (after Dugdale, 1965). Dots indicate points at which lattice constants were determined.

similar transition might occur in solid ³He, and in a study of the high pressure region Schuch and Mills (1961b) found that at 18.76°K and 1690 atm ³He is in fact cubic close packed, with $a = 4.242 \pm 0.016$ Å, as obtained from five observed lines on seven different photographs. (This form had already been termed γ-³He by Schuch and Mills, 1961a.) The point at which the β-³He → γ-³He transition, that is, hexagonal close packed → cubic close packed, joined the melting curve was not determined in this work. It was, however, reported that the β (hexagonal close packed) phase is still stable at 15.98°K and 1341 atm, with lattice constants of $a = 3.046*$, $c = 4.986$ Å, $c/a = 1.637$. The line of transition between the two solids was later studied by Franck (1961) and by Dugdale and Franck (1964); the phase line cuts the melting curve at 17.80°K and 1557 atm and moves to higher pressures at higher temperatures, in accord with the equation

$$p = 1557 + 1096 (T - 17.80) \text{ atm}$$

The phase diagram for ³He is presented in Figs. 2-3 and 2-4.

*The value in the paper of $a = 0.046$ Å is a misprint (A. F. Schuch, private communication.)

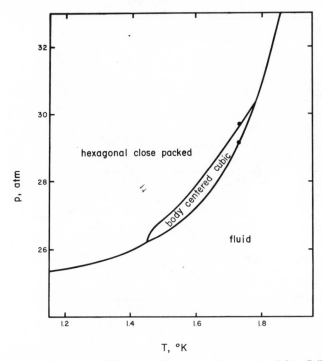

Fig. 2-2.   Phase diagram for ^{4}He in the low temperature region (after Grilly & Mills, 1962). Dots indicate points at which lattice constants were determined.

DISCUSSION

Nomenclature Problems

For chronological reasons a rather preposterous situation developed, as must be obvious from the foregoing discussion. The use of Greek letters in characterizing allotropes having simple structures, although economical of space, is not totally satisfactory. In the present case we find that α means hexagonal close packed in the case of ^{4}He, but body centered cubic (with two atoms per unit cell) in the case of ^{3}He, that β means cubic close packed (i.e., face centered cubic with four atoms per unit cell) in the case of ^{4}He, but hexagonal close packed in the case of ^{3}He, and that γ means body centered cubic (with two atoms per unit cell) in the case of ^{4}He, but cubic close packed in the case of ^{3}He. An attempt to rectify this mess, which requires unnecessary mental gymnastics on the part of those interested in the phase behavior of the solid heliums, was made by Daunt, Schuch, and Mills (1964). They proposed that the previous Greek letter system by abolished and be replaced by hcp, ccp,

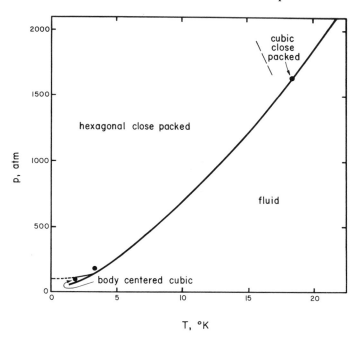

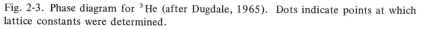

Fig. 2-3. Phase diagram for ^{3}He (after Dugdale, 1965). Dots indicate points at which lattice constants were determined.

and bcc. Their preference for ccp (cubic close packed) over fcc (face centered cubic) was made on the valid grounds that the term fcc is ambiguous, referring to a cubic close packed structure only when the fcc unit cell contains four atoms. However, the same objection may be raised with regard to the term bcc, which, as sometimes used, refers to a body centered cubic unit cell with two atoms per unit cell, in disregard of an infinity of other possible body centered cubic structures with *more* than two atoms per unit cell. The nomenclature of Daunt et al. should accordingly be emended to "bcc(2)" to avoid any ambiguity.

Some Volume and Distance Relationships

All of the values reported for the lattice constants, and the conditions under which they were determined, are presented in Table 2-1. Because these were made under widely varying pressures and temperatures, it is not possible to make from them direct comparisons among the atomic volumes and interatomic distances in the various forms. The functions given by Grilly and Mills were used, therefore, to calculate lattice constants and other quantities at the

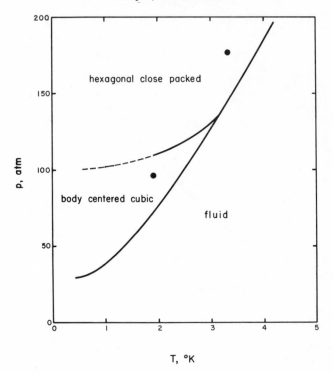

Fig. 2-4. Phase diagram for ^{3}He in the low temperature region (after Dugdale, 1965). Dots indicate points at which lattice constants were determined.

triple points, hcp–bcc(2)–fluid in both ^{3}He and ^{4}He; these results are presented in Table 2-2. (In the case of the hcp forms, the axial ratio was assumed to be 1.633.) At the transitions hcp → bcc(2) there are volume increases of 0.7 to 1.0%, while the nearest neighbor distances decrease by 2.5 to 2.6%

In case of the transition hcp → bcc(2) in ^{3}He, Grilly and Mills (1959) stated that their data extrapolated to a volume change of zero at 0°K and 99 atm. The nearest neighbor distances in the bcc(2) form would then be 2.8% shorter than those in the hcp form.

Direct observation of the volume changes at the hcp → ccp transitions does not appear to have been made. From thermal measurements on ^{4}He Dugdale and Simon (1953) estimated that there was an increase in the molar volume of 4×10^{-4} cm^3, and Dugdale and Franck (1964), by the same method, estimated the corresponding increase in the case of ^{3}He was 1.4×10^{-4} cm^3. These amounts are much too small to be detected directly by diffraction methods.

Table 2-1.   Directly Determined Lattice Constants of Helium Allotropes

	a,Å	c,Å	c/a	$T,°K$	p,atm	Atomic Volume, Å³	Reference
³He							
bcc(2)	4.01	—	—	1.9	97	32.24	a
	3.963	—	—	2.88	125	31.12	b
hcp	3.46	5.60	1.619	3.3	177	29.03	a
	3.501	5.721	1.634	3.48	163	30.36	b
	3.046	4.986	1.637	15.98	1341	20.03	c
ccp	4.242	—	—	18.76	1690	19.08	c
⁴He							
bcc(2)	4.110	—	—	1.73	29.01	34.71	d
hcp	3.531	5.693	1.612	1.15	66	30.74	e
	3.577	5.842	1.633	1.45	37	32.37	f
	—	—	1.594	1.7	175	—	g
	3.650	5.945	1.629	1.73	29.70	34.30	d
	3.555	5.798	1.631	2.51	61.2	31.73	b
ccp	4.240	—	—	16.0	1255	19.06	b

[a]Schuch, Grilly, & Mills, 1958
[b]Schuch & Mills, 1961a
[c]Schuch & Mills, 1961b
[d]Schuch & Mills, 1962
[e]Donohue, 1959a
[f]Keesom & Taconis, 1938
[g]Schuch, 1958

Comparisons between hcp ³He and hcp ⁴He under the same conditions can also be made by the use of the functions of Grilly and Mills, assuming that the axial ratios have the ideal value of 1.633. These comparisons are made in Table 2-3, where it may be seen that the atomic volume of ³He varies between 9.8 and 3.2% larger than that of ⁴He over the range 169 to 3438 atm, at temperatures lying on the melting curve of ³He.

Table 2-2. Lattice Constants and Other Quantities
as Calculated from PVT Data, for ^{3}He and ^{4}He at
the Triple Points hcp—bcc(2)—fluid

	a, Å	c, Å	Atomic Volume, Å^3	d, Å[a]
^{3}He at 3.15°K and 136 atm				
hcp	3.536	5.744	31.27	3.536
bcc(2)	3.978		31.49	3.445
^{4}He at 1.44°K and 25.94 atm				
hcp	3.663	5.982	34.75	3.663
bcc(2)	4.123		35.06	3.571
^{4}He at 1.76°K and 29.67 atm				
hcp	3.649	5.959	34.37	3.649
bcc(2)	4.110		34.70	3.559

[a]d, the nearest neighbor distance, 12 of each in hcp, 8 of each in bcc(2).

Table 2-3. Lattice Constants and Other Quantities
as Calculated from PVT Data, for ^{3}He and ^{4}He

	a, Å	c, Å	Atomic Volume, Å^3	d, Å
At 3.735°K and 169 atm				
^{3}He	3.488	5.696	30.01	3.488
^{4}He	3.382	5.523	27.33	3.382
At 30.184°K and 3438 atm				
^{3}He	2.862	4.674	16.57	2.862
^{4}He	2.832	4.625	16.06	2.832

NEON

Diffraction studies on solid neon between 4.2 and 24°K, beginning with the first determination by de Smedt, Keesom, and Mooy (1930), agree that the substance is cubic close packed. No phase changes have been reported, even on cold working (Barrett, Meyer, and Wasserman, 1966), but Goringe and

Valdre (1964) observed a poorly defined ring on an electron diffraction transmission photograph of neon deposited on carbon, which they thought most likely was to be explained by "the presence of stacking faults of both extrinsic and intrinsic nature".

The various determinations of the lattice constant are presented in Table 10-1, from which it may be seen that the agreement is not very good. The values of Mauer and Bolz (1961) and of Bolz and Mauer (1963), $a = 4.462_2 \pm 0.001$ Å for natural neon, $4.462_4 \pm 0.001$ Å for ^{20}Ne, and $4.454_0 \pm 0.001$ Å for ^{22}Ne, all at 4.2°K, appear to be the most reliable.

The thermal expansion in the range 4.2 to 24.5°K has been reported graphically by Mauer and Bolz (1961) and by Bolz and Mauer (1963), whose re-

Table 10-1. Lattice Constants of Neon

a, Å	T, °K	Diffraction Method	Reference
Natural			
4.515 ± 0.017[a]	ca. 4.2	x-ray, 4 lines	de Smedt, Keesom, & Mooy, 1930
4.428 ± 0.009[b]	4.2	neutron, 7 lines	Henshaw, 1958b
4.479 ± 0.004[a]	4.2	x-ray	Kogan, Lazarev, & Bulatova, 1961a
4.462 ± 0.001	4.2	x-ray, 2 lines	Mauer & Bolz, 1961
$4.461_6 \pm 0.001$[c]	4.2	x-ray, 2 lines	Bolz & Mauer, 1963
4.54	ca. 10	electron	Goringe & Valdre, 1964
4.467	10	x-ray, 2 lines	Mauer & Bolz, 1961
4.48	16	x-ray	Barrett, Meyer, & Wasserman, 1966
4.486 ± 0.001	16	x-ray, 2 lines	Bolz and Mauer, 1963
4.46377 ± 0.00008	4.25	x-ray	Batchelder, Losee, & Simmons, 1967
^{20}Ne			
4.480 ± 0.004	4.2	x-ray	Kogan, Lazarev, & Bulatova, 1961a
$4.462_4 \pm 0.001$	4.2	x-ray, 2 lines	Bolz & Mauer, 1963
^{22}Ne			
4.464 ± 0.004	4.2	x-ray	Kogan, Lazarev, & Bulatova, 1961a
$4.454_0 \pm 0.001$	4.2	x-ray, 2 lines	Bolz & Mauer, 1963

[a]From kX, calculated from the published spacings.
[b]Calculated from the published spacings.
[c]Calculated by Vegard's rule from the values given for ^{20}Ne and ^{22}Ne, assuming 91% ^{20}Ne and 9% ^{22}Ne.

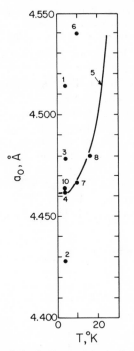

Fig. 10-1.　The variation of the lattice constant of natural neon with temperature. The numbers refer to the references in the same order as in Table 10-1.

sults on natural neon are presented in Fig. 10-1. The extrapolated values of the lattice constant at $0°K$ is 4.462 Å.

The results of Batchelder, Losee, and Simmons (1967) in the range 4.2 to $20.6°K$ are in essential agreement with those of Bolz and Mauer, but the former obtained a slightly higher extrapolated lattice constant at $0°K$ of $4.46368 ± 0.00009$ Å.

ARGON

Early and recent crystal structure determinations agree that solid argon is cubic close packed. The results of various determinations of the lattice constant are presented in Table 18-1 and in Fig. 18-1. The agreement leaves much to be desired, as can be seen, but it seems likely that the best values are the averages between those of Dobbs et al. (1956), Barrett and Meyer (1964), and Peterson et al. (1966), whose results are in close agreement.

A metastable hexagonal close packed phase was discovered by Meyer, Barrett, and Haasen (1964). It was observed to occur from just below the

Table 18-1. Lattice Constants of Argon

Cubic Close Packed a, Å	T, °K	Reference
5.435[a]	40	Simon & von Simson, 1924
5.411[a]	20	de Smedt & Keesom, 1925a, b
5.256	4.2	Henshaw, 1958b
5.312	4.2	Barrett & Meyer, 1964
5.318	21	Meyer, Barrett, & Haasen, 1964
5.319	21	Barrett & Meyer, 1964
5.348[b]	40	Barrett & Meyer, 1964
5.318[c]	20	Dobbs, Figgins, Jones, Piercey, & Riley, 1956
5.345	40	Dobbs et al., 1956
5.3109	0	Peterson, Batchelder, & Simmons, 1966

Hexagonal a, Å	c, Å	c/a	
3.760[d]	6.141[d]	1.633	Barrett & Meyer, 1964

[a]From kX.
[b]Interpolated; for results at other temperatures, see Fig. 18-1.
[c]For results at other temperatures, see Fig. 18-1.
[d]At 21°K.

melting point of 84 down to 2°K, but it converted to the cubic close packed structure on plastic deformation. The atomic volume at 21°K was found to be identical with that of the cubic form, and the axial ratio had the ideal value of 1.633 expected for close packed spheres.

KRYPTON

Early work on the structure of solid krypton (Keesom and Mooy, 1930a,b,c; Natta and Nasini, 1930b; Ruhemann and Simon, 1931) established that it is cubic close packed, but the lattice constants reported are in rather poor agreement with those obtained more recently by Dobbs and Luszczynski (1955), Cheesman and Soane (1957), and Figgins and Smith (1960), as well as some unpublished work of Bolz and Mauer cited by Pollack (1964). Smoothed values for the lattice constants from 4.2 to 115°K are presented in Fig. 36-1. The extrapolated lattice constant at 0°K is 5.644 Å.

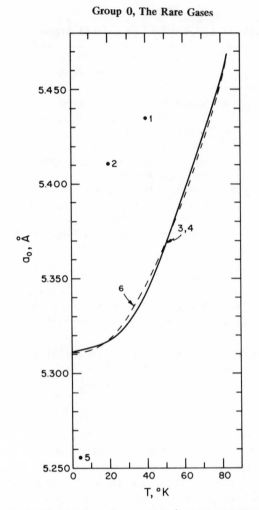

Fig. 18-1.   The lattice constant of argon at various temperatures. (1) Simon & von Simson, 1924; (2) de Smedt & Keesom, 1925a; (3) Dobbs et al., 1956; (4) Barrett & Meyer, 1964; (5) Henshaw, 1958b; (6) Peterson, Batchelder, & Simmons, 1966.

The results of Losee and Simmons (1968) show positive deviations from the previous results, as shown in Fig. 36-1. Their value for the extrapolated lattice constant at $0°K$ is 5.64587 ± 0.00010 Å.

XENON

As in the case of krypton, early work on solid xenon (Natta and Nasini, 1930a; Ruhemann and Simon, 1931) established the cubic close packed structure for

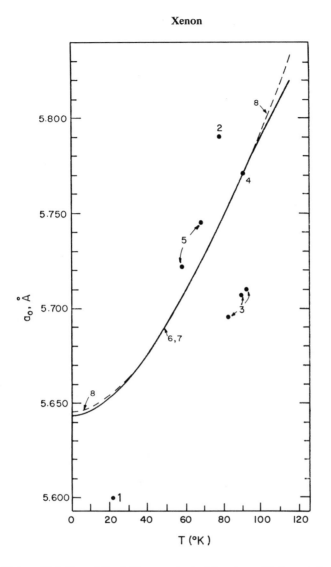

Fig. 36-1.   Variation of the lattice constant of krypton with temperature. (1) Keesom & Mooy, 1930a, b; (2) Natta & Nasini, 1930b; (3) Ruhemann & Simon, 1931; (4) Dobbs & Luszczynski, 1955; (5) Cheesman & Soane, 1957; (6) Figgins & Smith, 1960; (7) Bolz & Mauer, 1964; (8) Losee & Simmons, 1968.

this substance, but the lattice constants found are not in good agreement with more recent work. Values for the lattice constant between 5.5 and 160°K have been measured directly by x-ray diffraction (Cheesman and Soane, 1957;

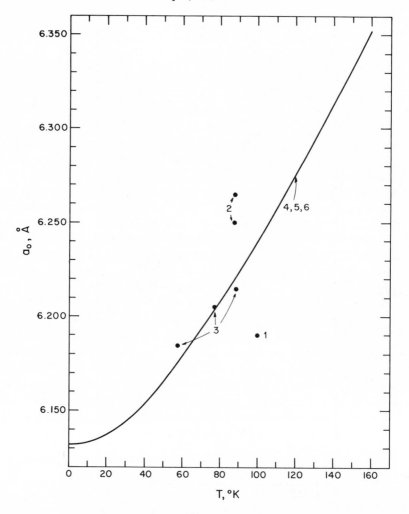

Fig. 54-1.   Variation of the lattice constant of xenon with temperature. (1) Natta &
Nasini, 1930a; (2) Ruhemann & Simon, 1931; (3) Cheesman & Soane, 1957; (4) Sears &
Klug, 1962; (5) Packard & Swenson, 1963; (6) Bolz & Mauer, 1964.

Sears and Klug, 1962) and indirectly by density measurements (Eatwell and
Smith, 1961; Packard and Swenson, 1963). The variation of the lattice constant with temperature is shown in Fig. 54-1. The extrapolated value at $0°K$
is 6.132 Å.

RADON

The crystal structure of solid radon has not been determined.

DISCUSSION

The extrapolated lattice constants of the rare gases at $0°K$ are compared in Table 86-1. Because helium does not exist as a solid below a pressure of ~ 34 atm for ^{3}He and ~ 25 atm for ^{4}He data for it are not included.

Theoretical attempts to explain the crystal structures of the rare gases have so far not been completely successful. From the simplest treatment (Jones and Ingham, 1925), which assumes interatomic interactions are of the van der Waals type only, the prediction results that *all* of the rare gases should be hexagonal close packed. Further considerations on the differences between hexagonal and cubic close packed lattices (Barren and Domb, 1955) led to the prediction that at elevated temperatures a transition hcp → ccp should take place, a result in agreement with the behavior of the heliums but not of the other rare gases. Cuthbert and Linnett (1958) suggested that the cause of the difference between these is that the helium atom has an outer shell of two electrons in a spherically symmetrical orbital, while the others have outer shells of eight electrons which tend to pair at the corner of a regular tetrahedron. Treatment of the bcc(2) → hcp transitions which occur at low temperatures in the heliums does not appear to have been attempted. More complete surveys of the situation may be found in the reviews by Dugdale and Franck (1964) on the heliums, and by Pollack (1964) on the other rare gases.

Table 86-1.   Some Properties of the Rare Gases at Absolute Zero and Atmospheric Pressure

	a, Å	Atomic Volume, Å^3	Density, g cm^{-3}	d, Å[a]
Ne	4.462	22.21	1.509	3.155
Ar	5.311	37.45	1.771	3.755
Kr	5.644	44.95	3.095	3.991
Xe	6.132	57.64	3.783	4.336

[a]d, the nearest neighbor distance.

Chapter 3

Group I, The Alkali Metals

LITHIUM

The structure of lithium at room temperature was first determined by Hull (1917c) to be body centered cubic, with two atoms in a unit cell having $a = 3.51$ Å (from kX), a value in excellent agreement with recent high precision values, which are presented in Table 3-1.

The thermal expansion of lithium from 77 to 300°K has been studied by Pearson (1954); expansion from 79 to 293°K has been studied by Owen and Williams (1954b); in addition, lattice constant measurements were made at 90°K by Lonsdale and Hume-Rothery (1945) and at 78°K by Barrett (1956b). Some of the values reported are included in Table 3-1.

A transition at low temperature was first discovered by Barrett (1947), who found that bcc(2) lithium if cold worked at about 78°K partially transformed to the cubic close packed structure. Barrett and Trautz (1948) then verified this finding and stated that it was accompanied with a volume change of zero; they also reported that a spontaneous partial transformation to the hexagonal close packed structure occurred at 72°K and below, and gave tentative lattice constants of $a = 3.14$, $c = 4.70$ Å, $c/a = 1.50$ at 78°K. Similar results were reported by Barrett (1950), who reported that bcc(2) lithium, when cooled to 63°K and exposed to white x-radiation, partially transformed to hexagonal close packed, with revised lattice constants of $a = 3.086$, $c = 4.828$ Å, $c/a = 1.564$; it was further stated that plastic deformation at or below liquid nitrogen temperature of bcc(2) lithium led to the appearance of a cubic close packed phase; the proportions depended on the temperature and the extent of cold working. At 78°K, the lattice constant of the cubic close packed phase was found to be 4.40 Å·

Table 3-1. The Lattice Constant of bcc(2) Lithium

a, Å	Reference
At 20°C	
3.5088[a]	Perlitz & Aruja, 1940
3.5093[a]	Lonsdale & Hume-Rothery, 1945
3.5087[a]	Pearson, 1954
3.5096[a]	Owen & Williams, 1954b
3.5093[b]	Covington & Montgomery, 1957
3.5092[c]	Nadler & Kempter, 1959b
3.5087[c]	Vogel & Kempter, 1959
av. 3.5091 ± 0.0004	
At 90°K	
3.464[d]	Simon & Vohsen, 1928a
3.4832[a]	Lonsdale & Hume-Rothery, 1945
3.4829[a]	Pearson, 1954
3.4834[a]	Owen & Williams, 1954
3.4925[e]	Barrett, 1956b
av. 3.4832 ± 0.0002	
At 4°K	
3.4795 ± 0.0002	Pearson, 1954

[a]From kX.
[b]Calculated by Vegard's law from the values given for ^{6}Li and ^{7}Li, 3.5107 and 3.5092 Å, respectively, assuming 7.4% ^{6}Li and 92.6% ^{7}Li.
[c]Corrected to 20°C from 25°C, and a thermal expansion coefficient of 4.7×10^{-5} deg^{-1}.
[d]From kX, recalculated from data given in the paper.
[e]Corrected to 90°K with the value of 3.491 Å given for 78°K and a thermal expansion coefficient of 3.5×10^{-5} deg^{-1}; omitted from the average.

More extensive data on the low temperature hexagonal close packed form were reported by Barrett (1956a,b): the previous tentative lattice constants were revised to the values $a = 3.111$, $c = 5.093$ Å, $c/a = 1.637$ at 78°K. At the same temperature the lattice constant of the bcc(2) phase was found to be 3.491 Å. Both forms coexist at 78°K, the hexagonal phase resulting from partial transformation on cold working. It was also stated that the cubic close packed form was produced by cold working, but no lattice constants were given.

Selected data obtained at low temperatures are presented in Table 3-2.

From electrical resistance measurements Stager and Drickamer (1963b) re-

Table 3-2. Lattice Constants and Other Properties of Lithium Allotropes
at 78°K

	a, Å	c, Å	c/a	Atomic Volume, Å^3	Density g cm^{-3}	d, Å
bcc(2)	3.482[b]	—	—	21.10	0.5459	3.016(8)
hcp	3.103[c]	5.080[c]	1.637	21.18	0.5440	3.103(6), 3.108(6)
ccp	4.388[d]	—	—	21.16	0.5445	3.103(12)

[a]d, the nearest neighbor distance; number of nearest neighbors given in parentheses.

[b]Calculated from the value of Table 3-1 for 90°K and a thermal expansion coefficient of 3.5 X 10^{-5} deg^{-1}.

[c]Converted from the lattice constants given by Barrett (1956b), using the lattice constant (b) of the coexisting bcc(2) phase as an internal standard.

[d]Calculated with the lattice constant of the bcc(2) phase and assuming that these two forms have the same atomic volume, as observed by Barrett and Trautz (1948).

ported a transition at 70 kbar and 23°C. The structure of the high pressure form is unknown.

SODIUM

The structure of sodium at room temperature was first determined by Hull (1917c), who found that it is body centered cubic with two atoms in a unit cell having a = 4.31 Å (from kX). From data given by Simon and Vohsen (1928a), a lattice constant of 4.248 Å (from kX) at liquid air temperature is obtained. In a somewhat later precision determination, Aruja and Perlitz (1938) found a = 4.2906 ± 0.0005 Å (from kX) at 20°C.

In 1948 Barrett found that samples of sodium which, after cold working at 20°K, had been allowed to warm up to 78°K gave one extra powder line at d = 3.083 Å; he tentatively suggested that this was caused by the appearance of a cubic close packed phase having a = 5.339 Å. Although the occurrence of a partial phase transformation was confirmed by a micrographic examination (Barrett, 1955), the observation of additional extra lines in powder patterns of sodium which had been cooled below 36°K or cold worked below 51°K (Barrett, 1956b) showed that the transformation was from body centered cubic to hexagonal close packed rather than to cubic close packed. [The single strong line previously observed proved to be (002) of hcp rather than (111) of ccp.]

Lattice constants and other properties are summarized in Table 11-1.

Table 11-1. Lattice Constants and Other Properties of Sodium

	a, Å	c, Å	c/a	T	Atomic Volume, Å³	Density, g cm⁻³	d, Åᵃ	Reference
bcc(2)	4.2906	–	–	20°C	39.49	0.9666	3.716(8)	Aruja & Perlitz, 1938
	4.235	–	–	78°K	37.98	0.9950	3.668(8)	Barrett, 1956b
	4.225	–	–	5°K	37.71	1.0122	3.658(8)	Barrett, 1956b
hcp	3.767	6.154	1.634	5°K	37.81	1.0095	3.767(6), 3.768(6)	Barrett, 1956b

$^a d$, nearest neighbor distance; number of nearest neighbors given in parentheses.

POTASSIUM

After some early work (Hull, 1917c; McKeehan, 1922a) in which it was reported that potassium was amorphous at room temperature, because no diffraction pattern could be detected, Posnjak (1928) established the structure as body centered cubic, two atoms per unit cell, $a = 5.344$ Å (from kX). At about the same time Simon and Vohsen (1928a) reported their work on the lattice constants of the alkali metals at liquid air temperatures. The value of the lattice constant as calculated from their data is 5.261 Å (from kX). This value is considerably higher than the value 5.21 Å (from kX) reported by McKeehan, at the same temperature. The value 5.321 Å (from kX) was found by Böhm and Klemm (1939) at 20°C; an identical value may be calculated from the density of 0.862 g cm⁻³ reported by Richards and Brink (1907). The values of a are presented in Fig. 19-1.

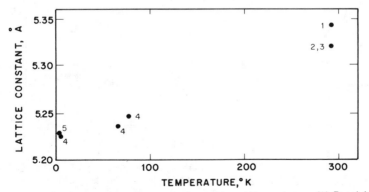

Fig. 19-1. The lattice constant of potassium at various temperatures. (1) Posnjak, 1928; (2) Böhm & Klemm, 1939; (3) Richards & Brink, 1907 (calculated from the density); (4) Simon & Vohsen, 1928; (5) Werner, Gürmen, & Arrott, 1969.

As opposed to the behavior of both lithium and sodium at low temperature, Barrett (1956b) found that cooling to 5°K did not induce any transformation, nor did cold working at 5°K. Lattice constants found are: at 78°K, 5.247 ± 0.002 Å; at 66°K, 5.236 ± 0.004 Å; at 5°K, 5.225 ± 0.002 Å.

Stager and Drickamer (1963b) found two transitions at 77°K, one at 280 kbar and the other at 360 kbar, from electrical resistance measurements. The structures of the new forms were not determined.

RUBIDIUM

The structure of rubidium was first determined by Simon and Vohsen (1928a), who found that at liquid air temperature it is body centered cubic with two atoms per unit cell, a = 5.636 Å (from kX, recalculated from the data given in the paper). Böhm and Klemm (1939) found a = 5.67 Å (from kX) at -10°C. The density of 1.5324 g cm^{-3} at 20°C reported by Richards and Brink (1907) corresponds to a lattice constant of 5.700 Å. Hume-Rothery and Lonsdale (1945) reported a = 5.711 at 18.5°C and 5.635 Å at -183°C (both from kX), while Kelly and Pearson (1955) found a = 5.610 Å at 77°K, 5.618 Å at 90°K, and 5.699 Å at 297°K. Barrett (1956b) reported that no transformation was induced by cooling rubidium to 5°K or by cold working at that temperature. He reported the following lattice constants: 5.605 ± 0.001 Å at 78°K; 5.585 ± 0.001 Å at 5°K. Values for a are presented in Fig. 37-1. The three room temperature values average to a = 5.703 ± 0.007 Å at 20.8 ± 2.8°C.

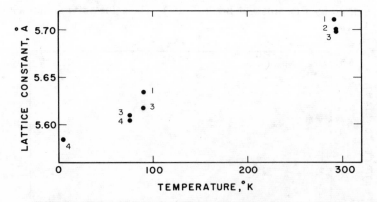

Fig. 37-1. The lattice constant of rubidium at various temperatures. (1) Hume-Rothery & Lonsdale, 1945; (2) Richards & Brink, 1907 (calculated from the density); (3) Kelly & Pearson, 1955; (4) Barrett, 1956b.

The high pressure studies of Bundy (1959) and Stager and Drickamer (1963) reveal two high pressure forms, both of unknown structure. The phase diagram is shown in Fig. 37-2.

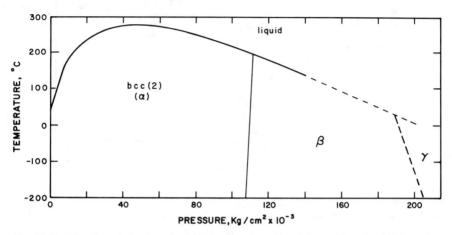

Fig. 37-2. The phase behavior of rubidium (from combined data of Bundy, 1959, and Stager & Drickamer, 1963).

CESIUM

The structure of cesium was first determined by Simon and Vohsen (1928a). At liquid air temperature the substance is body centered cubic, two atoms per cell, a = 6.070 Å (from kX recalculated from data in the paper). Böhm and Klemm (1939) found a = 6.14 Å (from kX) at -10°C, while Brauer (1947) reported a = 6.091 Å (from kX) at -100°C. The density of 1.873 g cm^{-3} determined by Richards and Brink (1907) corresponds to a = 6.176 Å at 20°C. This is considerably higher than the directly measured value of 6.141 ± 0.007 Å reported by Weir, Piermarini, and Block (1971). In a low temperature study Barrett (1956b) found that neither cooling to 5°K nor cold working at that temperature induced transformation. He found that at 78°K, a = 6.067 ± 0.002 Å and at 5°K, a = 6.045 ± 0.002 Å. The lattice constants are presented in Fig. 55-1.

Two transitions in cesium at high pressures were first discovered by Bridgman (1938) by displacement and electrical resistance measurements. A third transition, indicating the existence of a fourth polymorph stable over a

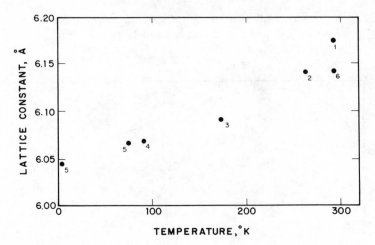

Fig. 55-1.   The lattice constant of cesium at various temperatures. (1) Richards &
Brink, 1907 (calculated from the density); (2) Böhm & Klemm, 1939; (3) Brauer, 1947;
(4) Simon & Vohsen, 1928a; (5) Barrett, 1956b; (6) Weir, Piermarini, & Block, 1971.

narrow pressure range, was discovered by Hall (1960). These phases were
studied by x-ray diffraction by Hall, Merrill, and Barnett (1964). Between 26
and 31°C the transitions are as follows:

$$\text{Cs(I)} \xrightarrow{23.7 \text{ kbar}} \text{Cs(II)} \xrightarrow{42.2 \text{ kbar}} \text{Cs(III)} \xrightarrow{42.7 \text{ kbar}} \text{Cs(IV)}$$

Below 26°C the Cs(III) phase is bypassed. This probably explains why it was
missed by Bridgman. Hall et al. found that both Cs(II) and Cs(III) were cubic
close packed. The respective lattice constants are

Cs(II) at $\sim$ 20°C and $\sim$ 24 kbar: $a = 6.465 \pm 0.015$ Å   (Weir et al., 1971)
        at 27°C and 41 kbar: $a = 5.984 \pm 0.014$ Å   (Hall et al., 1964)

Cs(III) at 27°C and 42.5 kbar: $a = 5.800 \pm 0.007$ Å   (Hall et al., 1964)

These data correspond to a volume decrease of 8.9 ± 0.8% for the transition
II→III, Hall, Merrill, and Barnett were unable to index their powder pattern
from Cs(IV) but estimated a volume decrease of about 2.4% for the transition
III→IV. The pressure dependence of the atomic volume is shown in Fig. 55-2.

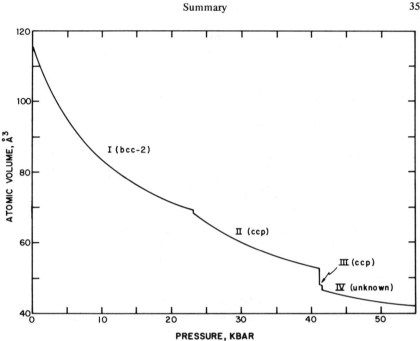

Fig. 55-2.   The variation of the atomic volume of cesium with pressure (calculated from the data of Hall, Merrill, & Barnett, 1964).

FRANCIUM

The crystal structure of francium has not been determined. The longest lived isotope has a half-life of 22 minutes.

SUMMARY

Lattice constants and interatomic distances for the alkali metals at various temperatures are presented in Table 87-1. The various observed allotropes are summarized in Table 87-2. The total number of forms observed so far for each alkali metal is:

lithium	four
sodium	two
potassium	three
rubidium	three
cesium	four

Table 87-1. Lattice Constants, Atomic Volumes, and Nearest Neighbor Distances for the bcc(2) Forms of the Alkali Metals

	At 5°K			At 20°C		
	a, Å	V, Å^3	d, Å	a, Å	V, Å	d, Å
Lithium	3.480	21.07	3.014	3.509	21.60	3.039
Sodium	4.225	37.71	3.659	4.291	39.50	3.716
Potassium	5.225	71.32	4.525	5.321	75.33	4.608
Rubidium	5.585	87.10	4.837	5.700	92.59	4.937
Cesium	6.045	110.45	5.235	6.176	117.79	5.348

Table 87-2. Observed Allotropy in the Alkali Metals

	bcc(2)	ccp	hcp	High Pressure
Li	5–300°K, at atmospheric pressure	By cold working below 78°K (partial)	Spontaneous below 72°K, or exposure to white x-rays below 63°K (both partial)	Transition at 23°C and 70 kbar
Na	5–293°K, at atmospheric pressure	Not observed after cooling and cold working at 5°K	Spontaneous below 36°K, or by cold working below 51°K (both partial)	No high pressure studies reported
K	5–293°K, at atmospheric pressure	No transitions on cooling to 5°K or cold working at 5°K		Two high pressure transitions at 77°K, at 280 kbar and 360 kbar
Rb	5–293°K, at atmospheric pressure	No transitions on cooling to 5°K or cold working at 5°K		Two high pressure transitions, see Fig. 37-2
Cs	5–293°K, at atmospheric pressure	No transitions on cooling to 5°K or cold working at 5°K		Three high pressure transitions, see Fig. 55-2

Chapter 4

Group II, The Alkaline Earth Metals

BERYLLIUM

The structure of beryllium was first determined by Meier (1921). She found the substance to be hexagonal close packed, with a = 2.291, c = 3.626 Å (from kX), c/a = 1.583. Subsequent determinations of the lattice constant are presented in Table 4-1.

The question of allotropy in beryllium has been the subject of a long controversy. Lewis (1929), on the basis of measurements of the temperature coefficient of resistance of a sample of 99.5+% purity beryllium, stated, "These results quite strongly indicate that within the temperature range -190°C to 700°C the sample passes through two allotropic transformations"; the first of these was at about -50°C, the second at about 450°C. Jaeger and Zanstra (1933) reported the "β-Be" forms when beryllium is heated to 500 to 700°C, and that the reverse transformation is very slow at room temperature. The "β-Be" was stated to have a hexagonal unit cell with a = 7.1, c = 10.8 Å, c/a = 1.52, containing 60 atoms. On the other hand, Owen and Richards (1936) observed no transformation between room temperature and 550°C, nor did Gordon (1949), who measured lattice constants between 24 and 1013°C. Sidhu and Henry (1950) reported that beryllium which had been heated to 1400 to 1480°C gave a powder pattern which showed 13 lines which were not attributable to the hexagonal close packed form. Nine of these lines could be indexed on a hexagonal cell having a = 6.93, c = 11.35 Å, c/a = 1.638. No structure was proposed. (It is probably a coincidence that the volume of this cell, 472 Å^3, is close to the volume of 471 Å^3 of the cell proposed by Jaeger and Zanstra.) The existence of this high temperature form was questioned by Seybolt, Lukesh, and White (1951), who pointed out that

Table 4-1.   Lattice Constants of Beryllium at $25°C$[a]

a,Å	c, Å	c/a	Reference
2.2726[b]	3.6016[b]	1.5848	Neuberger, 1935b
2.2860[c]	3.5854[c]	1.5684	Owen, Pickup, & Roberts, 1935
2.2858[c]	3.5845[c]	1.5682	Owen & Pickup, 1935
2.2856[d]	3.5809[e]	1.5667	Kossalopov & Trapeznikov, 1936
2.2860[c]	3.5846[c]	1.5681	Owen & Richards, 1936
2.2859	3.5843	1.5680	Gordon, 1949
2.2856	3.5831	1.5677	Kaufman, Gordon, & Lillie, 1950
2.2854	3.5829	1.5677	Kaufman, Gordon, & Lillie, 1950
2.2853	3.5829	1.5678	Kaufman, Gordon, & Lillie, 1950
2.2852	3.5830	1.5679	Kaufman, Gordon, & Lillie, 1950
2.2866	3.5833	1.5671	Schwarzenberger, 1959
2.2851[c]	3.5849[c]	1.5688	Amonenko et al., 1962
2.2859[f]	3.5845[f]	1.5681	Mackay & Hill, 1963
av. 2.2857	3.5839	1.5680	
± 0.0004	±0.0009	± 0.0004	

[a]All values corrected to 25°C with $\alpha_{\parallel} = 10.6 \times 10^{-6}$ and $\alpha_{\perp} = 8.9 \times 10^{-6}$ deg^{-1}.
[b]From kX and 20°C; not included in the average.
[c]From kX and 20°C.
[d]From kX and 18°C.
[e]From kX and 18°C; not included in the average.
[f]From 20.5°C.

seven of the extra lines were attributable to beryllium oxide and three to $AuBe_5$, although it was not explained why a sample of spectroscopically pure beryllium would contain gold.

A transition in beryllium was at last directly observed at high temperature by Martin and Moore (1959), who measured the lattice constants of the hexagonal close packed form from -196 to about 1250°C and found a transition to the body centered cubic structure with two atoms per unit cell at about 1250°C. This form apparently persists up to the melting point of about 1290°C. From the data in the paper it is found that at about 1250°C, a = 2.343, c = 3.659 Å, c/a = 1.562 for the hexagonal form and a = 2.550 Å for the cubic form. The transition was also observed by Amonenko et al. (1961). Values read from their graphs give, for the hexagonal form, a = 2.342,

c = 3.662 Å (from kX), c/a = 1.547, at 1200°C; for the cubic form at 1254°C they found a = 3.5515 Å (from kX).

Marder (1963) observed a transition in beryllium at 25°C and 93 kbar from electrical resistance measurements. The structure of this new phase was not determined.

MAGNESIUM

The structure of magnesium was first determined by Hull (1917), who found that the substance is hexagonal close packed, with a = 3.23, c = 5.13 Å (from kX), c/a = 1.61. The results of later measurements are presented in Table 12-1.

No change in structure was detected in high temperature measurements of the lattice constants from 10 to 597°C (Raynor and Hume-Rothery, 1939; Busk, 1952) or at pressures up to 300 kbar (Clendenen and Drickamer, 1964b). At 300 kbar a 30.5% decrease in the atomic volume was observed, and an increase of c/a to 1.673 (Stager and Drickamer, 1963a).

Finch and Quarrell (1933) found that magnesium deposited in thin films on platinum is also hexagonal close packed, with lattice constants a = 3.26,

Table 12-1.   Lattice Constants of Magnesium at 25°C

a, Å	c, Å	c/a	Reference
3.2091[a]	5.2104[a]	1.6236	Stenzel & Weerts, 1932
3.2095[b]	5.2107[b]	1.6235	Jette & Foote, 1935
3.2091[a]	5.2117[a]	1.6240	Owen, Pickup, & Roberts, 1935
3.2093[b]	5.2103[b]	1.6235	Ievins, Straumanis, & Karlsons, 1938
3.2096[c]	5.2113[c]	1.6237	Raynor & Hume-Rothery, 1939
3.2094[a]	5.2109[a]	1.6236	Raynor, 1940
3.2093[b]	5.2103[b]	1.6235	Hume-Rothery & Boultbee, 1949
3.2088[b]	5.2099[b]	1.6236	Busk, 1950, 1952
3.2094	5.2103	1.6234	Swanson & Tatge, 1953
3.2094	5.2107	1.6236	von Batchelder & Raeuchle, 1957
3.2099	5.2108	1.6234	Hardie & Parkins, 1959
av. 3.2093	5.2107	1.6236	
± 0.0003	± 0.0005	± 0.0002	

[a]From kX, corrected to 25°C with α_a = 27 × 10^{-6} and α_c = 29 × 10^{-6} (values from Hanawalt & Frevel, 1937).
[b]From kX.
[c]From kX, average of values given for 10 and 30°.

$c = 5.36$ Å (from kX), $c/a = 1.64$. In this condition the atomic volume is about 6% larger than that of bulk magnesium.

CALCIUM

The structure of calcium at room temperature was first determined by Hull (1920), who found that it is cubic close packed with $a = 5.57$ Å (from kX), a value in good agreement with the later, more precise measurements included in Table 20-1.

Table 20-1.   Lattice Constant of ccp Calcium at Room Temperature

	a, Å	Reference
	5.576[a]	Klemm & Mika, 1941
	5.582	Smith, Carlson, & Vest, 1956
	5.601	Schottmiller, King, & Kanda, 1958
	5.5884	Smith & Bernstein, 1959b
	5.592	Peterson & Fattore, 1961
av.	5.588	
	± 0.010	

[a]From kX.

   Transitions in calcium at elevated temperature have been a subject of much confusion. Ebert, Hartmann, and Peisker (1933) reported a transition at ca. 450°C to a hexagonal close packed form having $a = 3.99$, $c = 6.53$ Å (from kX), $c/a = 1.638$, while Graf (1933) reported a transition at 450°C to the bcc(2) structure. Somewhat later Graf (1934) reported a second transition, at 300°C, to a complex structure, which transformed to the hexagonal close packed structure at 450°C. Graf examined five different preparations of calcium, including the sample used by Ebert et al. In 1956, Dunsmore, Calvert, and Alexander reported the existence of *four* phases in calcium, but no details were given. Melsert, Tiedma, and Burgers (1956) found that at 250°C there is a transition to the hexagonal close packed structure, which is stable up to 450°C, at which temperature there is a transition to the bcc(2) structure. They gave the following lattice constants:

at	20°C	ccp	$a = 5.56$ Å
at	300°C	hcp	$a = 3.94$, $c = 6.44$ Å, $c/a = 1.635$
at	500°C	bcc(2)	$a = 4.38$ Å.

In a study of the calcium-strontium system, Schottmiller, King, and Kanda (1958) reported the following transitions:

$$ccp \xrightarrow{300°C} complex\ form \xrightarrow{344°C} hcp \xrightarrow{610°C} bcc(2)$$

They referred to their sample as pure calcium, although it contained 0.1% magnesium. The following lattice constants were reported:

at 25°C	ccp	$a = 5.601$ Å
at 415°C	hcp	$a = 4.00, c = 6.50$ Å, $c/a = 1.625$
at 615°C	bcc(2)	$a = 4.488$ Å.

This contradictory situation was clarified by the work of Smith, Carlson, and Vest (1956), who studied three samples of calcium of different purity. Their results are summarized in Table 20-2 and may be expressed schematically as follows:

Sample A: $ccp \xrightarrow{300°C} complex\ form \xrightarrow{375°C} hcp \xrightarrow{500°C} bcc(2)$

Sample B: $ccp \xrightarrow{300°C} complex\ form \xrightarrow{\hspace{1.5cm}450°C\hspace{1.5cm}} bcc(2)$

Sample C: $ccp \xrightarrow{\hspace{2.5cm}464°C\hspace{2.5cm}} bcc(2)$

Table 20-2. Transitions in Calcium Found by Smith, Carlson, and Vest

Sample[a]	Behavior
A	ccp to 300°C, $a = 5.612$ Å at 18°C
B	ccp to 300°C
C	ccp to 464°C, $a = 5.582$ Å at 18°C
A	Complex form, 300° to 375°C
B	Complex form, 300° to 450°C
C	Complex form not observed
A	hcp, 375° to 500°C, $a = 3.97, c = 6.49$ Å, $c/a = 1.635$ at 500°C
B	hcp form not observed
C	hcp form not observed
A	bcc(2), 500°C to melting point, $a = 4.474$ Å at 500°C
B	bcc(2), 450°C to melting point
C	bcc(2), 464°C to melting point, $a = 4.477$ Å at 500°C

[a]Analytical data:

	A	B	C
Mg	0.300%	0.110%	0.010%
N	0.025	0.022	0.011
Fe	0.006	0.010	0.010
Al	0.001	0.001	0.001
Mn	0.004	0.005	0.005

Because sample C was of the highest purity, Smith et al. concluded that *pure* calcium exhibited only the bcc(2) structure at high temperatures. Additional work by Smith and Bernstein (1959a) showed that the presence of small amounts of nitrogen or carbon induces the formation of the complex form, while the presence of a small amount of hydrogen induces the formation of the hexagonal close packed form, and neither oxygen nor boron is responsible for the appearance of either of these forms. The powder pattern of the complex form (which may or may not be a single phase) was the same as that reported by Graf (1934); it has not been indexed.

The occurrence of only one transformation was confirmed by Peterson and Fattore (1961), who stated that the ccp→bcc(2) transition takes place at 448°C. Their sample was 99.94% pure, the largest impurities being 0.03% of magnesium and 0.022% of hydrogen. They also found that an intermediate phase, probably hcp, stable between 320 and 600°C, was produced by hydrogen.

Smith and Bernstein (1959b) measured the lattice constants of the ccp form from 26 to 371°C and of the bcc(2) form at 467 and 523°C; at 26°C they found a = 5.5884 Å for the former, and at 467°C a = 4.480 Å for the latter.

In a high pressure study Jayaraman, Klement, and Kennedy (1963) also observed only two forms. Their phase diagram is presented in Fig. 20-1. However, working at much higher pressures, Stager and Drickamer (1963a) found two transitions at 25°C, one at about 150 kbar, the other at about 300 kbar. Both forms are of unknown structure.

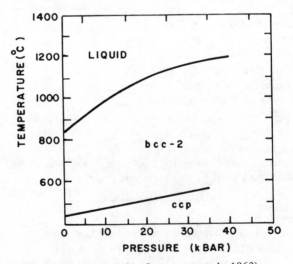

Fig. 20-1.   Phase diagram for calcium (after Jayaraman et al., 1963).

STRONTIUM

The structure of strontium was first determined by Simon and Vohsen (1928b who found that it is cubic close packed with a = 6.04 Å (from kX). Subsequent determinations of the lattice constant with higher precision are included in Table 38-1.

**Table 38-1.   Lattice Constant of Strontium
at Room Temperature**

a, Å	Reference
6.087[a]	King, 1929
6.06[b]	Ebert & Hartmann, 1929
6.061[b]	Klemm & Mika, 1941
6.088[a]	King, 1942
6.085[a]	Sheldon & King, 1953
6.088[a]	Hirst, King, & Kanda, 1956
6.084	Schottmiller, King, and Kanda, 1958
av. 6.086	
± 0.002	

[a]From kX.
[b]From kX, omitted from the average.

Evidence for phase changes at high temperatures was found by Cubicciotti and Thurmond (1949), who observed a pause in the cooling curve at 589°C, and by Rinck (1952), who found transitions at 235 and 540°C from dilatometric, electrical resistance, and thermal measurements. Diffraction data collected with a sample of 99.5% pure strontium yielded the following results (Sheldon and King, 1953; all lattice constants from kX):

$$\text{ccp} \xrightarrow{215 \pm 10°C} \text{hcp} \xrightarrow{605 \pm 10°C} \text{bcc}(2)$$

ccp	hcp	bcc(2)
a = 6.085 Å	a = 4.32 Å	a = 4.85 Å
at 25°C	c = 7.06 Å	at 614°C
	c/a = 1.634	
	at 248°C	

Hirst, King, and Kanda (1956) also observed three forms of strontium. Their results may be summarized as follows:

$$\text{ccp} \xrightarrow{\;213 \pm 3^{\circ}\text{C}\;} \text{hcp} \xrightarrow{\;602 \pm 8^{\circ}\text{C}\;} \text{bcc(2)}$$

$a = 6.076$ Å	$a = 4.28$ Å	$a = 4.87$ Å
at 25°C	$c = 7.05$ Å	at 614°C
	$c/a = 1.647$	
	at 225°C	

These results are in good agreement with those of Schottmiller, King, and Kanda (1958):

$$\text{ccp} \xrightarrow{\;213 \pm 3^{\circ}\text{C}\;} \text{hcp} \xrightarrow{\;621 \pm 6^{\circ}\text{C}\;} \text{bcc(2)}$$

$a = 6.084$ Å	$a = 4.33$ Å	$a = 4.87$ Å
at 25°C	$c = 7.05$ Å	at 614°C
	$c/a = 1.628$	
	at 415°C	

In view of the behavior of impure calcium (see Table 20-2), the existence of a hexagonal close packed form of strontium should be accepted with caution until measurements are carried out on samples of high purity.

McWhan and Jayaraman (1963) studied strontium at room temperature at pressures up to 83 kbar. They found one transition, at 35 kbar, at which point the structure changed from cubic close packed to the bcc(2) structure. From the data collected at 42 kbar given in their paper, a lattice constant of 4.437 ± 0.002 Å may be calculated.

In a study of the high pressure behavior Jayaraman, Klement, and Kennedy (1963) also observed only two forms. Their phase diagram is presented in Fig. 38-1.

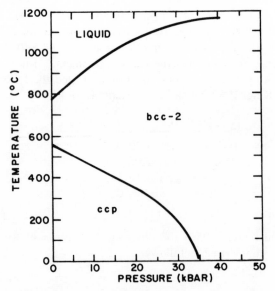

Fig. 38-1.   Phase diagram for strontium (after Jayaraman et al., 1963).

BARIUM

The structure of barium at room temperature was first determined by Clark, King, and Hyde (1928), who found that it has the bcc(2) structure, with $a = 5.05$ Å (from kX). Later determinations are presented in Table 56-1.

Table 56-1.   Lattice Constant of Barium at Room Temperature

a, Å	Reference
5.05[a]	Clark, King, & Hyde, 1929
5.025[b]	King & Clark, 1929
5.02[a]	Ebert & Hartmann, 1929
5.019[b]	Klemm & Mika, 1941
5.025	Swanson, Fuyat, & Ugrinic, 1955
5.023[b]	Hirst, King, & Kanda, 1956
5.0227[b]	King, 1956
av. 5.023 ± 0.002	

[a]From kX, omitted from the average.
[b]From kX.

Barrett (1956c) found no change in structure down to 5°K, even after cold working. The lattice constant at 5°K was reported to be 5.000 Å.

In a high pressure study, Barnett, Bennion, and Hall (1963a) found one transition between atmospheric pressure and 62 kbar (at room temperature). It occurs at 59 kbar, at which point the structure changes to hexagonal close packed. At the highest pressure it was found that $a = 3.901$, $c = 6.154$ Å, $c/a = 1.578$. These correspond to a decrease in volume of 35.9%. The transition pressure was later revised to 53.3 ± 1.2 kbar by Jeffery, Barnett, Vanfleet, and Hall (1966). In studies at much higher pressures by Stager and Drickamer (1963a) a third form of barium, of unknown structure, was discovered. The phase diagram is shown in Fig. 56-1.

RADIUM

The crystal structure of radium was not determined until 1968, by Weigel and Trinkl. It has the bcc(2) structure, with $a = 5.148$ Å at room temperature.

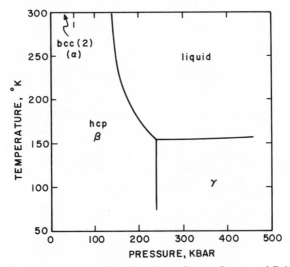

Fig. 56-1.   The phase behavior of barium (according to Stager and Drickamer, 1963a).

SUMMARY

Lattice constants, atomic volumes, and interatomic distances in the alkaline
earth metals at 1 atm and 25°C are presented in Table 88-1. The various
allotropes observed are summarized in Table 88-2. The total number of forms
observed so far is:

beryllium	two
magnesium	one
calcium	four
strontium	two (three?)
barium	three
radium	one

**Table 88-1.   Lattice Constants, Atomic Volumes, and Nearest Neighbor
Distances for the Alkaline Earth Metals at 25°C and 1 atm**

	Structure	a, Å	c, Å	V, Å^3	d, Å
Be	hcp	2.2857	3.5839	8.108	2.286(6), 2.225(6), av. 2.256
Mg	hcp	3.2093	5.2107	23.239	3.209(6), 3.197(6), av. 3.203
Ca	ccp	5.588		43.62	3.951(12)
Sr	ccp	6.086		56.35	4.303(12)
Ba	bcc(2)	5.023		63.36	4.350(8)
Ra	bcc(2)	5.148		68.22	4.458(8)

**Table 88-2. Observed Allotropy in the Alkaline Earth Metals
(atmospheric pressure unless otherwise stated)**

	hcp	ccp	bcc(2)	Other
Be	–196 to 1250°C	Not observed	Above 1250°C	Transition at 93 kbar and 25°C
Mg	10 to 597°C	Not observed	Not observed	No change below 600 kbar
Ca	Below 448°C	Observed only in impure samples	Above 448°C	Transitions at ~ 150 kbar and ~ 300 kbar (both at 25°C)
Sr	Below 213°C	From 213 to 621°C (purity unknown)	Above 621°C	Same bcc(2) phase at 25°C above 35 kbar
Ba	Above 53 kbar at 25°C	Not observed	5°K to 25°C	Transition at 240 kbar and 25°C
Ra	–	–	Room temperature (not studied under other conditions)	–

Chapter 5

Group III

BORON

Boron has the second most complicated structural chemistry among the elements, being surpassed in this regard only by sulfur. There may be as many as 16 distinct allotropes, but the evidence for some is rather flimsy. The situation is further complicated because there is a large number of very boron-rich borides such as $NiB_{\sim 50}$, $YB_{\sim 66}$, and $PuB_{\sim 100}$, and there has been controversy concerning whether or not some preparations described as "boron" were not in fact borides having boron frameworks stable only in the presence of very small amounts of a metal.

Structural data reported so far for elemental boron are summarized in Table 5-1. The forms listed are discussed in the same order in the text. The following convention will be used to designate the forms: a capital letter for the crystal system (R = rhombohedral, T = tetragonal, C = cubic, O = ortho=rhombic, M = monoclinic, H = hexagonal, U = unknown), followed by a number denoting the number of atoms in the unit cell, where known. In cases where this number is large, it must be considered approximate. The references for each of these forms may be found in the sections in which they are discussed.

The structural chemistry of elemental boron is dominated by the regular icosohedron. Three views of this Platonic solid are shown in Fig. 5-1.

Tetragonal-50 Boron

T-50 boron was first characterized by Laubengayer, Hurd, Newkirk, and Hoard (1943), who described the needle crystals from their preparations as tetragonal with $a = 8.95$, $c = 5.07$ Å (from kX) and about 50 atoms per unit

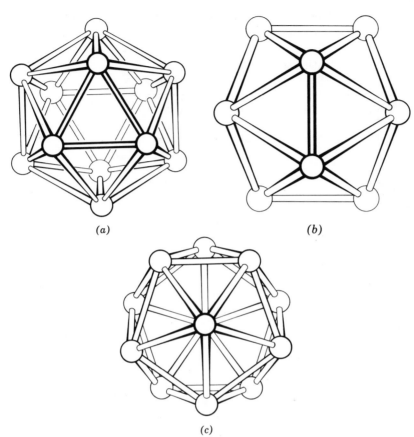

Fig. 5-1.   Three views of the regular icosahedron: (a) face first; (b) edge first; (c) vertex first.

cell. The observed density was 2.310 g cm^{-3} (calc., 2.30). No structure was proposed.

In 1951 Hoard, Geller, and Hughes gave a preliminary description of the same material. They reported a unit cell with a much smaller value of a, 8.73 Å, with c only slightly smaller, 5.03 Å. These values correspond to 50 atoms per unit cell (calculated density, 2.34 g cm^{-3}), which were placed in position 2(b), and six sets of position 8(i) of space group $P\bar{4}n2$–D_{2d}^8. Of the other two possible space groups, $P4/mnm$–D_{4h}^{14} was said to be eliminated, and the same arrangement, with a different grouping of equivalent atoms, was said to be obtainable from $P4nm$–C_{4v}^4. The idealized values of the 18 parameters gave a structure consisting of nearly regular icosahedra, centered at $(\frac{\bar{1}}{4}\frac{1}{4}\frac{1}{4})$, $(\frac{\bar{1}}{4}\frac{3}{4}\frac{3}{4})$, $(\frac{3}{4}\frac{\bar{1}}{4}\frac{3}{4})$, $(\frac{3}{4}\frac{3}{4}\frac{\bar{1}}{4})$, of the 48 boron atoms in positions 8(i), cross linked by

Table 5-1.   Summary of Structural Results on Boron Allotropes

Allotrope[a]	Material Studied[b]	Diffraction Results
T-50	s	Complete structure
R-12	s	Complete structure
R-105	s	Complete structure
T-192	s	Unit cell and space group
T-100	s	Unit cell and space group
M(?)-200	s	Unit cell
O-288	s	Unit cell
C-1708	s	Unit cell
T-78	p	Unit cell
H-90	p	Unit cell
H-108	p	Unit cell
H-154	p	Unit cell
U(i)	p	Not indexed[d]
U(ii)	p	Not indexed[d]
U(iii)	p	Not indexed[d]
U(iv)	p	Not indexed[d]

[a]Crystal system and number of atoms per unit cell; see text.
[b]s = single crystals; p = powder.
[c]Calculated value.
[d]Also not identifiable with any of the other known forms.

Table 5-1. (continued)

Allotrope[a]	Method of Preparation	Observed Density g cm^{-3}	Page
T-50	BBr$_3$ +H$_2$ on W or Ta at ~1300°C	2.310	48
R-12	From B-Pt melts at 830°C; BX$_3$ +H$_2$ at 750–1200°C	2.46	57
R-105	From melted B	2.35	61
T-192	BBr$_3$ +H$_2$ on W or Re at 1550–1280°C; from R-105 B in argon plasma	2.366	78
T-100	BBr$_3$ +H$_2$ on W or Ta at 1300°C	2.27	79
M(?)-200	BBr$_3$ +H$_2$ on W or Ta at ~1300°C	2.21^{c}	79
O-288	Condensed vapor at 1600°C	2.38	80
C-1708	From R-105 B in argon plasma	2.367	80
T-78	BBr$_3$ on W or Mo at 1000–1300°C	2.33	80
H-90	BBr$_3$ arced	2.39^{c}	80
H-108	From R-105 B in argon plasma	2.367	81
H-154	BBr$_3$ on W or Mo at 1500–1600°C	2.33	81
U(i)	BBr$_3$ +H$_2$ on hot W	—	81
U(ii)	BCl$_3$ +H$_2$ on Ti at 1075–1125°C	—	81
U(iii)	BCl$_3$ +H$_2$ on graphite at 1075–1125° C	—	81
U(iv)	Quenched B from 100–150 kbar and 1500–1200°C	2.49	82

the two atoms in position $2(b)$. Because this structure was redetermined later with new data, it will not be described in detail here. Hoard et al. apparently did not notice that the ideal coordinates they gave as being expressible in both space group $P\bar{4}n2$ and P4nm corresponded, in fact, to a structure having space group $P4_2/nnm-D_{4h}^{12}$. (This space group was later found to be the correct one, as described in the next paragraph.)

Hoard, Hughes, and Sands (1958) collected data from crystals of widely dissimilar habit which had been prepared earlier by Laubengayer et al. (1943). The fundamental arrangement of Hoard, Geller, and Hughes (1951) was confirmed. The lattice constants, as can be seen in Table 5-2, were found to vary from specimen to specimen. This effect may arise from a large and variable degree of internal disorder. The space group $P4_2/nnm-D_{4h}^{12}$ was now chosen as correct. A few weak reflections which had been observed in the earlier work of Hoard et al. (1951), used by them to reject $P4_2/nnm$, were shown to be artifacts, that is, Renninger, or double, reflections. Spectrometric data were collected for both a needle and a plate crystal, and the parameters were refined with both sets by Fourier methods. The results are presented in Table 5-3. The atoms lie in the following positions: two B(1) in $2(b)$ at $(00\frac{1}{2})$, $(\frac{1}{2}\frac{1}{2}0)$; eight B(2) and eight B(3) in two sets of $8(m)$, at (xxz), $(\bar{x}\bar{x}z)$, $(\bar{x}xz)$, $(x\bar{x}\bar{z})$, $(\frac{1}{2}+x\ \frac{1}{2}+x\ \frac{1}{2}-z)$, $(\frac{1}{2}-x\ \frac{1}{2}-x\ \frac{1}{2}-z)$, $(\frac{1}{2}-x\ \frac{1}{2}+x\ \frac{1}{2}+z)$, $(\frac{1}{2}+x\ \frac{1}{2}-x\ \frac{1}{2}+z)$; and 16 B(4) and 16 B(5) in two sets of $16(n)$, at (xyz), $(\bar{x}\bar{y}z)$, $(\bar{x}y\bar{z})$, $(x\bar{y}\bar{z})$, $(\bar{y}x\bar{z})$, $(y\bar{x}\bar{z})$, (yxz), $(\bar{y}\bar{x}z)$, $(\frac{1}{2}+x\ \frac{1}{2}+y\ \frac{1}{2}-z)$, $(\frac{1}{2}-x\ \frac{1}{2}-y\ \frac{1}{2}-z)$, $(\frac{1}{2}-x\ \frac{1}{2}+y\ \frac{1}{2}+z)$, $(\frac{1}{2}+x\ \frac{1}{2}-y\ \frac{1}{2}+z)$, $(\frac{1}{2}-y\ \frac{1}{2}+x\ \frac{1}{2}+z)$, $(\frac{1}{2}+y\ \frac{1}{2}-x\ \frac{1}{2}+z)$, $(\frac{1}{2}+y\ \frac{1}{2}+x\ \frac{1}{2}-z)$, $(\frac{1}{2}-y\ \frac{1}{2}-x\ \frac{1}{2}-z)$.

Table 5-2. Lattice Constants of Tetragonal-50 Boron

a, Å	c, Å	Remarks	Reference
8.95[a]	5.07[a]	Needle	Laubengayer, Hurd, Newkirk, & Hoard, 1943
8.73	5.03	Needle	Hoard, Geller, & Hughes, 1951
8.743	5.03	Needle I[b]	
8.740	5.068	Needle III[c]	
—	5.090	Needle IV	Hoard, Hughes, & Sands, 1958
8.771	5.088	Plate[d]	
8.73	5.06	"Typical"	
8.756	5.078	Average of c and d; values adopted here	

[a]From kX.
[b]Used in the original photographic work of Hoard et al., 1951.
[c]Used in spectrometric measurements of Hoard et al., 1958.
[d]Used in spectrometric measurements.

Table 5-3.   Positional Parameters in T-50 Boron, According to Hoard, Hughes, and Sands, 1958

Atom	Parameter	Needle	Plate	Average[a]
B(2)	x	0.2425	0.2490	0.2548
	z	0.5815	0.5865	0.5840
B(3)	x	0.1195	0.1245	0.1220
	z	0.3780	0.3870	0.3825
B(4)	x	0.2272	0.2265	0.2269
	y	0.0805	0.0825	0.0815
	z	0.0865	0.0895	0.0880
B(5)	x	0.3253	0.3255	0.3254
	y	0.0883	0.0875	0.0879
	z	0.3985	0.4005	0.3995

[a]Values adopted here.

Although there are differences between the needle and the plate that are probably significant, both in the lattice constants and positional parameters, the averages are adopted here. The differences are partly due to irregular occupancy of natural holes in the 50-atom framework. Hoard et al. (1958) obtained evidence which showed partial filling of such holes by extra boron atoms, but in variable concentrations between specimens. The average parameters used here, then, correspond to a disordered structure which is the average of two of many possible disordered structures. In this average there are, roughly, $\frac{1}{16}$ B atom in position 2(a) at (000), $(\frac{1}{2}\frac{1}{2}\frac{1}{2})$, $\frac{3}{32}$ B atom in position 4(c) at $(0\frac{1}{2}0)$, $(\frac{1}{2}00)$, $(0\frac{1}{2}\frac{1}{2})$, $(\frac{1}{2}0\frac{1}{2})$, and $\frac{1}{16}$ B atom in position 4(g) at $\pm(00z)$ $(\frac{1}{2}\ \frac{1}{2}\ \frac{1}{2}\ +z)$ with $z \leqslant 0.183$. These disordered atoms are not considered below in the discussion of the interatomic distances in the main framework.

The fifty atoms in the unit cell consist of four nearly regular icosahedra plus two more atoms (type 1), those in position 2(b), needed to complete the framework. Each atom in an icosahedron forms five intraicosahedral and one extraicosahedral bond, while each of the two atoms of type 1 form four bonds to different icosahedra. The bond distances for the five different kinds of boron atom are presented in Table 5-4, and those within an icosahedron are shown in Fig. 5-2. The icosahedra lie on points having crystal symmetry $2/m$–C_{2h}, but the actual symmetry may well be regular icosahedral. It is possible, on the other hand, that some of the differences between the individual bond distances of Table 5-4 are truly significant, but there does not appear to be a ready explanation for the fact that whereas the four different kinds of atoms in an icosahedron form five intraicosahedral bonds of equal average length, the extraicosahedral bonds which these atoms form range from 1.624 to 1.847 Å.

Table 5-4.   Bond Distances in Tetragonal-50 Boron

From Atom Type	To Atom Type	Distance, Å	Number	Average
Bonds within an icosahedron				
2	4	1.763	2 ⎫	
	5	1.810	2 ⎬	1.798 ± 0.035
	3	1.843	1 ⎭	
3	4	1.790	2 ⎫	
	5	1.808	2 ⎬	1.808 ± 0.022
	2	1.843	1 ⎭	
4	2	1.763	1 ⎫	
	3	1.790	1 ⎪	
	4	1.800	1 ⎬	1.794 ± 0.019
	5	1.803	1 ⎪	
	5	1.816	1 ⎭	
5	4	1.803	1 ⎫	
	3	1.808	1 ⎪	
	2	1.810	1 ⎬	1.819 ± 0.023
	4	1.816	1 ⎪	
	5	1.860	1 ⎭	
			av.	1.806 ± 0.025
Extraicosahedral bonds				
1	3	1.624	4	
2	2	1.689	1	
3	1	1.624	1	
4	4	1.684	1	
5	5	1.847	1	

The structure is viewed in projection down the c-axis in Fig. 5-3, and the contents of one and one-half unit cells are viewed along the b-axis in Fig. 5-4. The observed axial ratio of $c/a = 0.5799$ is very nearly equal to $1/\sqrt{3} = 0.5773$, the value expected if the icosahedra in the x-z plane are bonded to form close packed layers, as in Fig. 5-5. Distortions from this ideal structure result from the fact that icosahedra are not spheres, but only fairly close approximations thereof.

According to Hughes (1970) the framework in T-50 boron is based on a metallic boride structure $M_x B_{25}$, which is probably stable over the entire range $0 \leqslant x \leqslant 1$. Two examples are BeB_{12} (Becher, 1960) and NiB_{25} (Decker and Kasper, 1960).

Uno (1958) prepared boron by use of the method of Laubengayer et al. (1943) and presented a powder pattern which, according to Hoard and Newkirk (1960), indicated that boron T-50 and R-12 were both present.

Following the discovery of T-192 boron the designation of α-tetragonal boron has sometimes been applied to T-50 boron.

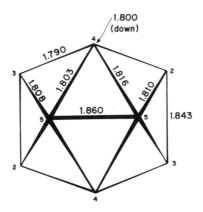

Fig. 5-2.   Bond distances within the icosahedra of tetragonal-50 boron.

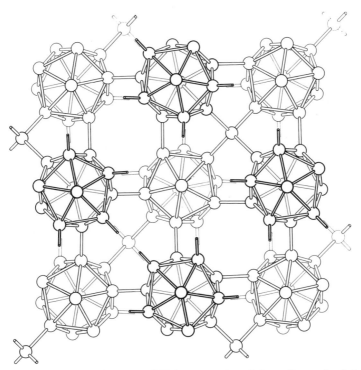

Fig. 5-3.   The structure of tetragonal-50 boron, projected down the *c*-axis. Intericosahedral bonds formed up and down by the atoms at the tops and bottoms of the icosahedra are not shown.

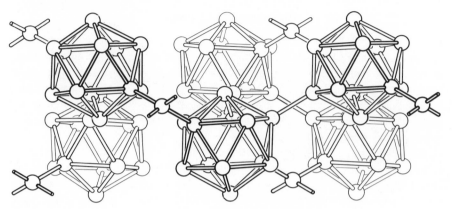

Fig. 5-4.   The structure of tetragonal-50 boron, viewed along the b-axis.

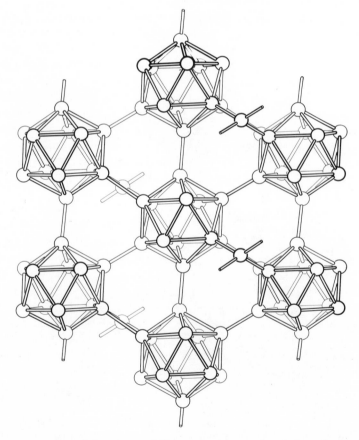

Fig. 5-5.   A portion of the tetragonal-50 boron structure, viewed along the b-axis.

56

Rhombohedral-12 Boron

In 1958 McCarty, Kasper, Horn, Decker, and Newkirk announced in a brief communication the discovery of a hitherto unreported crystalline modification of boron. It had been obtained by pyrolitic decomposition of BI_3 on tantalum, tungsten, or boron nitride surfaces heated to 800 to 1000°C. The crystals are a beautiful, clear red color. Details of the experimental method of preparation were published later by McCarty and Carpenter (1960), and the refined structure was described by Decker and Kasper (1959).

It is interesting that R-12 boron had previously been synthesized, but unrecognized as such, by Robertson (1943) and by Uno (1958) according to Hoard and Newkirk (1960). Somewhat different preparative methods were reported by Horn (1959b), who found it in boron-platinum melts at 830°C, and by Robb and Landau (1959), who decomposed boron hydrides and BI_3 on surfaces heated to 800 to 1100°C.

The single crystal study of Decker and Kasper was based on complete data accessible with MoK_α radiation. A total of 112 reflections was observed. The crystals are rhombohedral with $a = 5.057$ Å, $\alpha = 58.06°$; the corresponding hexagonal axes are $a = 4.908$, $c = 12.567$ Å. The observed density of 2.46 g cm^{-3} gives 12.0 as the number of boron atoms per (rhombohedral) unit cell. Of the three possible space groups consistent with the intensity data, R32–D_3^7, R3m–C_{3v}^5, and R$\bar{3}$m–D_{3d}^5, only the centric R$\bar{3}$m was considered, and it was found to meet all demands of the data. The 12 atoms were assigned to two sets of position 6h, $\pm(xxz)$, (xzx), (zxx); the corresponding equivalent positions in the hexagonal cell are position 18h at (000), $(\frac{1}{3}\frac{2}{3}\frac{2}{3})$, $(\frac{2}{3}\frac{1}{3}\frac{1}{3}) \pm (x\bar{x}z)$, $(x\,2x\,z)$, $(\overline{2x}\,\bar{x}\,z)$. The final parameters, as obtained from Fourier analyses, are presented in Table 5-5. These give an R value of 11.8%.

The structure may be described as composed of units of nearly regular icosahedra of boron atoms in a slightly deformed cubic close packed arrangement: perfect cubic close packing would correspond to a rhombohedral angle α of 60°. One of the close packed layers perpendicular to the c-axis is shown in Fig. 5-6. Each B(2) atom, in addition to forming five bonds within an icosahedron, forms two bonds to two neighboring icosahedra in the same layer. Bonds between the layers are formed by the B(1) atoms, one per atom. An icosahedron in the second layer thus forms three bonds to different icosahedra in the first layer, as shown in Fig. 5-7. The second type of close packed layer is shown in Fig. 5-8. It is defined by the plane through icosahedra centered at (000), (100), and $(\frac{2}{3}\frac{1}{3}\frac{1}{3})$. Figure 5-9 indicates how these layers stack together in this direction. Two other layers of this type occur in the crystal: they are defined by planes through (000), (110), $(\frac{2}{3}\frac{1}{3}\frac{1}{3})$ and through (100), (110), $(\frac{2}{3}\frac{1}{3}\frac{1}{3})$. In true cubic close packing all four layers are equivalent.

The individual bond distances are presented in Table 5-6 and shown in Fig. 5-10.

Table 5-5.   Positional Parameters in R-12
Boron (Decker & Kasper, 1959)

Atom	Rhombohedral cell	Hexagonal cell[a]
B(1)		
x	0.0104	0.1177
z	−0.3427	−0.1073
B(2)		
x	0.2206	0.1961
z	−0.3677	0.0245

[a]The two sets of parameters are related by the equations $x_R = x_H + z_H$ and $z_R = z_H - 2x_H$.

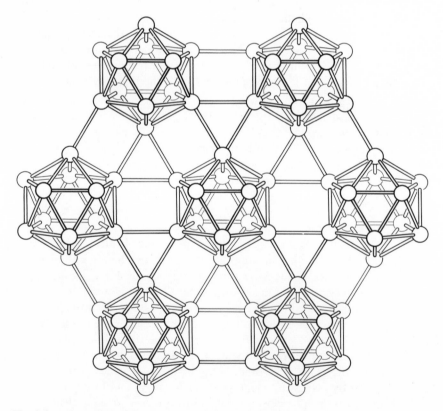

Fig. 5-6.   A portion of the structure of rhombohedral-12 boron; one layer viewed along the hexagonal c-axis.

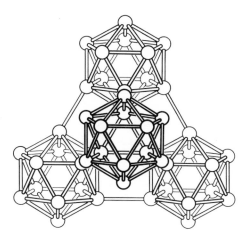

Fig. 5-7.   Part of the layer of Fig. 5-6, plus an additional icosahedrom from the second layer.

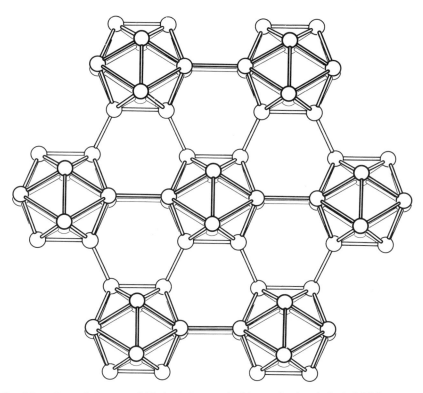

Fig. 5-8.   Part of the second kind of close packed layers in rhombohedral-12 boron.

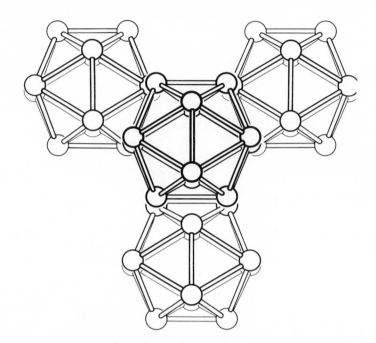

Fig. 5-9. Part of the layer of Fig. 5-8, plus an additional icosahedron in the next layer up.

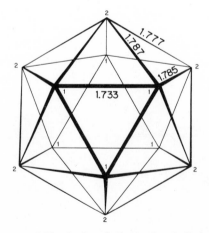

Fig. 5-10. Bond distances within the icosahedra in rhombohedral-12 boron.

Table 5-6.   Bond Distances in Rhombohedral-12 Boron

From Atom Type	To Atom Type	Distance, Å	Number	Average
Bonds within an icosahedron				
1	1	1.733	2	
	2	1.785	1	1.765 ± 0.029
	2	1.787	2	
2	2	1.777	2	
	1	1.785	1	1.783 ± 0.004
	1	1.787	2	
				av. 1.774 ± 0.030
Bonds between icosahedra				
1	1	1.709	1	
2	2	2.021	2	

Rhombohedral-105 Boron

The space group and unit cell of R-105 boron (sometimes termed β-rhombohedral, or complex rhombohedral boron) were first determined by Sands and Hoard (1957), who reported a rhombohedral unit cell with $a = 10.12$ Å, $\alpha = 65°28'$, accurate to 0.2%; the corresponding hexagonal lattice constants are $a = 10.95$, $c = 23.73$ Å. The observed density of 2.35 ± 0.01 g cm^{-3} corresponds to 107.5 ± 0.5 atoms per rhombohedral unit cell. Single crystals had been obtained by allowing melted boron to crystallize in a helium atmosphere. (The melting point of boron is approximately 2250°C.) No structure was proposed.

This variety had been obtained before but not recognized as such, as pointed out by Hoard and Newkirk (1960), who compared the powder pattern as calculated from single crystal intensity data with various previously published powder patterns. Thus the preparation of Newkirk (1940), who reduced BBr$_3$ with hydrogen on tungsten in the range 1550 to 1280°C, was said to be a mixture of R-105 and T-192 borons; the pattern which Godfrey and Warren (1950) obtained from a sample prepared by an unspecified method was essentially that expected for R-105 boron; and the eight samples of Lagrenaudie (1953), six of which had been obtained from the melt and two by arcing BBr$_3$, gave strong indications that each was largely R-105 boron (Lagrenaudie's own interpretation of these patterns was stated to be "wholly unacceptable".) Finally, the boron sample obtained by Stern and Lynds (1958) by reduction of BCl$_3$ with hydrogen on titanium at 1100 to 1200°C gave a powder pattern in excellent agreement with that expected for R-105 boron. Hoard and Newkirk found it surprising that R-105 boron was formed at such a low temperature.

It was pointed out shortly after by Parthé and Norton (1958, 1960) that the pattern of one form of "AlB_{12}" indexed as monoclinic by Halla and Weil (1939) could be explained on a simpler rhombohedral cell having $a = 10.16$ Å (from kX) and $\alpha = 65°$. Parthé and Norton obtained the same pattern from 99% pure boron which had been melted under argon. Kohn, Katz, and Giardini (1958) reached the identical conclusion that the AlB_{12} of Halla and Weil was in fact the rhombohedral allotrope of boron described by Sands and Hoard. It is rather ironic that this preparation, originally described as AlB_{12}, proved to be elemental boron—after many years other preparations which had been described as "pure crystalline boron" proved to be AlB_{12}! (See Laubengayer et al., 1943.)

It has since been well established that boron crystals obtained from the melt are exclusively the R-105 allotrope (Horn, 1959a; Tavadze, Bairamashvili, Tsagareishvili, Tsomaya, and Zoidze, 1965). These crystals are black, with a metallic luster (Parthé, 1971).

An approximate structure for this modification was described very briefly by Kolakowski (1962), who reported lattice constants of $a = 10.96$, $c = 23.78$ Å. The structure was based on space group $R\bar{3}m$–D_{3d}^5, with atoms in four sets of position $12i$ and ten sets of position $6h$, the rhombohedral unit cell thus containing 108 atoms. Several cycles of least squares refinement reduced the R value to only 33%; it thus appears likely that although this structure may contain some features which are approximately correct, there may be others which are grossly incorrect, and it will not, accordingly, be examined in detail here.

The following year Hughes solved the structure by the stochastic method (Hughes, Kennard, Sullenger, Weakliem, Sands, and Hoard, 1963). A brief description was given; revised lattice constants are $a = 10.145 \pm 0.015$ Å, $\alpha = 65°17 \pm 8'$, and the unit cell was now described as containing 105 atoms; the calculated density is thus 2.29 ± 0.01 g cm^{-3}. In this preliminary report no parameters were presented, and the R value for all reflections out to $\sin\theta/\lambda < 0.71$ was given as 15%.

Seven years later the results of two independent refinements of this structure appeared almost simultaneously (Hoard, Sullenger, Kennard, and Hughes, 1970; Geist, Kloss, and Follner, 1970).* Some details of these two investigations are presented in Table 5-7. Except for the partial occupancy of two of the sets of position $6h$ found by Hoard et al., as opposed to the full occupancy of one of these sites and zero occupancy of the other found by Geist et al., the two determinations are in essential agreement. This anomaly may possibly be due to the somewhat lower purity of the sample used by Hoard

*The paper of Hoard et al. was received August 15, 1969, and published January 1970, while that of Geist et al. was received September 6, 1968, in revised form July 25, 1969, and published November 1970.

Table 5-7.   Two Independent Refinements of the Structure of R-105 Boron

	Hoard et al.	Geist et al.
Sample purity	99.4–99.5%	99.999%
Radiation	MoKα, CuKα	AgKα
Number of reflections	1445, of which the 921 innermost were refined (unobs. F's not used)	501 used in refinement, plus 361 unobserved
Maximum sin θ/λ	1.32 Å^{-1}, but only those up to 0.81 Å^{-1} refined	0.70 Å^{-1}
Final R value	9.8% for refined F's, 12.9% for all F's	10.6% for obs. F's, 51.9% for unobs. F's
Average standard error on parameters		
Positional	0.0034 Å	0.011 Å
Thermal	0.07 Å^2	0.23 Å^2
Space group	R$\bar{3}$m–D$_{3d}^5$[a]	R$\bar{3}$m–D$_{3d}^5$
Atom positions[b]	Four sets of 12i	Four sets of 12i
	Eight sets of 6h	Nine sets of 6h
	One set of 2c	One set of 2c
	One set of 1b	One set of 1b
	One set of 6h, $\frac{2}{3}$ occupied	
	One set of 6h, $\frac{1}{3}$ occupied	

[a]The two other space groups of lower symmetry compatible with the diffraction pattern, R3m–C$_{3v}^5$ and R32–D$_3^7$, were eliminated by the absence of the slightest piezoelectric signal in a capacitance measurement.
[b]See Table 5-9 for the list of equivalent positions.

et al. It is also difficult to understand why the standard errors of Hoard et al. are only one-third, on the average, those of Geist et al. The fewer number of reflections used by the latter by no means can account for this discrepancy. There is excellent agreement between the individual values of the parameters, as shown in Table 5-8. The average difference in the positional parameters is 0.013 Å, and in the thermal parameters (excluding atom 13), 0.22 Å^2; both of these numbers are close to the average standard errors given by Geist et al. and therefore about three times the average standard errors of Hoard et al.

The various determinations of the lattice constants which have been reported are presented in Table 5-10. In the discussion which follows the lattice constants of Hughes et al. (1963) and the parameters of Hoard et al. (1970)

Table 5-8. Parameters in R-105 Boron (Hexagonal Coordinate System[a]); Upper Numbers, Hoard et al.; Lower Numbers, Geist et al.

x	y	z	B	Atom Number
Atoms in position 36i[b]				
0.1730	0.1742	0.1768	0.49	1
0.1712	0.1728	0.1761	0.85	5
0.3189	0.2966	0.1291	0.54	2
0.3183	0.2968	0.1295	0.58	6
0.2607	0.2172	0.4199	0.48	3
0.2621	0.2181	0.4205	0.73	8
0.2351	0.2506	0.3473	0.62	4
0.2342	0.2494	0.3470	0.54	9
Atoms in position 18h				
0.0549		−0.0560	0.55	5
0.0552		−0.0577	0.58	1
0.0867		0.0130	0.61	6
0.0878		0.0130	0.78	2
0.1091		−0.1140	0.47	7
0.1078		−0.1145	0.43	3
0.1703		0.0281	0.62	8
0.1703		0.0280	0.55	7
0.1285		−0.2338	0.57	9
0.1296		−0.2328	0.61	4
0.1022		−0.3020	0.35	10
0.1023		−0.3014	0.96	10
0.0562		0.3267	0.41	11
0.0571		0.3272	0.55	13
0.0894		0.3990	0.35	12
0.0906		0.3989	0.60	12
0.0574		−0.4460	0.78	13[c]
0.0606		−0.4479	2.54	11
0.0546		0.1166	1.94	16[d]
—		—	—	—
Atoms in position 6c				
(0)	(0)	0.3852	0.27	14
(0)	(0)	0.3848	0.61	14
Atoms in position 3b				
(0)	(0)	$(\frac{1}{2})$	1.32	15
(0)	(0)	$(\frac{1}{2})$	0.68	15

[a]Rhombohedral coordinates of Geist et al., converted to hexagonal coordinates by use of the relations $x_H = \frac{1}{3}(2x_R - y_R - z_R)$, $y_H = \frac{1}{3}(x_R + y_R - 2z_R)$, $z_H = \frac{1}{3}(x_R + y_R + z_R)$, and changed to the appropriate equivalent position for comparison purposes.
[b]See Table 5-9 for the list of equivalent positions.
[c]Two-thirds occupancy.
[d]One-third occupancy.

Table 5-9.   Equivalent Positions in $R\bar{3}m-D_{3d}^5$

Position	Rhombohedral	Hexagonal
b	$(\frac{1}{2}\,\frac{1}{2}\,\frac{1}{2})$	$(00\frac{1}{2})$, $(\frac{1}{3}\frac{2}{3}\frac{1}{6})$, $(\frac{2}{3}\frac{1}{3}\frac{5}{6})$
c	$\pm(xxx)$	$(H)^a\pm(00z)$
h	$\pm(xxz)$, (xzx), (zxx)	$(H)\pm(x\bar{x}z)$, $(x\;2x\;z)$, $(\overline{2x}\;\bar{x}\;z)$
i	$\pm(xyz)$, (zyx), (yzx)	$(H)\pm(xyz)$, $(\bar{y}\;x\text{-}y\;z)$, $(y\text{-}x\;\bar{x}\;z)$
	(yxz), (zyx), (xzy)	$(\bar{y}\bar{x}z)$, $(x\;x\text{-}y\;z)$, $(y\text{-}x\;y\;z)$

[a]H stands for hexagonal centering: (000), $(\frac{1}{3}\frac{2}{3}\frac{2}{3})$, $(\frac{2}{3}\frac{1}{3}\frac{1}{3})$.

Table 5-10.   The Lattice Constants of R-105 Boron

Rhombohedral Cell		Hexagonal Cell		
a, Å	α	a, Å	c, Å	Reference
10.12	65°28′	10.95	23.73	Sands & Hoard, 1957
10.16	65°	10.92	23.90	Parthé & Norton, 1958, 1960
10.14	65°25′	10.96	23.78	Kolakowski, 1962
10.145	65°17′	10.944	23.81	Hughes et al., 1963
10.17	65°12′	10.96	23.89	Geist et al., 1970

were used in all calculations, since these values are the most precise ones available.

The structure consists of an enormously complicated framework of boron atoms which contains, however, some simple and beautiful features which would have made Kepler leap with joy. Because this framework is three dimensional there is more than one way of describing its components. In identifying the atoms the numbering scheme of Hoard et al. is used, and the description is based on the hexagonal unit cell.

Atoms 5 and 6 form nearly regular icosahedra centered at (000), $(\frac{1}{3}\frac{2}{3}\frac{2}{3})$, and $(\frac{2}{3}\frac{1}{3}\frac{1}{3})$. These icosahedra, termed type A, have crystal symmetry $\bar{3}m-D_{3d}$. The four different intraicosahedral bond lengths, which average 1.767 ± 0.017 Å, are shown in Fig. 5-11. Each atom also forms one extra-icosahedral bond, six with length 1.624 Å (6-8) and six with length 1.722 Å (5-7). The disposition of these is discussed below.

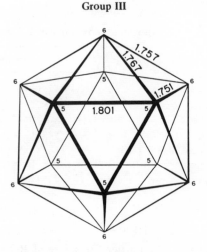

Fig. 5-11.   Bond lengths in the icosahedra of type A of rhombohedral-105 boron.

Atoms 1, 2, 7, and 9 also form nearly regular icosahedra, centered at (000), $(\frac{1}{3}\frac{2}{3}\frac{2}{3})$, $(\frac{2}{3}\frac{1}{3}\frac{1}{3}) + (\frac{1}{2}0\frac{1}{2})$, $(0\frac{1}{2}\frac{1}{2})$, $(\frac{1}{2}\frac{1}{2}\frac{1}{2})$. These icosahedra, termed type B, have crystal symmetry $2/m$–C_{2h}. The nine different intraicosahedral bond lengths, which average 1.850 ± 0.028 Å, are shown in Fig. 5-12. Each atom also forms one extraicosahedral bond; the individual lengths are 1–1, 1.880 Å; 2–3, 1.715 Å; 7–5, 1.722 Å; and 9–10, 1.700 Å. The disposition of these is discussed below.

Atoms 3, 4, 8, 10, 11, 12, 13, and 14 form icosahedra, termed type C, which are somewhat more distorted than the icosahedra of types A and B. The 17 different intraicosahedral bond lengths are shown in Fig. 5-13; the average

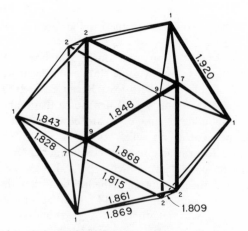

Fig. 5-12.   Bond lengths in the icosahedra of type B of rhombohedral-105 boron.

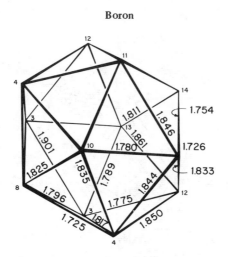

Fig. 5-13.   Bond lengths in the icosahedra of type C of rhombohedral-105 boron.

is 1.805 ± 0.046 Å. Unlike the icosahedra of types A and B, which form only
radially directed extraicosahedral bonds, the icosahedra of type C form trimers
in which atoms 11 and 12 are common to two icosahedra and atom 14 common
to all three. A view of one of these trimeric groups, C_3, which consists of 28
atoms, is shown in Fig. 5-14. These trimers have crystal symmetry $3m-C_{3v}$
and are centered at approximately $\pm(00\frac{1}{3})$, $(\frac{1}{3}\frac{2}{3}0)$, $(\frac{1}{3}\frac{2}{3}\frac{1}{3})$. Extraicosahedral
bonds are: 3–2, 1.715 Å; 4–4, 1.705 Å; 8–6, 1.624 Å; 10–1, 1.699 Å; and
13–15, 1.684 Å. Atom 13 forms, in addition, two intratrimer bonds to
equivalent atoms 13 of length 1.885 Å.

Atoms 15, at $(00\frac{1}{2})$, $(\frac{1}{3}\frac{2}{3}\frac{1}{6})$, $(\frac{2}{3}\frac{1}{3}\frac{5}{6})$, are not part of any of the icosahedral
systems but form connective links between two C_3 groups, thru atoms 13.
Each atom 15 forms six bonds, at 1.684 Å, three above and three below, directed toward the vertices of a trigonal antiprism. Hoard et al., while assigning a $\frac{2}{3}$ occupancy factor to atoms 13, noted the shortness of this bond distance and remarked that it seems unlikely that a boron atom would form six
short octahedral bonds. In their scheme then, each atom 15 forms, on the
average, four bonds at 1.684 Å. This distance is longer than that observed for
the four-coordinated boron atoms in tetragonal-50 boron, where the bonds are
of length 1.624 Å. Geist et al., on the other hand, did not observe partial
occupancy of the atom 13 sites. They did find, however, a much larger thermal parameter for these atoms (see Table 5-8). The possibility that this effect
is due to positional disorder and/or thermal anisotropy rather than occupa-
tional disorder was not tested by either group of investigators.

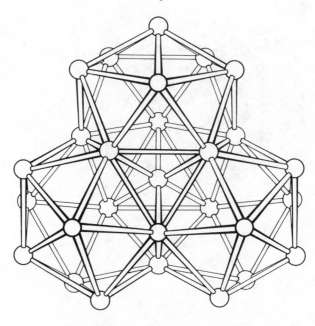

Fig. 5-14.   The condensation of three icosahedra to give the C_3 trimeric groups in rhombohedral-105 boron. The central atom is of type 14 (cf. Fig. 5-13).

In the above description the hexagonal unit cell is seen to contain three icosahedra of type A, nine icosahedra of type B, six trimeric C_3 units, and three atoms of type 15, for a total of 315 (=3×105) atoms.

Individual bond distances for the various atoms are presented in Table 5-11. Atoms 5, 6, 1, 2, 7, 9, 3, 4, 8, and 10 all form six bonds, five within an ico-sahedron plus one radially directed to adjacent icosahedra. Atoms 11 and 12, which are shared by two icosahedra of the C_3 units, are eight-coordinated. Atom 13 is also eight-coordinated, but in a different way: six of the bonds are similar in nature to those formed by atoms 5,6, . . ., with two more to the other two icosahedra of the C_3 unit. Atom 14, which is shared by all three icosahedra in a C_3 unit, is nine-coordinated; six of these bonds are di-rected toward the vertices of a trigonal antiprism, with three more out from alternate equatorial faces. The coordination of atom 14, viewed down the c-axis, is shown in Fig. 5-15. It is remarkable that this projection is virtually identical to the vertex figure of the semi-regular four-dimensional polytope $s\{3,4,3\}$ (Coxeter, 1948). A second view of this group of atoms, perpendicular to the first, is shown in Fig. 5-16. These ten atoms comprise the "B_{10} sub-units" in the discussion of Hoard et al.

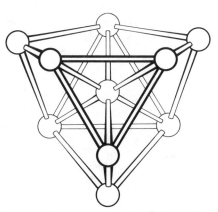

Fig. 5-15. The coordination of atoms of type 14 in rhombohedral-105 boron, viewed in the *c*-direction.

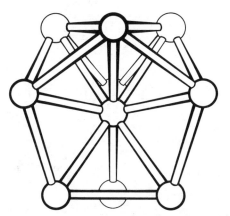

Fig. 5-16. A second view of the coordination of atoms of type 14 in rhombohedral-105 boron, viewed in a direction normal to Fig. 5-15.

Icosahedra of type A are bonded to six icosahedra of type B, three above and three below, as shown schematically in Fig. 5-17. The equator of the A type is girdled by six C_3 groups, as shown schematically in Fig. 5-18. Each A type is thus bonded to 12 atoms, each in a different icosahedron. Each of these 12 atoms is bonded to a pentagon of atoms in its icosahedron, and each of these pentagons, in turn, is bonded to five adjacent pentagons in five different neighboring icosahedra, the interpentagonal bonds forming nearly regular hexagons. The resulting figure which surrounds an A type is thus a truncated icosahedron, one of the Archimedean semi-regular solids (Fig. 5-19), with added atoms dropped inward from each of the pentagonal faces. The grouping consisting of the central A icosahedron, its 12 immediate legands,

Table 5-11.   Bond Distances in Rhombohedral-105 Boron

From Atom Type	To Atom Type	Distance, Å	Number	Average
Bonds within an icosahedron				
5[a]	6	1.751	1	1.777 ± 0.023
	6	1.767	2	
	5	1.801	2	
6[a]	5	1.751	1	1.760 ± 0.007
	5	1.767	2	
	6	1.757	2	
1[b]	1	1.920	1	1.864 ± 0.035
	2	1.861	1	
	2	1.869	1	
	7	1.828	1	
	9	1.843	1	
2[b]	1	1.861	1	1.844 ± 0.030
	1	1.869	1	
	2	1.809	1	
	7	1.815	1	
	9	1.868	1	
7[b]	1	1.828	2	1.827 ± 0.014
	2	1.815	2	
	9	1.848	1	
9[b]	1	1.843	2	1.854 ± 0.013
	2	1.868	2	
	7	1.848	1	
3[c]	3	1.901	1	1.816 ± 0.016
	4	1.817	1	
	8	1.796	1	
	12	1.775	1	
	13	1.789	1	
4[c]	3	1.817	1	1.814 ± 0.051
	8	1.725	1	
	10	1.835	1	
	11	1.844	1	
	12	1.850	1	
8[c]	3	1.796	2	1.773 ± 0.046
	4	1.725	2	
	10	1.825	1	
10[c]	4	1.835	2	1.811 ± 0.029
	8	1.825	1	
	11	1.780	2	

Table 5-11. Bond Distances in Rhombohedral-105 Boron (Continued)

From Atom Type	To Atom Type	Distance, Å	Number	Average
Bonds within an icosahedron				
11^c	4	1.844	1	
	10	1.780	1	
	11	1.846	1	1.811 ± 0.042
	12	1.833	1	
	14	1.754	1	
12^c	3	1.775	1	
	4	1.850	1	
	11	1.833	1	1.809 ± 0.057
	13	1.861	1	
	14	1.726	1	
13^c	3	1.789	2	
	12	1.861	2	1.822 ± 0.037
	14	1.811	1	
14^c	11	1.754	2	
	12	1.726	2	1.754 ± 0.035
	13	1.811	1	
Extraicosahedral bonds				
5	7	1.722	1	
6	8	1.624	1	
1	1	1.880	1	
2	3	1.715	1	
7	5	1.722	1	
9	10	1.700	1	
3	2	1.715	1	
4	4	1.705	1	
8	6	1.624	1	
10	9	1.699	1	
11	4^d	1.844	1	
11	10^d	1.780	1	
11	11^d	1.846	1	
12	3^d	1.775	1	
12	4^d	1.850	1	
12	13^d	1.861	1	
13	13^d	1.885	2	
13	15	1.684	1	
14	11^d	1.754	1	
14	12^d	1.726	1	
14	13^d	1.811	2	
15	13	1.684	6	

[a]In icosahedron A. [c]In icosahedron C.
[b]In icosahedron B. [d]In another part of a C_3 group.

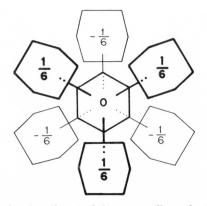

Fig. 5-17.   Schematic drawing of part of the surroundings of a type A icosahedron (center) in rhombohedral-105 boron, showing the six ligands of type B (outer). The relative positions of the B icosahedra should be compared with the positions of the six innermost atoms of Fig. 5-1*a*. Numbers give heights as fractions along the *z*-axis.

and the 60 atoms at the vertices of the truncated icosahedron comprise the "B_{84} unit" of Hoard et al. Part of one of these units is shown in Fig. 5-20.

The three-dimensional network is generated by completing the half-icosahedra of Fig. 5-20 and continuing to add those groups to which these are bonded.

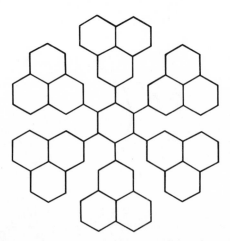

Fig. 5-18.   Schematic drawing of part of the surroundings of a type A icosahedron (center) in rhombohedral-105 boron, showing the six ligands of type C (outer). All icosohedra centered roughly in the same *x-y* plane. The relative positions of the C icosahedra should be compared with the positions of the six outermost atoms of Fig. 5-1*a*.

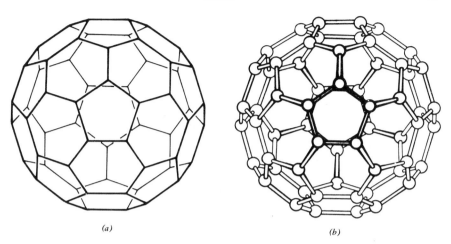

Fig. 5-19. The truncated icosahedron, pentagon first, (a) as a polyhedron and (b) as a complex of bonded atoms.

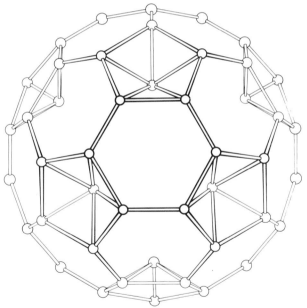

Fig. 5-20. Part of the B_{84} unit in rhombohedral-105 boron. Only the upper part of truncated icosahedron is shown, and the central icosahedron, which is bonded to the atoms in from the pentagonal faces, has been omitted.

Each icosahedron of type B is also bonded to 12 adjacent icosahedra, two of type A, four of type B, and eight of type C. The disposition of the individual atoms in these icosahedra is shown in Fig. 5-21.

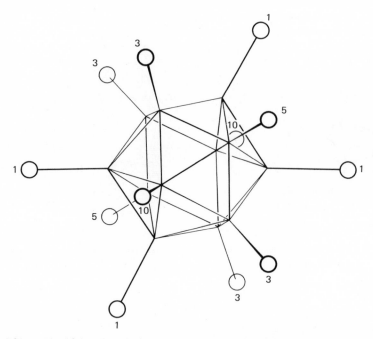

Fig. 5-21.   The 12 intericosahedral ligands of a type B icosahedron (cf. Fig. 5-12).

The 156-atom grouping of a central icosahedron bonded to 12 surrounding icosahedra occurs as a discrete packing unit in the YB_{66} structure (Richards and Kasper, 1969); the situation is more complicated in the R-105 boron structure where there are two different kinds of the 156-atom complexes, neither of which is a discrete group, and which differ, moreover, in the way the 12 outer icosahedra are linked to each other.

An alternate way of discussing the three-dimensional network may be done

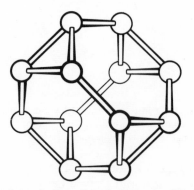

Fig. 5-22.   The truncated tetrahedron, viewed along one of its fourfold inversion axes.

in terms of another semiregular solid, the truncated tetrahedron (Fig. 5-22). If 20 slightly distorted truncated tetrahedra are condensed so that each shares three of its hexagonal faces with adjoining truncated tetrahedra, then in the resulting complex six of the 12 vertices of each truncated tetrahedron are shared among five tetrahedra and six vertices between two tetrahedra. The total number of atoms in a complex obtained by placing an atom at each vertex is thus $(6 \times \frac{1}{5} + 6 \times \frac{1}{2}) \times 20 = 84$, and the complex is identical to the B_{84} units of Fig. 5-20. The central icosahedron is formed by the 20 innermost triangular faces of the truncated tetrahedra. The formation of one kind of truncated tetrahedron in the R-105 boron structure is presented in Fig. 5-23, which shows parts of one type A and three type B icosahedra; the formation of a second kind is presented in Fig. 5-24, which shows parts of one type A, two type B, and three type C icosahedra. The beginning of the way in which these truncated tetrahedra merge to give the B_{84} units (and the resulting truncated icosahedron) is seen in Fig. 5-25. A third kind of truncated tetrahedron is formed of parts of one type A, one type B, and two type C icosahedra, as in Fig. 5-26. These share one hexagonal face with each other and merge with other two kinds, as in Fig. 5-27.

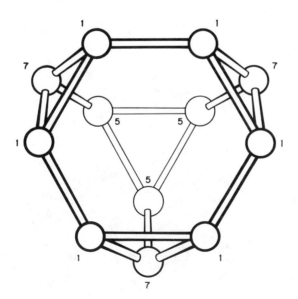

Fig. 5-23. The formation of one kind of truncated tetrahedron in rhombohedral-105 boron by part of a type A icosahedron (atoms 5; cf. Fig. 5-11) and part of three type B icosahedra (atoms 1 and 7; cf. Fig. 5-12).

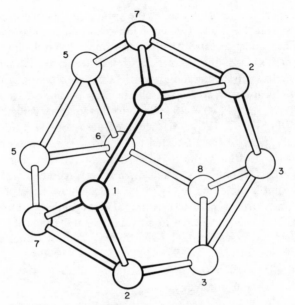

Fig. 5-24. The formation of the second kind of truncated tetrahedron in rhombo-hedral-105 boron by part of a type A icosahedron (atoms 5 and 6; cf. Fig. 5-11), parts of two type B icosahedra (atoms 1, 2, 7, and 9; cf. Fig. 5-12), and part of a type C ico-sahedron (atoms 3 and 8; cf. Fig. 5-13).

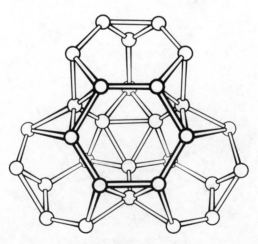

Fig. 5-25. Union of one truncated tetrahedron of the first kind (central atoms; cf. Fig. 5-23) with three truncated tetrahedra of the second kind (outer atoms; cf. Fig. 5-24).

76

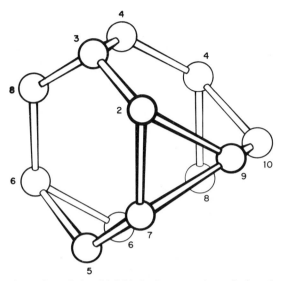

Fig. 5-26. The formation of the third kind of truncated tetrahedron in rhombohedral-105 boron by part of a type A icosahedron (atoms 5 and 6; cf. Fig. 5-11), part of a type B icosahedron (atoms 2, 7, and 9; cf. Fig. 5-12), and parts of two type C icosahedra (atoms 4, 8, and 10; cf. Fig. 5-13).

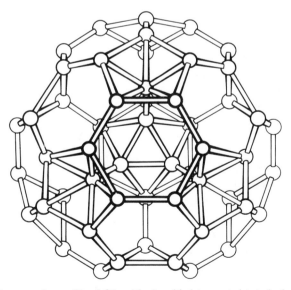

Fig. 5-27. The grouping as Fig. 5-25, with six added truncated tetrahedra of the third kind (cf. Fig. 5-20).

The holes in the centers of the truncated tetrahedra are large enough to accommodate atoms having a diameter of about 2.1 Å, and are thus large enough to accommodate boron atoms, or, for that matter, numerous metallic ions. Hoard et al. found evidence that one-third of the tetrahedra of the kind shown in Fig. 5-23 were occupied by boron atoms (their type 16), and a detailed exploration of all other holes in the structure did not reveal any major degree of occupancy. Geist et al. apparently did not find any such interstitial atoms in their electron density maps.

Tetragonal-192 Boron

A second tetragonal form of boron was first reported by Talley, LaPlaca, and Post (1960). As opposed to T-50 boron, it has sometimes been termed β-tetragonal boron (e.g., by Hoard and Hughes, 1967). T-192 boron was prepared by reduction of BBr_3 by H_2 on incandescent tungsten or rhenium, deposited at about 1270°. It is black when viewed by reflected light, but red by transmitted light. Powder data (35 published spacings, plus additional unlisted lines) were consistent with a tetragonal cell with $a = 10.12$, $c = 14.14$ Å, $c/a = 1.397$. The observed density of 2.364 g cm^{-3} corresponds to 190.6 atoms per unit cell, and Talley et al. speculated that the unit cell contained 16 icosahedra, or 192 atoms. No structure was proposed.

According to Hoard and Newkirk (1960) this form of boron had been previously synthesized by Newkirk (1940), whose preparations, made by reduction of BBr_3 by H_2 on hot tungsten at 1550 to 1280°C, consisted of a mixture of T-192 boron and R-105 boron, as deduced from powder data.

This same form of boron was observed by Sullenger, Phipps, Seabaugh, Hudgens, Sands, and Cantrell (1969), who dropped finely powdered boron R-105 through an argon plasma. Along with three other boron polymorphs, single crystals of T-192 boron were recovered. The observed density was 2.367 ± 0.002 g cm^{-3}. Lattice constants $a = 10.061$, $c = 14.210$ Å, $c/a = 1.412$ were reported, and the space group was given as $P4_1 2_1 2$-D_4^4, and the enantiomorphic $P4_3 2_1 2$-D_4^8. The number of atoms per unit cell was stated to be 190 (calc. 189.5) but, because this is impossible for these two space groups, in which the number of atoms in the unit cell must be a multiple of four, we assume here that the cell contains 192 atoms. Future work may show that occupancy disorder leads to the approximation of the smaller number. The calculated density for 192 atoms per cell is 2.398 g cm^{-3}.

The structure of T-192 boron has not yet been worked out, but, because of a general similarity of both intensities and lattice constants, it must be closely related to two metallic borides, tetragonal AlB_{12}, with $a = 10.30$, $c = 14.33$ Å (from kX), $c/a = 1.391$ (Halla and Weil, 1939), or $a = 10.17$, $c = 14.28$ Å, $c/a = 1.404$ (Yannoni, 1961), and tetragonal BeB_6, with $a =$

10.16, c = 14.28 Å, c/a = 1.406 (Sands, Cline, Zalkin, and Hoenig, 1961). Both of these borides are also of unknown structure.

When T-192 boron is melted and cooled the R-105 modification is formed.

Tetragonal-100 Boron

Hoard (1953) was reported as finding that some weak ($hh\ell$) reflections, observed from specimens of T-50 boron, indicated that the length of the c-axis must be doubled. This apparently would correspond to a unit cell having a = 8.73, c = 10.06 Å. He did not, however, consider that this constituted evidence of a distinct boron allotrope, but that weak reflections could be the result of variable amounts of extra boron atoms in the ideal T-50 boron structure.

This same doubling of the c-axis was later reported by Gorski (1965), who found a = 8.75, c = 10.15 Å for a tetragonal modification of boron. He also stated that this variety had the same space group, $P4_2/nnm$–D_{4h}^{12}, as the T-50 variety. In this case the doubling of the c-axis was based on evidence on rotation photographs of single crystals, but no further details were given.

The classification of T-100 boron as a valid allotrope should be considered unsettled until additional data are obtained.

Monoclinic(?)-200 Boron

Laubengayer et al. (1943) obtained two types of crystals in their preparations, made by reduction of BBr_3 at about 1300°C. The needlelike crystals were later identified with T-50 boron, as mentioned previously. The platelike crystals were reported as requiring doubling of two of the axes, and the values a = 10.15, b = 8.95, c = 17.90 Å (from kX), $\alpha = \beta = \gamma = 90°$. Sketchy optical examination indicated a symmetry not higher than monoclinic. Laubengayer et al. believed that the structures of the two crystalline forms are not fundamentally different, but said the doubling of two of the translations in the needles "cannot be explained at this time".

Hoard et al. (1951) alluded to the T-200 boron of Laubengayer et al. as a distinct modification, and Hoard (1953) gave lattice constants a = 10.0, b = 8.9, c = 17.8 Å, $\beta = 90°$, close to the values given above, for a monoclinic form of boron, but he added that this might not be the primitive cell.

Hoard et al. (1958), on the other hand, examined in detail both the needle and plate crystals, and concluded that both were T-50 boron.

The extra reflections requiring the doubling of two of the axes for some samples of what is ostensibly T-50 boron thus remain unexplained.

Orthorhombic-288 Boron

Gorski (1963) examined single crystals of boron obtained by crystallization from the gaseous state at about 1600°C by Niemyski and Olempska (1962). He found that the small (2 mm) steel grey crystals were orthorhombic, with $a = 8.73$, $b = 11.87$, $c = 20.89$ Å. The observed density of 2.38 g cm^{-3} corresponds to 288 atoms per unit cell (calc. 286.8). This could mean that the cell contents consist of 24 icosahedra.

Cubic-1708 Boron

In the argon plasma experiments of Sullenger et al. (1969) single crystals of a cubic allotrope were isolated. The lattice constant was found to be 23.472 ± 0.008 Å, and observed density, 2.367 ± 0.002 g cm^{-3}; the number of atoms in the unit cell is thus 1705 ± 3. For an unstated reason, this number was chosen as 1708. The space group was given as Pn3n–O_h^2. The number of atoms in the unit cell, $61 \times 7 \times 2 \times 2$, does not appear to be numerologically related to the icosahedron. The calculated powder pattern of C-1708 boron agrees well with the powder pattern observed by Eick (1965) for PuB_{100} but does not agree with that of YB_{66} observed by Richards and Kasper (1969). All three substances have similar lattice constants: for PuB_{100}, $a = 23.43$ Å, and for YB_{66}, $a = 23.440$ Å. Sullenger et al. accordingly concluded that there are two distinctive cubic frameworks for boron with large, nearly identical lattice translations.

Tetragonal-78 Boron

Náray-Szabó and Tobias (1949) reported a powder pattern obtained from "graphitic" boron, prepared by thermal decomposition of BBr_3 at 1000 to 1300°C on W or Mo. This pattern of 25 lines was indexed on a tetragonal unit cell with $a = 8.59$, $c = 8.15$ Å (from kX). The observed density of 2.33 g cm^{-3} corresponds to 78 (calc. 78.0) atoms per unit cell.

Hexagonal-90 Boron

Rollier (1953) obtained a powder pattern of 21 lines from a sample of boron prepared by the arcing of BBr_3. It was indexed on a hexagonal cell with $a = 8.93$, $c = 9.80$ Å. With 90 atoms per unit cell the calculated density is 2.39 g cm^{-3}. According to Hoard and Newkirk (1960) this powder pattern cannot be interpreted in terms of known polymorphs of boron.

Hexagonal-108 Boron

Sullenger et al. (1969) obtained, in their argon plasma experiments, a powder pattern of a form of boron not reconcilable with any published pattern of any form of boron or boride. It was compatible with a hexagonal cell having $a = 9.755$, $c = 10.016$ Å. These values, combined with the observed density of 2.367 g cm^{-3}, yield 108 (calc. 108.8) for the number of atoms in the unit cell.

Hexagonal-154 Boron

Náray-Szabó and Tobias (1949) reported a powder pattern of 30 lines obtained from crystalline boron prepared by thermal decomposition of BBr$_3$ on W or Mo at 1500 to 1600°C. It was indexed on a hexagonal cell with $a = 12.00$, $c = 9.56$ Å (from kX). The observed density of 2.33 g cm^{-3} gives 154 (calc. 154.6) atoms per unit cell. Náray-Szabó and Tobias stated that the unit cell contained 180 atoms, but this value leads to an impossibly high value of 2.71 g cm^{-3} for the density; it appears that they neglected the factor $\sqrt{3}/2$ in calculating the volume of the unit cell.

According to Hoard and Newkirk (1960) most of the lines in this pattern are in fair agreement with the pattern of R-105 boron, but, since the correlation is not fully satisfactory, significant admixture of an unidentified phase is required.

Unknown Boron (i)

Mellor, Cohen, and Underwood (1936) prepared boron by reduction of BBr$_3$ with hydrogen on hot tungsten. They were unable to index the powder pattern (of 22 lines). Hoard and Newkirk (1960), on the basis of that pattern, were unable to establish that any of the known polymorphs of boron was present.

Unknown Borons (ii and iii)

Stern and Lynds (1958) prepared three forms of boron by reduction of BCl$_3$ with hydrogen. The first, deposited on titanium at 1100 to 1200°C, was R-105 boron, according to Hoard and Newkirk (1960), on the basis of the 16-line pattern of Stern and Lynds. Hoard and Newkirk were unable to interpret the powder patterns of the other two preparations on the basis of known boron allotropes. One of these was a 15-line pattern obtained from boron deposited on titanium at 1075 to 1125°C, the other a 12-line pattern obtained from boron deposited on graphite, also at 1075 to 1125°C.

Unknown Boron (iv)

Wentorf (1965) subjected ordinary forms of pure boron to pressures exceeding 100 kbar at temperatures between 1500 and 2000°C, then cooled to 25°C and reduced the pressure to 1 atm. The powder pattern of 21 lines showed some weak lines corresponding to some of the known forms of boron but also showed a strong pattern which could not be attributed to any combination of the known forms of boron. It is not known whether this product is a single crystalline phase or a mixture of new and previously recognized forms of boron

The foregoing complexities may be drastically simplified if the suggestion of Amberger and Ploog (1971) turns out to be correct. They pyrolyzed mixtures of boron tribromide and hydrogen containing "no foreign atoms" on tantalum in the temperature range 900 to 1300°C and observed but three products: (1) glassy boron; (2) R-12 boron; and (3) T-192 boron. They never obtained T-50 boron and concluded that it is not a modification of pure boron but a boron-rich boride, and conjectured that R-105 boron is also a boron. rich boride. The "convolution molecule method" on the diffraction data from T-192 boron showed the presence of nearly regular icosahedra.

If the above results are substantiated, then the number of crystalline allotropes of boron is reduced to two, and explanation of the contradictory results on the other 14 forms described in this chapter will be required.

ALUMINUM

The structure of aluminum was first determined by Hull (1917b), who reported that it is cubic close packed, with a = 4.06 Å (from kX). There have been an unusually large number of precision determinations of this lattice constant. These are included in Table 13-1.

No change in structure was detected down to –196° (Figgins, Jones, and Riley, 1956) nor up to 650° (Wilson, 1942). Finch and Quarrell (1933), however, observed two different forms of aluminum in thin films deposited on platinum. One was cubic close packed, with a lattice constant of 4.02 Å (from kX), the other face centered tetragonal (four atoms per unit cell) with a = 3.91, c = 4.03 Å (from kX), c/a = 1.03.

At 205 kbar and room temperature aluminum is partially transformed to an allotrope having the hexagonal close packed structure, according to Roy and Steward (1969). From the data given in their paper, and assuming an ideal value for the axial ratio, the following quantities for hexagonal close packed aluminum at 205 kbar may be calculated: a = 2.693, c = 4.398 Å, c/a = 1.633.

Structural data for the various forms of aluminum are summarized in Table 13-2.

Table 13-1.

Table 13-1. Lattice Constant of Aluminum at 25°C[a]

a, Å		a, Å	
4.04945[b]	Owen & Iball, 1932	4.04785[f]	Zhmudski, 1949
4.04947[b]	Owen & Yates, 1933a	4.04954	Sully & Hardy, 1949
4.04955[c]	Jette & Foote, 1935	4.04948[g]	Kochanovska, 1949
4.04962	Ieviņš & Straumanis, 1936a	4.04944[h]	Dorn, Pietrowsky, &
4.04959	Ieviņš & Straumanis, 1936a		Tietz, 1950
4.04961	Ieviņš & Straumanis, 1936b	4.04938[c]	Poole & Axon, 1952
4.04928	Miller & DuMond, 1940	4.04953[c]	Černohorsky, 1952
4.04973[d]	Lu & Chang, 1941	4.04940	Swanson & Tatge, 1953
4.04954[d]	van Bergen, 1941	4.04935	Hill & Axon, 1953
4.04959[e]	Wilson, 1942	4.04952[c]	Pearson, 1954
4.04963[d]	Siebel, 1943	4.04960	Smakula & Kalnajs,
4.04950[c]	Axon & Hume-Rothery, 1948		1955
4.04946[b]	Owen, Liu, & Morris, 1948	4.04963	Figgins, Jones & Riley,
4.04965[b]	Goniche & Graf, 1948		1956
4.04973[c]	Ellwood & Silcock, 1948	4.04958[c]	Weyerer, 1956a
4.04958[c]	Hume-Rothery & Boultbee,	4.04954	Straumanis, 1959
	1949	4.04961[i]	Cooper, 1962
4.04963[c]	Straumanis, 1949	4.04934[j]	Otte, Montague, &
			Welch, 1963

av. 4.04953 ± 0.00011

[a]Values not determined at 25°C corrected to that temperature with a thermal expansion coefficient of 23.4×10^{-6} deg^{-1}.
[b]From kX, from 18°C.
[c]From kX.
[d]From kX, from 20°C.
[e]From kX, from 0 and 50°C.
[f]From kX, from 19°C, omitted from the average.
[g]From kX, from 22°C.
[h]From kX, average value from two different samples.
[i]From kX, from 24.8°C.
[j]From 24.2°C.

Table 13-2.   Structural Data for Forms of Aluminum

Structure	Atomic Volume, Å^3	Interatomic Distances, Å
At 25°C and atmospheric pressure		
Cubic close packed	16.602	2.863 (12)
In thin film on platinum		
Cubic close packed	16.22	2.84 (12)
Face centered tetragonal	15.36	2.81 (8), 2.76 (4)
At room temperature and 205 kbar		
Hexagonal close packed	13.81	2.693 (12)

SCANDIUM

The structure of scandium was first studied by Meisel (1939), who reported that it is dimorphous, having both the cubic close packed structure, with a = 4.541 Å (from kX), and the hexagonal close packed structure, with a = 3.31, c = 5.24 Å (from kX), c/a = 1.583. On the other hand, only the hexagonal form was observed by Bommer (1939b), who found a = 3.309, c = 5.256 Å (from kX), c/a = 1.588. The existence of the cubic close packed form was questioned by Klemm (1949), who suggested that it might be scandium nitride. Other determinations of the lattice constants of the hexagonal form have been reported by Spedding, Daane, and Herrmann (1956), Spedding, Daane, Wakefield, and Dennison (1960), Spedding, Hanak, and Daane (1961), and Beaudry and Daane (1964). Spedding and Beaudry (1971), who examined nine high purity group III metals having the hexagonal close packed structure, found that various contaminants, especially hydrogen, lead to high results.

No allotropic change between 20 and 1009°C was observed by Spedding et al. (1961). The various determinations of the lattice constants are presented in Table 21-1.

YTTRIUM

The structure of yttrium was first determined by Quill (1932a), who found, for a 99.5% pure sample at room temperature, that it is hexagonal close packed with a = 3.670, c = 5.826 Å (from kX), c/a = 1.587. Somewhat different lattice constants were later reported by Bommer (1939b), who found a = 3.636, c = 5.762 Å (from kX), c/a = 1.585. Later precision measurements

Table 21-1.   The Lattice Constants of Scandium at
Room Temperature

a, Å	c, Å	c/a	Reference
3.31[a]	5.24[a]	1.58	Meisel, 1939
3.309[a]	5.256[a]	1.588	Bommer, 1939b
3.3090	5.2733[b]	1.5936	Spedding et al., 1956
3.308[b]	5.267[b]	1.592	Spedding et al., 1960
3.3085	5.2669	1.5919	Spedding et al., 1961
3.313[b]	5.276[b]	1.593	Beaudry & Daane, 1964
3.3088	5.2680	1.5921	Spedding & Beaudry, 1971
av. 3.3088	5.2675	1.5920	
±0.0002	±0.0006	±0.0001	

[a]From kX, omitted from the average.
[b]Omitted from the average.

were made by Spedding et al. (1956), Spedding, Hanak, and Daane (1961), Finkel and Vorobiev (1968), and Spedding and Beaudry (1971). These are presented in Table 39-1.

No change in structure up to 987°C was detected by Spedding, Hanak, and Daane nor down to 77°K by Finkel and Vorobiev.

Table 39-1.   Lattice Constants of Yttrium at
Room Temperature

a, Å	c, Å	c/a	Reference
3.670[a]	5.826[a]	1.587	Quill, 1932a
3.636[a]	5.762[a]	1.585	Bommer, 1939b
3.647$_4$	5.730$_6$	1.571	Spedding et al., 1956
3.645$_7$	5.733$_0$	1.572	Spedding et al., 1961
3.650[b]	5.741[b]	1.573	Finkel & Vorobiev, 1968
3.6482	5.7318	1.5711	Spedding & Beaudry, 1971
av. 3.6471	5.7318	1.5716	
±0.0010	±0.0010	±0.0005	

[a]From kX, omitted from the average.
[b]Read from a graph, omitted from the average.

LANTHANIDES

LANTHANUM

The first studies on the structure of lanthanum (at room temperature) gave conflicting results: it was found to be hexagonal close packed by McLennon and McKay (1930b), Quill (1932c), and Rossi (1934), but cubic close packed by Zintl and Neumayer (1933), Klemm and Bommer (1937), and Young and Ziegler (1952). In a dilatometric study Trombe and Foëx (1943) reported that lanthanum lengthens reversibly between –195 and 152°C, and above 377°C, but that between 152 and 377°C two forms coexist. They termed the low temperature form α and the high temperature form β. They also stated that the $\alpha \rightarrow \beta$ transformation was accompanied with a 0.19% decrease in volume. Ziegler, Young, and Floyd (1953) studied the transition in detail. They reported that the transition hcp→ccp begins in the range 150 to 254°C but that even after four days at 400°C some of the hexagonal form was still present. The reverse transformation, moreover, was found to be very slow at room temperature, taking one year or more. It could be speeded up by cooling the metal to –195°C, or by severe deformation, such as filing.

In 1956, Spedding, Daane, and Herrmann reported that the low temperature form of lanthanum is not simply hexagonal close packed, but the length of the c-axis is doubled. This corresponds to a stacking sequence of · ·[abac] · · instead of · ·[ab] · ·. This observation was subsequently verified by Spedding, Hanak, and Daane (1961) and by Eliseev et al. (1964). Regarding the hexagonal⇌cubic transitions, the former investigators noted that surface impurities appeared to stabilize the cubic form below 293°C, whereas above that temperature it was the only form present. Calculated values of the lattice constants at 293°C are: hexagonal, $a = 3.775$, $c = 12.227$ Å, $c/a = 2 \times 1.619$; cubic, $a = 5.314$ Å. These correspond to a decrease in atomic volume of 0.6% in going from the hexagonal to the cubic form. Various values of the lattice constants as determined at room temperature are presented in Table 57-1. At this temperature the volume change for the hexagonal→cubic transition is $+0.13 \pm 0.19$ Å^3, or 0.4%.

A second transition in lanthanum at 864°C was discovered by Spedding, Daane, and Herrmann (1957). The high temperature phase was found by Spedding, Hanak, and Daane to be body centered cubic, with two atoms per unit cell. The lattice constant at 887°C, as obtained by measurement of one

Table 57-1.   Lattice Constants of Lanthanum at Room Temperature

a, Å	c, Å	c/a	Reference
		Hexagonal Form	
3.728[a]	6.072[a]	1.629	McLennan & McKay, 1930b
3.762[b]	6.075[c]	1.615	Quill, 1932
3.765[b]	6.061[c]	1.610	Rossi, 1934
3.755[d]	6.035[d]	1.607	Fox et al., 1952
3.748[b]	6.072[c]	1.620	Ziegler, Young, & Floyd, 1953
3.748[b]	6.072[c]	1.620	Stalinski, 1955
3.770	12.159	2×1.613	Spedding, Daane, & Herrmann, 1956
3.770	12.137	2×1.610	Spedding, Hanak, & Daane, 1961
3.755	12.024[e]	2×1.601	Eliseev et al., 1964
av. 3.760	12.143	2×1.615	
±0.009	±0.001	±0.004	
		Cubic Form (ccp)	
5.307[b]			Zintl & Neumayr, 1933b
5.305[b]			Klemm & Bommer, 1937
5.302[b]			Young & Ziegler, 1952
5.275[d]			Fox et al., 1952
5.296[a]			Ziegler, Young, & Floyd, 1953
5.303[b]			Stalinski, 1955
5.304			Spedding, Hanak, & Daane, 1961
5.291[e]			Eliseev et al., 1964
5.3058			Hill & Ellinger, 1971
av. 5.3045 ±0.0017			

[a]From kX, omitted from the average.
[b]From kX.
[c]From kX, multiplied by two in the averaging.
[d]Average of values from two different samples; omitted from the average.
[e]Omitted from the average.

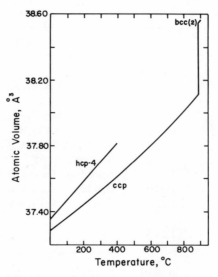

Fig. 57-1.   The atomic volume of lanthanum at various temperatures, at atmospheric pressure (calculated from the data of Spedding, Hanak, & Daane, 1961).

of their diagrams which gives the atomic volume at various temperatures, is 4.256 ± 0.007 Å. The extrapolated lattice constant of the cubic close packed form at that temperature is 5.341 Å   The variation of the atomic volume with temperature, as calculated from data given by Spedding, Hanak, and Daane, is presented in Fig. 57-1.

The cubic close packed structure may also be obtained at room temperature from the hexagonal form at high pressure. Piermarini and Weir (1964) found that the transition occurred at about 23 kbar. At this pressure the lattice constant is 5.17 Å. This value corresponds to a volume contraction of 7.0%. They also reported that at 40 kbar a = 5.02 Å. The overall contraction in this case is 14.9%. At the still higher pressure of 70 ± 10 kbar McWhan and Bond (1964) found a = 4.95 Å, a value which corresponds to an overall contraction of 18.0%. The atomic volumes at various pressures are presented in Fig. 57-2, and those found under several sets of conditions summarized in Table 57-2.

The phase behavior at high temperature and high pressure was studied by Jayaraman (1965b); the results are presented in Fig. 57-3.

CERIUM

The structure of cerium was first investigated by Hull (1921b), who reported that two forms exist at room temperature, one hexagonal close packed with

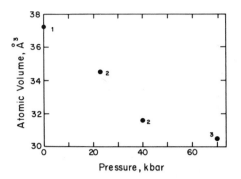

Fig. 57-2. The atomic volume of lanthanum at various pressures, at room temperature. (1) Table 57-1; (2) Piermarini & Weir, 1964; (3) McWhan and Bond, 1964. (The errors in the pressures of 2 and 3 were given as ± 10 kbar.)

Table 57-2.   The Atomic Volume of Lanthanum Under Various Conditions

Atomic Volume, Å^3	Structure	Temperature	Pressure
37.17 ± 0.18	hcp-4	Room	1 atm
37.30 ± 0.05	ccp	Room	1 atm
37.72	hcp-4	293°C	1 atm
37.51	ccp	293°C	1 atm
38.09	ccp	887°C	1 atm
38.55 ± 0.19	bcc(2)	887°C	1 atm
34.55	ccp	Room	23 kbar

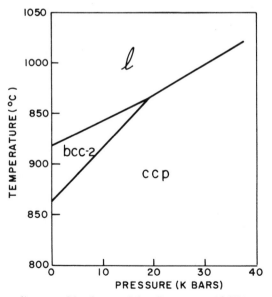

Fig. 57-3.   Phase diagram of lanthanum (after Jayaraman, 1965b).

89

a = 3.66, c = 5.92 Å (from kX), c/a = 1.62, the other cubic close packed with a = 5.13 Å (from kX). The hexagonal form was not observed later by Quill (1932c), Schuch and Sturdivant (1950), Dialer and Rothe (1955), Spedding, Daane, and Herrmann (1956), Weiner and Raynor (1959), nor by Spedding, Hanak, and Daane (1961); all of these investigators observed only the cubic close packed form.

By the use of dilatometric, magnetic, and electrical measurements, Trombe and Foëx (1943, 1944, 1947, 1952) discovered that cerium exists in four modifications, which they termed α, β, γ, and δ. The γ form is the one stable at room temperature. Cooling it to about –173°C gives the α form, the change being accompanied by about a 10% decrease in volume. Cycling the γ form between 20 and –195°C gives the β form and a volume decrease of 0.7%. The γ form changes to the δ form above 600 to 700°C. No structural studies were made. The identification of these forms is made below.

On the other hand, Klemm and Bommer (1937) observed hexagonal cerium but did not give any lattice constants, and Fox et al. (1952) observed both the hexagonal and cubic close packed forms coexisting in a sample which contained 35% of the former. McHargue and Yakel (1960) clarified the situation by studying cerium at low temperature. They found that at –10 ± 10°C cubic close packed cerium which had not been previously cooled begins to transform to a hexagonal close packed form having a c-axis double that previously reported. This corresponds to the stacking sequence · · [abac] · · as in hexagonal lanthanum. They also found that the reverse transformation does not begin until the sample is warmed to 100 ± 5°C and that it is not complete until the temperature reaches 147 ± 5°C. Additional transformations observed by McHargue and Yakel are discussed below. Gschneider, Elliott, and McDonald (1962) obtained the hexagonal form (with the doubled c-axis) by cycling the cubic close packed form from 23 to –198°C five to ten times. They observed that cubic to hexagonal transition began at 1°C, while the reverse transformation began at 81°C, in fair agreement with the values found by McHargue and Yakel.

The so-called "collapsed" form of cubic close packed cerium was discovered at high pressure (15,000 atm) and room temperature by Lawson and Tang (1949) and at atmospheric pressure and low temperature (–183°C) by Schuch and Sturdivant (1950); the respective values for the lattice constant reported are 4.84 and 4.82 Å. Studies of this transition at various temperatures and pressures by Herman and Swenson (1958) linked the high pressure-room temperature and atmospheric pressure-low temperature transitions. No lattice constants were reported. Piermarini and Weir (1964) found a = 4.82 Å at room temperature and 15 kbar; Davis and Adams (1964) showed that at about 350 to 400°C and 20 to 22 kbar there was a critical end point for the boundary between the two phases, as shown in Fig. 58-1.

An even more collapsed form of cubic close packed cerium was discovered by Franceschi and Olcese (1969), who found that at room temperature and 50 kbar the atomic volume decreases abruptly by 4.38% without any change in structure. This "supercollapsed" form exists at least up to 82.5 kbar with a pressure-independent lattice constant. Figure 58-2 shows variation of the lattice constant with pressure.

The low temperature transitions were studies in detail by McHargue and Yakel. Their results are summarized in Fig. 58-3. This figure oversimplifies the transformation behavior of cerium: the transitions exhibit hysteresis, and some of them are suppressed or enhanced, depending on the previous history of the sample.

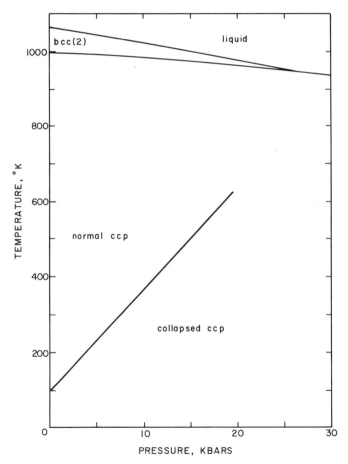

Fig. 58-1.   The phase diagram for cerium (after Jayaraman, 1965a), showing the critical end point in the boundary between the normal and collapsed forms.

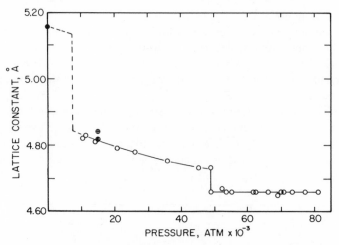

Fig. 58-2.   Pressure dependence of the atomic volume of cerium; ●Table 58-1;   Lawson & Tang; 1949; ⊕Piermarini & Weir, 1964; ○Franceschi & Olcese, 1969.

The existence of the high temperature transition first noted by Trombe and Foëx (1947, 1952) was verified by Spedding, Daane, and Herrmann (1957), who found a break in the electrical resistivity curve for cerium at 730°C. The structure of the high temperature form was determined to be body centered cubic, with two atoms per unit cell, by Spedding, Hanak, and Daane (1961). Their observed values of $\sin^2 \theta$ give 4.103 ± 0.007 Å for the lattice constant at 757°C. The change of atomic volume with temperature, as calculated from their data, is presented in Fig. 58-4.

The various determinations of the lattice constants of three of the forms are presented in Tables 58-1 and 58-2, and the atomic volumes under various conditions are presented in Table 58-3.

The foregoing interpretation leads to the following identification of the forms of Trombe and Foëx:

α: collapsed cubic close packed
β: double hexagonal close packed
γ: normal cubic close packed
δ: body centered cubic, two atoms per cell

To complicate the situation further, a *fifth* form of cerium at room temperature was reported by Dialer and Rother (1955), Weiner and Raynor (1959), and Gschneider, Elliott, and McDonald (1962). This form was said to be cubic close packed, but with a lattice constant about 0.04 Å smaller than the normal cubic close packed form. The new form was obtained in various ways, including annealing the sample at over 500°C for 24 hours (Weiner and

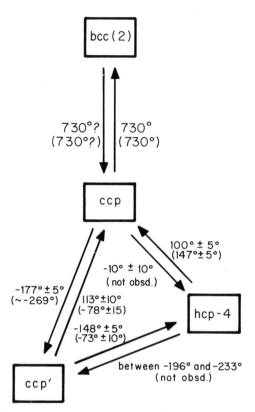

Fig. 58-3. Phase changes in cerium at atmospheric pressure. The collapsed cubic close packed phase is denoted by ccp′. The upper temperature is the observed start of the transition, the lower temperature (in parentheses) the point at which the transition is complete. All temperature in degrees centigrade. (Drawn from the data of McHargue & Yakel, 1960.)

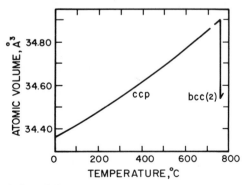

Fig. 58-4. The variation of the atomic volume of cerium with temperature (calculated from the data of Spedding, Hanak, & Daane, 1961).

Table 58-1.   Lattice Constants of Cubic Close Packed Cerium

a, Å	Reference
At room temperature and 1 atm	
5.13[a]	Hull, 1921b
5.153[a]	Quill, 1932c
5.150[a]	Klemm & Bommer, 1937
5.14[b]	Lawson & Tang, 1949
5.140[b]	Schuch & Sturdivant, 1950
5.152[b]	Dialer and Rothe, 1955
5.1612	Spedding, Daane, & Herrmann, 1956
5.1606[c]	Weiner & Raynor, 1959
5.15[b]	McHargue and Yakel, 1960
5.1611	Spedding, Hanak, & Daane, 1961
5.1601	Gschneider, Elliott, & McDonald, 1962
5.1612[c]	Harris & Raynor, 1964
5.1600	Gschneider & Waber, 1964
5.1612[c]	Speight, Harris, & Raynor, 1968
av. 5.1608 ± 0.0005	
At room temperature and 15,000 atm	
4.84	Lawson & Tang, 1949
4.82	Piermarini & Weir, 1964
At room temperature and 49,000 to 81,000 atm	
4.73	Franceschi & Olcese, 1969
4.66	Franceschi & Olcese, 1969
At –183°C and 1 atm	
4.82	Schuch & Sturdivant, 1950
At 757°C and 1 atm	
5.188[d]	Spedding, Hanak, & Daane, 1961

[a]From kX, not included in the average.
[b]Not included in the average.
[c]From kX.
[d]Extrapolated value.

Raynor) and quenching the sample in liquid hydrogen, then warming (Gschneider et al.). However, after numerous experiments in which cerium was heated in the presence of various gases, Gschneider and Waber (1964) concluded that this "form " is CeX, with the sodium chloride structure, where X is probably oxygen.

Table 58-2.   Lattice Constants of Hexagonal Cerium (at room temperature
and 1 atm)

a, Å	c, Å	c/a	
3.66[a]	5.92[a]	1.62	Hull, 1921b
3.62[b]	5.99[b]	1.65	Fox et al., 1952
3.680[c]	11.922[c]	2 × 1.620	McHargue & Yakel, 1960
3.673	11.802	2 × 1.607	Gschneider, Elliott, & McDonald, 1962
av. 3.677	11.862	2 × 1.613	
±0.005	±0.085	±0.012	

[a]From kX, omitted from the average.
[b]Omitted from the average.
[c]Recalculated from the spacings given in the paper.

Table 58-3.   The Atomic Volume of Cerium under Various Conditions

Atomic Volume, Å³	Structure	Temperature	Pressure
34.36 ± 0.01	ccp	Room	1 atm
28.17	ccp	Room	15,000 atm
28.00	ccp	–183°C	1 atm
25.30 ± 0.16	ccp	Room	48,900–81,400 atm
34.91	ccp	757°C	1 atm
34.54 ± 0.17	bcc(2)	757°C	1 atm
34.72 ± 0.27	hcp-4	Room	1 atm

PRASEODYMIUM

The structure of praseodymium was first reported by Rossi (1932) to be
hexagonal close packed, with $a = 3.664$ Å, $c = 5.936$ Å (from kX), $c/a =$
1.620. Klemm and Bommer (1939) reported lattice constants close to these,
but stated that the c-axis might have to be doubled to account for some extra
lines on their photographs. The double hexagonal close packed structure,
stacking sequence · ·[abac]· ·, was established by Spedding, Daane, and
Herrmann (1956), and by Spedding, Hanak, and Daane (1961). The determinations of the lattice constants are presented in Table 59-1.

A second form of praseodymium at room temperature was observed by

Klemm and Bommer (1937). It was stated to be cubic close packed, with a = 5.161 Å (from kX). This form, admixed with the hexagonal form, was also observed by Fox et al. (1952).

A transition at 792°C in praseodymium was detected by Spedding, Daane, and Herrmann (1957) by electrical resistivity measurements. The high temperature form was shown by Spedding, Hanak, and Daane (1961) to be body centered cubic, two atoms per unit cell, with lattice constant a = 4.13 Å at 821°C. The change of the atomic volume with temperature as calculated from their data is presented in Fig. 59-1.

A transition at about 40 kbar and room temperature to a collapsed cubic close packed form was discovered by Piermarini and Weir (1964); at that pressure the lattice constant is 4.88 Å. This represents a contraction of 15.9% from the volume of the double hexagonal close packed form at the same temperature.

Lattice constants of cubic praseodymium under various conditions are presented in Table 59-2.

The phase diagram of praseodymium as determined by Jayaraman (1965b) is presented in Fig. 59-2.

Table 59-1.   Lattice Constants of Hexagonal Praseodymium at Room Temperature

a, Å	c, Å	c/a	Reference
3.664[a]	5.936[a]	1.620	Rossi, 1932
3.659[a]	5.891[a]	1.610	Rossi, 1934
3.669[a]	5.920[a]	1.614	Klemm & Bommer, 1939
3.64[b]	5.89[b]	1.62	Fox et al., 1952
3.673	11.835	2 × 1.611	Spedding, Daane, & Herrmann, 1956
3.671	11.831	2 × 1.611	Spedding, Hanak, & Daane, 1961
3.6715	11.8354	2 × 1.612	Lundin, Yamamoto, & Nachmar, 1966
3.6714[c]	11.831[c]	2 × 1.611	Speight, Harris, & Raynor, 1968
av. 3.6717	11.833	2 × 1.6114	
±0.0008	±0.002	±0.0004	
3.684[d]	11.937[d]	2 × 1.620	Spedding, Hanak, & Daane, 1961

[a]From kX, not included in the average.   [c]From kX.
[b]Not included in the average.   [d]Extrapolated value for 821°C.

Table 59-2.   Lattice Constants of Praseodymium

Structure	a, Å	Temperature	Pressure	Reference
ccp	5.161	Room	1 atm	Klemm & Bommer, 1937
ccp	5.11	Room	1 atm	Fox et al., 1952
ccp	4.88	Room	40 kbar	Piermarini & Weir, 1964
bcc(2)	4.13	821°C	1 atm	Spedding, Hanak, & Daane, 1961

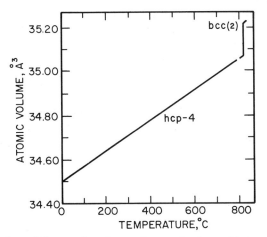

Fig. 59-1.   Variation of the atomic volume of praseodymium with temperature (calculated from the data of Spedding, Hanak, & Daane, 1961).

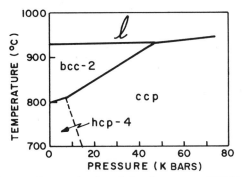

Fig. 59-2.   Phase diagram of praseodymium (after Jayaraman, 1965b). The position of the collapsed ccp′ form observed above 40 kbar at room temperature (see text) is not indicated.

97

NEODYMIUM

Quill (1932c) first studied the structure of neodymium; he reported that it is
hexagonal close packed, with $a = 3.664$, $c = 5.892$ Å (from kX), $c/a = 1.608$.
Klemm and Bommer (1937) found closely similar lattice constants but later
(1939) remarked that the c-axis might have to be doubled in length. Ellinger
later (1955) found that the c-axis is indeed doubled, an observation verified
by Spedding, Daane, and Herrmann (1956) and by Spedding, Hanak, and
Daane (1961). The metal, at room temperature, is thus double hexagonal
close packed, with stacking sequence $\cdot\cdot[abac]\cdot\cdot$. Determinations of the lattice
constants are presented in Table 60-1.

In electrical resistivity measurements Spedding, Daane, and Herrmann (1957)
detected a transition in neodymium at 862°C. The high temperature form was
shown by Spedding, Hanak, and Daane (1961) to be body centered cubic,
with two atoms per unit cell. At 883°C the lattice constant is 4.13 Å. The
change in atomic volume with temperature, as calculated from data given by
them, is presented in Fig. 60-1.

The hexagonal form transforms to a collapsed face centered cubic form,
with lattice constant 4.80 Å, at 50 kbar and room temperature (Piermarini and
Weir, 1964). The volume decrease is 19.2%. The phase behavior at high temperature and high pressure, as reported by Jayaraman (1965), is presented in
Fig. 60-2.

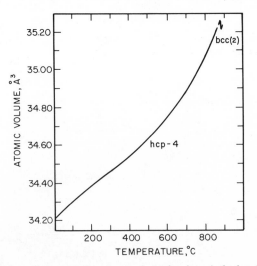

Fig. 60-1.   Variation of the atomic volume of neodymium (calculated from the data of
Spedding, Hanak, & Daane, 1961).

Table 60-1.   Lattice Constants of Neodymium at Room Temperature

a, Å	c, Å	c/a	Reference
3.664[a]	5.892[a]	1.608	Quill, 1932c
3.662[a]	5.892[a]	1.609	Klemm & Bommer, 1937
3.657[b]	5.902[c]	1.614	Klemm & Bommer, 1939
3.59[d]	5.83[d]	1.62	Fox et al., 1952
3.655	11.796	2 × 1.614	Ellinger, 1955
3.658	11.799	2 × 1.613	Spedding, Daane, & Herrmann, 1956
3.659	11.805	2 × 1.613	Spedding, Hanak, & Daane, 1961
3.6566	11.7983	2 × 1.6133	Lundin, Yamamoto, & Nachman, 1966
3.6564[b]	11.7948[b]	2 × 1.6129	Speight, Harris, & Raynor, 1968
av.3.6570	11.7995	2 × 1.6133	
±0.0011	±0.0038	±0.0007	
3.689[e]	11.961[e]	2 × 1.621	Spedding, Hanak, & Daane, 1961

[a]From kX, not included in the average.   [d]Not included in the average.
[b]From kX.   [e]Extrapolated value for 883°C.
[c]From kX, doubled in the averaging.

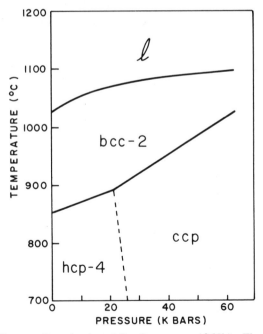

Fig. 60-2.   Phase diagram of neodymium (after Jayaraman, 1965b). The position of the collapsed ccp′ form observed at 50 kbar and room temperature is not indicated.

PROMETHIUM

The structure of promethium at room temperature was determined by Pallmer and Chikalla (1971), who studied the isotope ^{147}Pm, which has a half-life of 2.6 years. The metal is double hexagonal close packed, stacking sequence $\cdot\cdot[abac]\cdot\cdot$, with $a = 3.65 \pm 0.01$, $c = 11.65 \pm 0.09$ Å, $c/a = 2 \times 1.596$.

SAMARIUM

The structure of samarium was first reported by Daane, Dennison, and Spedding (1953), who stated that it is rhombohedral with approximate lattice constants $a \approx 8$ Å and $\alpha \approx 23.5°$, but gave no further structural information. A detailed structure, based on 38 powder lines, was soon thereafter given by Ellinger and Zachariasen (1953). They also reported that samarium is rhombohedral, with lattice constants $a = 8.982$ Å, $\alpha = 23.31°$. This unit cell contains three samarium atoms, one type I at (000) and two type II at (xxx), $(\bar{x}\,\bar{x}\,\bar{x})$. The space group is D_{3d}^5-R$\bar{3}$m. Intensity considerations lead to a value of x of 0.222 ± 0.003, or $\frac{2}{9}$, within the limits of error of the determination. The alternate description of this structure in the more easily visualized hexagonal lattice is $a = 3.629$, $c = 26.20$ Å, $c/a = 7.220$, nine atoms per cell, three type I at (000), $(\frac{2}{3}\frac{1}{3}\frac{1}{3})$, $(\frac{1}{3}\frac{2}{3}\frac{2}{3})$, and six type II at $(0\ 0\ \pm z)$, $(\frac{2}{3}\ \frac{1}{3}\ \frac{1}{3}\pm z)$, $(\frac{1}{3}\ \frac{2}{3}\ \frac{2}{3}\pm z)$, with $z = \frac{2}{9}$. This structure may be described as consisting of close packed hexagonal layers perpendicular to the c-axis, with stacking sequence $\cdot\cdot[ababcbcac]\cdot\cdot$. The axial ratio is 4.5×1.604, or slightly smaller than the ideal value of 1.633, as in the other hexagonal rare earth elements. The stacking sequence consists of a layer of atoms of type I followed by two layers of type II. If the axial ratio were ideal the environment of each of the type I atoms would be identical to that found in cubic close packing; the environment of each of the type II atoms is the same as that which occurs in hexagonal close packing. The stacking sequence might thus be alternately described as $\cdot\cdot[CHH]\cdot\cdot$.

The above structure for samarium was confirmed by Daane, Rundle, Smith, and Spedding (1954), who worked with single crystals. Their values for the lattice constants are included in Table 62-1. The single crystal work also showed that the structure was either disordered or submicroscopically twinned.

A transition in samarium at 917°C was included in a table by Spedding and Daane (1956). Evidence for this transformation was based on thermal analyses. Details of this work was reported later by Spedding, McKeown, and Daane, (1960), but no diffraction studies were made, so the structure of the high temperature form was not determined.

Table 62-1.   Lattice Constants of Samarium at Room Temperature

Rhombohedral Cell		Hexagonal Cell			
a, Å	α,°	a, Å	c, Å	c/a	Reference
Normal Form					
~8[a]	~23.5[a]	~3.3[a]	~23[a]		Daane, Dennison, & Spedding, 1953
8.982	23.31	3.629	26.20	4.5 × 1.604	Ellinger & Zachariasen, 1953
8.996	23.22	3.621	26.25	4.5 × 1.611	Daane, Rundle, Smith, & Spedding, 1954
8.991	23.28	3.6281[b]	26.231[b]	4.5 × 1.607	Speight, Harris, & Raynor, 1968
8.974	23.31	3.626[c]	26.18	4.5 × 1.605	Mardon & Koch, 1970
av. 8.988	23.28	3.626	26.222	4.5 × 1.607	
±0.009	±0.04	±0.003	±0.027	±0.002	
Metastable Form[d]					
		3.618	11.66	2 × 1.611	Jayaraman & Sherwood, 1964b

[a]Not included in the average.
[b]From kX.
[c]Read from a graph.
[d]Double hexagonal close packed, obtained at
high temperature and pressure, and quenched.

A high pressure-high temperature transition was observed by Jayaraman and Sherwood (1964b) to take place at about 40 kbar and 300°C. The new form was found to be metastable at room temperature; it is double hexagonal close packed, with lattice constants $a = 3.618$, $c = 11.66$ Å, $c/a = 1.611$. The atomic volume of 33.05 Å^3 does not differ significantly from that of the normal form, 33.17 ± 0.11 Å^3.

In his study of the phase behavior of samarium Jayaraman (1965b) assumed that the high temperature form is body centered cubic. His phase diagram is presented in Fig. 62-1.

A third form of samarium was discovered by Kumar and Srivastava (1969), who examined thin films deposited on nitrocellulose. They found these films were hexagonal close packed, with $a = 3.65$, $c = 5.86$ Å, $c/a = 1.605$. These give an atomic volume of 33.81 Å^3, a value higher, but probably not significantly, than those of the other two forms. This near equality of the volumes is, however, unimportant, for Boulesteix, Caro, Gasgnier, Henry la Blanchetais, Pardo, and Valiergue (1970) showed that thin films of B-Sm_2O_3 gave diffraction patterns identical to those reported by Kumar and Srivastava for "samarium".

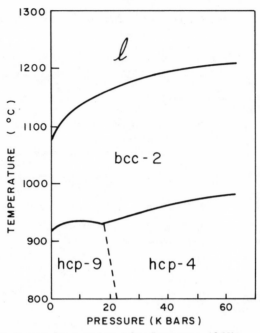

Fig. 62-1.   Phase diagram for samarium (after Jayaraman, 1965b).

EUROPIUM

The structure of europium was first determined by Klemm and Bommer (1937), who found that it is body centered cubic with two atoms per unit cell, and a lattice constant of 4.582 Å (from kX). Subsequent work has confirmed this structure. Various determinations of the lattice constant are presented in Table 63-1.

Table 63-1.   Lattice Constant of Europium at Room Temperature

a, Å	Reference
4.582[a]	Klemm & Bommer, 1937
4.606[b]	Spedding, Daane, & Herrmann, 1956
4.578[b]	Barrett, 1956
4.5820	Spedding, Hanak, & Daane, 1958
4.5822	Spedding, Hanak, & Daane, 1961
av. 4.5821 ± 0.0001	

[a]From kX, not included in the average.   [b]Not included in the average.

Barrett (1956) found no transformation occurred, even with cold working, at 5°K. At that temperature, as well as at 78°K, the lattice constant is 4.551 Å. No change in structure between room temperature and 352°C was observed by Spedding, Hanak, and Daane (1961). The change in atomic volume with temperature, as calculated from their data, is presented in Fig. 63-1.

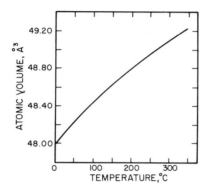

Fig. 63-1.   Variation of the atomic volume of europium with temperature (calculated with the data of Spedding, Hanak, & Daane, 1961).

GADOLINIUM

Gadolinium was found by Klemm and Bommer (1937) to be hexagonal close packed, with a = 3.629, c = 5.760 Å (from kX), c/a = 1.587. This structure was confirmed by subsequent work; the various determinations of the lattice constants are presented in Table 64-1. The structure does not change at 5°K, even after cold working (Barrett, 1958), nor on heating up to 1200°C (Spedding, Hanak, and Daane, 1961). Above 1262°C, however, some evidence was obtained for the existence of a body centered cubic structure, and from measurement of four diffraction maxima an approximate lattice constant of 4.06 Å was obtained. Assuming an error on this value of ± 1%, the change in atomic volume for the transition hcp→bcc is -(0.75 ± 0.99) Å^3, or -(2.2 ± 2.8)%, where the volume of the hexagonal phase is extrapolated with the function given by Spedding et al.

The c-axis exhibits anomalous expansion when cooled below 298°K (Banister, Legvold, and Spedding, 1954; Darnell, 1963a; Vorobiev, Smirnov, and Finkel, 1966). The temperature dependence of the lattice constants is presented in Fig. 64-1. Darnell and Vorobiev et al. discuss these in terms of the magnetic properties of the metal. The change in atomic volume with temperature is presented in Fig. 64-2.

Table 64-1. Lattice Constants of Gadolinium at Room Temperature

a, Å	c, Å	c/a	Reference
3.622[a]	5.760[a]	1.590	Klemm & Bommer, 1937
3.62[b]	5.76[b]	1.59	Fox et al., 1952
3.635	5.776	1.589	Banister, Legvold, & Spedding, 1954
3.6360	5.7826	1.5903	Spedding, Daane, & Herrmann, 1956
3.632	5.778	1.591	Spedding, Hanak, & Daane, 1961
3.6308[c]	5.7790[c]	1.5917	Darnell, 1963a
3.634	5.785	1.592	Beaudry & Daane, 1964
3.6331[c]	5.7766[c]	1.5900	Vorobiev, Smirnov, & Finkel, 1966
3.6336	5 7810	1.5910	Spedding & Beaudry, 1971
3.6315	5.7810	1.5908	Gupty & Anantharaman, 1971
av. 3.6333	5.7794	1.5907	
±0.0018	±0.0032	±0.0012	

[a]From kX, omitted from the average.
[b]Omitted from the average.
[c]Read from a graph.

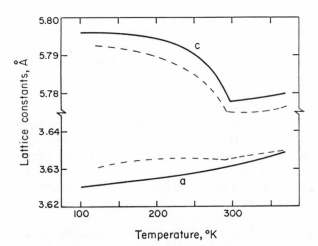

Fig. 64-1. Variation of the lattice constants of gadolinium with temperature. Solid lines after Darnell, 1963a; dashed lines after Vorobiev et al., 1966. There is no change in structure in this temperature range.

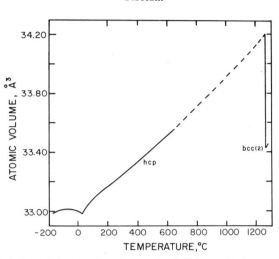

Fig. 64-2.   Variation of the atomic volume of gadolinium with temperature; below 20°C
from data of Darnell, 1963a; above 20°C from data of Spedding, Hanak, & Daane, 1961.
The discontinuity at 20°C is not accompanied by a change in phase.

A transition in gadolinium at high pressure and high temperature was discovered by Jayaraman and Sherwood (1964a). This form was stated to be isostructural with the room temperature form of samarium, that is, stacking sequence of close packed layers.··[ababcbcac]··. At about 40 kbar and 400°C
the lattice constants are $a = 3.61$, $c = 26.03$ Å, $c/a = 4.5 \times 1.60$. This same form
was observed at 35 kbar and room temperature by McWhan and Stevens (1965).
Under these conditions the lattice constants are $a = 3.49$, $c = 25.6$ Å, $c/a =
4.5 \times 1.63$. The values correspond to a contraction of 9.2% under the volume
of the hexagonal close packed form. More extensive studies of the phase behavior were made by Jayaraman (1965b); the resulting phase diagram is shown in
Fig. 64-3.

TERBIUM

Terbium was discovered to be hexagonal close packed by Klemm and Bommer
(1937). Their values for the lattice constants, plus those of later investigators,
all of whom confirmed the structure, are presented in Table 65-1.
      A low temperature transition in terbium was discovered by Darnell (1963b).
At 220°K transformation to an orthorhombic form takes place. The hexagonal close packed structure may be alternately described, as in Fig. 65-1, with

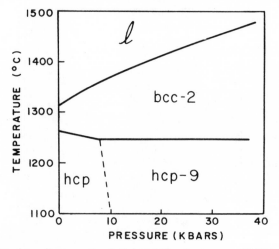

Fig. 64-3.   The phase diagram of gadolinium (after Jayaraman, 1965b).

Table 65-1.   Lattice Constants of Terbium at Room Temperature

a, Å	c, Å	c/a	Reference
3.592[a]	5.675[a]	1.580	Klemm & Bommer, 1937
3.6010	5.6936	1.5811	Spedding, Daane, & Herrmann, 1956
3.600[b]	5.698[b]	1.583	Spedding, Hanak, & Daane, 1961
3.608[c]	5.701[c]	1.580	Darnell, 1963b
3.6091[d]	5.6970[d]	1.5785	Finkel, Smirnov, & Vorobiev, 1967
3.6059[e]	5.6973[e]	1.5800	Speight, Harris, & Raynor, 1968
3.6055	5.6966	1.5800	Spedding & Beaudry, 1971
3.5990	5.6960	1.5827	Gupty & Anantharaman, 1971
av. 3.6041 ± 0.0041	5.6961 ± 0.0015	1.5805 ± 0.0022	

[a]From kX, omitted from the average.
[b]Omitted from the average.
[c]Read from a graph, omitted from the average.
[d]Read from a graph.
[e]From kX.

106

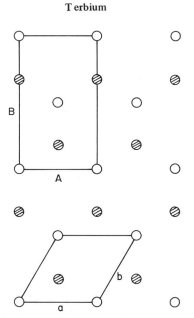

Fig. 65-1. Two layers of a hexagonal close packed structure: open circles, layer at $z = 0$; shaded circles, layer at $z = \frac{1}{2}$. The hexagonal unit cell shown by a, b, and the orthorhombic unit cell by A, B; c and C both perpendicular to projection plane.

an end-centered orthorhombic unit cell having $b = \sqrt{3}a$, with atoms at (000), $(\frac{1}{2}\frac{1}{2}0)$, $(\frac{1}{2}\frac{1}{6}\frac{1}{2})$, $(0\frac{2}{3}\frac{1}{2})$; on moving the origin to the centers of symmetry, one finds that these are positions 4c of space group D_{2h}^{17}–Cmcm, $(000\,\frac{1}{2}\frac{1}{2}0) \pm (0y\frac{1}{4})$, with the y parameter having the ideal value of $\frac{1}{6}$. As the temperature is lowered from 220°K the axial ratio of $b:a$ increases over the ideal value of $\sqrt{3}$, and the symmetry is no longer hexagonal, but orthorhombic. It is not known whether the value of the y-parameter changes from the ideal value of $\frac{1}{6}$, this value no longer being fixed by symmetry when the structure is orthorhombic.

The pressure-volume-temperature studies of Monfort and Swenson (1965) show a discontinuity in the volume change with temperature at zero pressure at about 230°K, but their V/V_0 data, normalized to the appropriate atomic volume at room temperature, are in only fair agreement with those of Darnell. Finkel, Smirnov, and Vorobiev (1967), in contrast to Darnell, observed only a small orthorhombic distortion below 223°K. The variation of the lattice constants with temperature is shown in Fig. 65-2; the corresponding change in the atomic volume is presented in Fig. 65-3.

A high temperature-high pressure transition to the double hexagonal close packed structure, such as occurs in samarium, was *not* observed at 40 kbar and

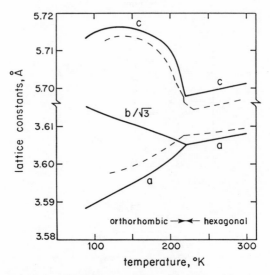

Fig. 65-2.   The variation of the lattice constants of terbium with temperature. Solid lines after Darnell, 1963b; dashed lines after Finkel et al., 1971, who observed only a very small orthorhombic distortion below 223° K.

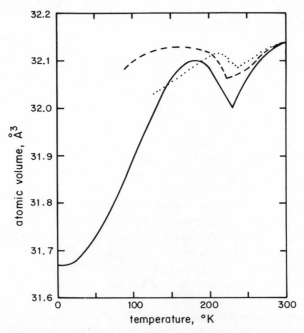

Fig. 65-3.   The variation of the atomic volume of terbium in the low temperature range. Solid line, data of Monfort & Swenson, 1965; dashed line, data of Darnell, 1963b; dotted line, data of Finkel et al., 1967.

108

350°C in terbium (Jayaraman and Sherwood, 1964b). On the other hand, at 60 kbar and room temperature McWhan and Stevens (1965) reported that terbium has the room temperature samarium structure with lattice constants $a = 3.41$, $c = 24.5$ Å, $c/a = 4.5 \times 1.60$. These correspond to a contraction of 14.4% from the volume of the hexagonal close packed form at atmospheric pressure.

No change in phase between room temperature and 950°C was observed by Spedding, Hanak, and Daane (1961); their data give the atomic volume versus temperature curve presented in Fig. 65-4. At 1316°C, however, they observed a discontinuity in the electrical resistance, and Jayaraman (1965b) assumed that the high temperature form is body centered cubic. The phase diagram is shown in Fig. 65-5.

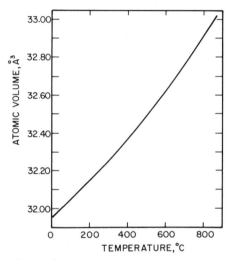

Fig. 65-4.   The variation of the atomic volume of terbium in the high temperature region (calculated from the data of Spedding, Hanak, & Daane, 1961).

DYSPROSIUM

The structure of dysprosium was found to be hexagonal close packed by Klemm and Bommer (1937), a result subsequently verified by a number of others. The various determinations of the lattice constants are presented in Table 66-1.

Dysprosium was stated by Banister, Legvold, and Spedding (1954) to retain the hexagonal close packed structure down to 49°K, and Spedding, Hanak, and Daane (1961) observed no transition between room temperature and 685°C. The atomic volume change with temperature as calculated from their

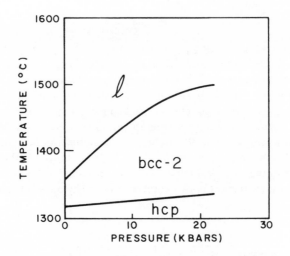

Fig. 65-5.   The phase diagram of terbium (after Jayaraman, 1965b). In addition to the phases shown the hcp-9 structure has been observed at room temperature and 60 kbar, and an orthorhombic structure has been observed below 220°K at atmospheric pressure.

Table 66-1.   Lattice Constants for Dysprosium at Room Temperature

a, Å	c, Å	c/a	Reference
3.585[a]	5.659[a]	1.579	Klemm & Bommer, 1937
3.595[b]	5.657[b]	1.574	Banister, Legvold, & Spedding, 1954
3.5903	5.6475	1.5730	Spedding, Daane, & Herrmann, 1956
3.5930	5.6567	1.5744	Spedding, Hanak, & Daane, 1961
3.594[c]	5.654[c]	1.574	Darnell & Moore, 1963; Darnell, 1963a
3.5917	5.6504	1.5732	Finkel & Vorobiev, 1967
3.5915	5.6501	1.5732	Spedding & Beaudry, 1971
3.5923	5.6545	1.5741	Gupty & Anantharaman, 1971
av. 3.5918	5.6518	1.5735	
±0.0010	±0.0037	±0.0011	

[a]From kX, not included in the average.
[b]Not included in the average.
[c]Read from a graph, not included in the average.

equation is shown in Fig. 66-1. Darnell (1963a) and Darnell and Moore (1963), on the other hand, observed a transition at 86°K to an orthorhombic form, similar to that found in terbium (see above). This transformation differs, however, in that there are abrupt changes in the three lattice constants, as shown in Fig. 66-2. At 86°K the lattice constants (as read from a figure presented by Darnell) are, for the hexagonal form, $a = 3.586$, $c = 5.666$ Å, $c/a = 1.580$, and for the orthorhombic form, $a = 3.595$, $b = 6.184$, $c = 5.678$ Å, $a:b:c = 1:1.720:1.579$. The volume change is zero within the accuracy of the measurements. The change in volume with temperature in the low temperature range is presented in Fig. 66-3.

Finkel and Vorobiev (1967) also observed discontinuities in the lattice constants at 85°K, but their identifications of the axes shown in in Fig. 66-2, differs from that of Darnell and Moore. Furthermore, the change in atomic volume at the transition given by Finkel and Vorobiev corresponds to +0.10 Å^3 (+0.32%) for orthorhombic→hexagonal, as opposed to the zero change of Darnell and Moore (see Fig. 66-3).

Observations on dysprosium at high pressure are not in complete agreement. Jamieson (1964) observed no structural transition up to 160 kbar at room temperature, nor did Jayaraman and Sherwood (1964b) up to 40 kbar at 350°C. McWhan and Stevens (1965), on the other hand, found that at 75 kbar and room temperature dysprosium has the room temperature samarium structure, that is, the c-axis 4.5 times the value for the hexagonal close packed structure. It is possible that the faint additional diffraction maxima caused by the longer c-axis were overlooked by the earlier workers. The variation of the atomic volume with pressure is presented in Fig. 66-4.

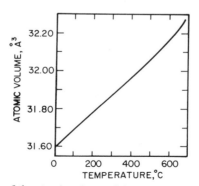

Fig. 66-1.   The variation of the atomic volume of dysprosium in the high temperature range (calculated from the data of Spedding, Hanak, & Daane, 1961).

112   Lanthanides

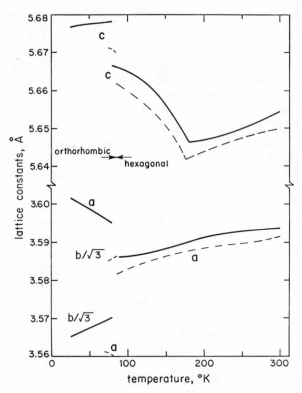

Fig. 66-2.   The variation of the lattice constants of dysprosium with temperature. Solid lines after Darnell & Moore, 1963; dashed lines after Finkel & Vorobiev, 1967.

HOLMIUM

The hexagonal close packed structure was first assigned to holmium by Bommer (1939). This and later determinations of the lattice constants are presented in Table 67-1. No change in structure up to 966°C at atmospheric pressure was observed by Spedding, Hanak, and Daane (1961) nor at 350°C and 40 kbar by Jayaraman and Sherwood (1964b). The data of Spedding et al. give the atomic volume versus temperature curve of Fig. 67-1. However, McWhan and Stevens (1965) observed that at 75 kbar and room temperature holmium had transformed to the room temperature samarium structure, with $a = 3.34$, $c = 24.5$ Å, $c/a = 4.5 \times 1.63$. Furthermore, no change in structure down to about 15°K was detected by Darnell (1963a), although discontinuities in the lattice constants were observed, as shown in Fig. 67-2. Similar, but not identical discontinuities were observed by Finkel and Palatnik (1971). The dependence of the atomic volume on the temperature in this range is presented in Fig. 67-3.

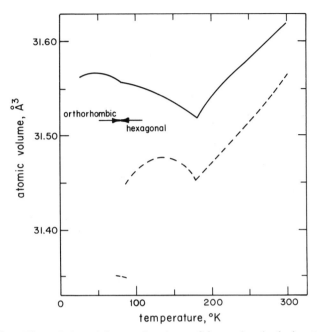

Fig. 66-3.   The variation of the atomic volume of dysprosium in the low temperature region. (Sources the same as Fig. 66-2)

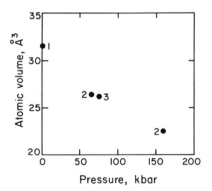

Fig. 66-4.   The atomic volume of dysprosium at various pressures and room temperature. (1) Table 66-1; (2) Jamieson, 1964; (3) McWhan & Stevens, 1965.

113

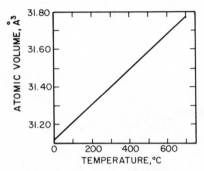

Fig. 67-1. The variation of the atomic volume of holmium in the high temperature range (calculated from the data of Spedding, Hanak, & Daane, 1961).

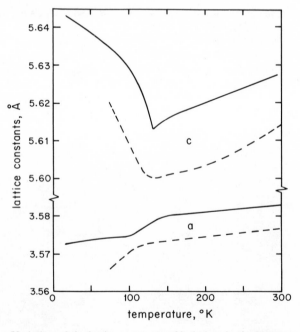

Fig. 67-2. Variation of the lattice constants of holmium in the low temperature range. Solid curves after Darnell, 1963a; dashed curves after Finkel & Palatnik, 1971.

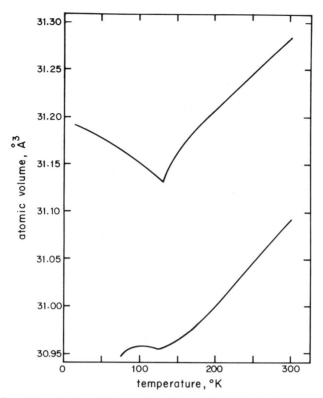

Fig. 67-3. The variation of the atomic volume of holmium in the low temperature range. Upper curve, data of Darnell, 1963a; lower curve, data of Finkel & Palatnik, 1971. There is no change in structure in this temperature range.

Table 67-1.   Lattice Constants of Holmium at Room Temperature

a, Å	c, Å	c/a	Reference
3.564[a]	5.631[a]	1.580	Bommer, 1939
3.5773	5.6158	1.5698	Spedding, Daane, & Herrmann, 1956
3.5768	5.6196	1.5711	Spedding, Hanak, & Daane, 1961
3.583[b]	5.628[b]	1.571	Darnell, 1963a
3.5778	5.6178	1.5702	Spedding & Beaudry, 1971
3.5764[c]	5.6140[c]	1.5697	Finkel & Palatnik, 1971
3.5761	5.6174	1.5708	Gupty & Anantharaman, 1971
av. 3.5769	5.6169	1.5703	
±0.0007	±0.0021	±0.0007	

[a]From kX, omitted from the average.   [c]Read from a graph.
[b]Read from a graph, omitted from the average.

ERBIUM

Erbium was reported by McLennan and Monkman (1929) to be hexagonal close packed, with $a = 3.75$, $c = 6.10$ Å (from kX), $c/a = 1.63$, but these lattice constants diverge so drastically from those reported in later studies that they must actually have examined some other substance. The various determinations of the lattice constants are presented in Table 68-1. No change in structure occurs in the range 20°K to 917°C (Banister, Legvold, and Spedding, 1954; Spedding, Hanak, and Daane, 1961; Darnell, 1963b) nor at 350°C and 40 kbar (Jayaraman and Sherwood, 1964b). The behavior of the lattice constants at low temperature is presented in Fig. 68-1, and the corresponding change in atomic volume is shown in Fig. 68-2. Finkel and Palatnik (1971) presented data collected in the range 77 to 300°K; these data give atomic volumes which are 0.24 ± 0.01 Å^3 smaller than those of Darnell in that range. The rather peculiar changes in a and c with temperature lead to an almost constant volume in the range 37 to 87°K. The volume change at high temperature is presented in Fig. 68-3.

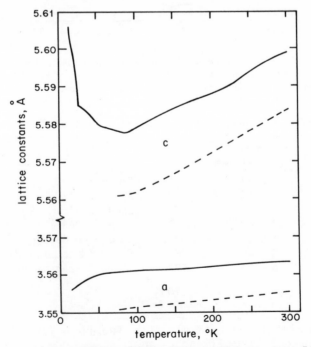

Fig. 68-1.   Variation of the lattice constants of erbium with temperature. Solid lines after Darnell, 1963b; dashed lines after Finkel & Palatnik, 1971.

Table 68-1.   Lattice Constants of Erbium at Room Temperature

a, Å	c, Å	c/a	Reference
3.75[a]	6.10[a]	1.63	McLennan & Monkman, 1929
3.539[a]	5.600[a]	1.583	Klemm & Bommer, 1937
3.562[b]	5.602[b]	1.573	Banister, Legvold, & Spedding, 1954
3.5588	5.5874	1.5700	Spedding, Daane, & Hermann, 1956
3.5596	5.5944	1.5716	Spedding, Hanak, & Daane, 1961
3.561[b]	5.594[b]	1.571	Smidt & Daane, 1963
3.563[c]	5.599[c]	1.571	Darnell, 1963b
3.557[b]	5.584[b]	1.570	Azarkh & Gavrilov, 1968
3.5592	5.5850	1.5692	Spedding & Beaudry, 1971
3.5551[c]	5.5839[d]	1.5707	Finkel & Palatnik, 1971
3.5590	5.5920	1.5700	Gupty & Anantharaman, 1971
av. 3.5592 ± ±0.0003	5.5885 ± ±0.0045	1.5702 ± ±0.0012	

[a]From kX, omitted from the average.   [c]Read from a graph, omitted from the av.
[b]Omitted from the average.   [d]Read from a graph.

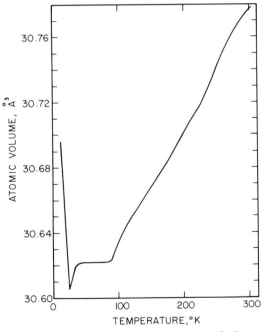

Fig. 68-2.   The variation of the atomic volume of erbium in the low temperature range (data of Darnell, 1963b).

117

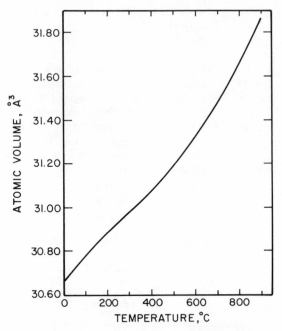

Fig. 68-3.   The variation of the atomic volume of erbium in the high temperature range (calculated from the data of Spedding, Hanak, & Daane, 1961).

THULIUM

The structure of thulium was first determined by Klemm and Bommer (1937), who found that it is hexagonal close packed. The values they reported for the lattice constants, together with the results from subsequent more precise de-terminations, are presented in Table 69-1. No change in structure between room temperature and 1004°C was observed by Spedding, Hanak, and Daane (1961). The change in atomic volume with temperature as reported by them is shown in Fig. 69-1. The thermal expansion in the range 90 to 300°K was reported by Singh, Khanduri, and Tsang (1971). Their results are presented in Fig. 69-2. No change in structure was observed in this temperature range.

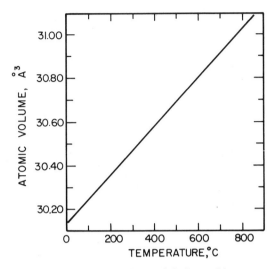

Fig. 69-1.   The variation of the atomic volume of thulium with temperature (calculated from the data of Spedding, Hanak, & Daane, 1961).

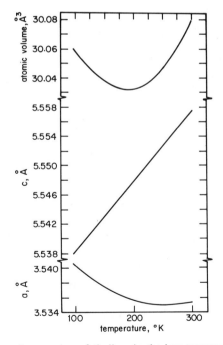

Fig. 69-2.   The thermal expansion of thulium in the low temperature range (after Singh et al., 1971).

119

Table 69-1.   Lattice Constants of Thulium at Room Temperature

a, Å	c, Å	c/a	Reference
3.530[a]	5.575[a]	1.579	Klemm & Bommer, 1937
3.5375	5.5546	1.570	Spedding, Daane, & Herrmann, 1956
3.5379	5.5644[b]	1.573	Spedding, Daane, & Herrmann, 1956
3.535[c]	5.558[c]	1.572	Singh, Khanduri, & Tsang, 1971
3.5375	5.5540	1.5700	Spedding & Beaudry, 1971
av. 3.5376	5.5543	1.5701	
±0.0003	±0.0003	±0.0001	

[a]From kX, omitted from the average.   [c]Read from a graph, omitted from the
[b]Omitted from the average.   average.

YTTERBIUM

Ytterbium was found by Klemm and Bommer (1937) to be cubic close packed.
Their and subsequent determinations of the lattice constant are presented in
Table 70-1.

Bucher et al. (1970) found that samples condensed from the vapor were a
mixture of the cubic close packed form together with a hexagonal close packed
form. The transition between the two was stated to occur in the range 100 to
360°K, but with a large degree of hysteresis. Their lattice constants for the
cubic close packed form are included in Table 70-1; those for the hexagonal
form are $a = 3.883$, $c = 6.328$ Å, $c/a = 1.629$. The hexagonal form was subsequently observed in high purity ytterbium by Kayser (1971), who stated
that it was thermodynamically stable below 270°K; the lattice constants reported are $a = 3.8799 \pm 0.0002$, $c = 6.3859 \pm 0.002$ Å, $c/a = 1.6459 \pm 0.0001$
at 23°C. These differ somewhat from those of Bucher et al., and are probably more accurate.

The atomic volume of the low temperature hexagonal close packed form is
41.63 Å^3, and that of the cubic close packed form, 41.24 Å^3, both at room
temperature. The transition thus is accompanied by a contraction of 0.39 Å^3,
or 0.9%. The corresponding interatomic distances are: 12 at 3.878 Å in the cubic
close packed form, and six at 3.880 and six at 3.900 Å, average, 3.890 Å, in
the hexagonal close packed form.

At 732°C the structure changes to body centered cubic, two atoms per unit
cell (Spedding, Hanak, and Daane, 1961). The dependence of the atomic
volume on the temperature is shown in Fig. 70-1. Lattice constants at 774°C
are: ccp, $a = 5.603$ Å (extrapolated); bcc(2), $a = 4.44$ Å. The change in
atomic volume for the transition ccp → bcc(2) is –0.2 Å^3, or –0.5%, with an
estimated error of about 0.3 Å^3.

Table 70-1.   Lattice Constant of Cubic Close Packed Ytterbium at Room
Temperature

a, Å	Reference
5.479[a]	Klemm & Bommer, 1937
5.460[b]	Daane, Dennison, & Spedding, 1953
5.4862	Spedding, Daane, & Herrmann, 1956
5.4838	Spedding, Hanak, & Daane, 1961
5.481	Hall, Barnett, & Merrill, 1963
5.486	Bucher et al., 1970
5.4847	Kayser, 1971
av. 5.4843 ± 0.0021	

[a]From kX, omitted from the average.
[b]Omitted from the average.

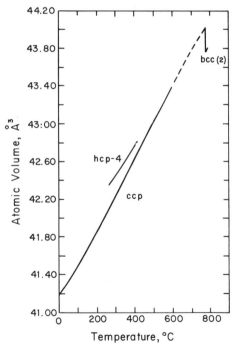

Fig. 70-1.   The change in atomic volume of ytterbium with temperature at atmospheric
pressure (after Spedding, Hanak, & Daane, 1961). (The hcp-4 form is observed only if small
amounts of atmospheric impurities are present.)

A high pressure transition at 25°C and 39.5 kbar to the body centered cubic structure with two atoms per unit cell was discovered by Hall and Merrill (1963). At the transition point the lattice constants are, for the cubic close packed form, a = 5.12 Å, and for the body centered cubic form, a = 4.02 Å. The volume change for the transition ccp → bcc(2) is thus –3.2%.

LUTECIUM

Lutecium was found to be hexagonal close packed by Klemm and Bommer (1937). Their and other determinations of the lattice constants are presented in Table 71-1. No change in structure between room temperature and 1400°C was observed by Spedding, Hanak, and Daane (1961). The change in atomic volume with temperature is presented in Fig. 71-1. The thermal expansion in the range 90 to 300°K was reported by Singh et al. (1971). No structural transition occurs in this range. Their results are presented in Fig. 71-2.

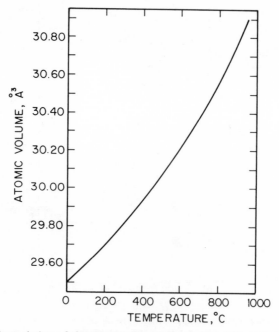

Fig. 71-1.   The variation of the atomic volume of lutecium with temperature (calculated from the data of Spedding, Hanak, & Daane, 1961).

Table 71-1.   Lattice Constants of Lutecium at Room Temperature

a, Å	c, Å	c/a	Reference
3.516[a]	5.570[a]	1.584	Klemm & Bommer, 1937
3.5031	5.5509	1.5846	Spedding, Daane, & Herrmann, 1956
3.5050	5.5509	1.5837	Spedding, Hanak, & Daane, 1961
3.516[b]	5.574	1.585	Singh, Khanduri, & Tsang, 1971
3.5052	5.5494	1.5832	Spedding & Beaudry, 1971
av. 3.5044	5.5504	1.5838	
±0.0005	±0.0007	±0.0004	

[a]From kX, omitted from the average.

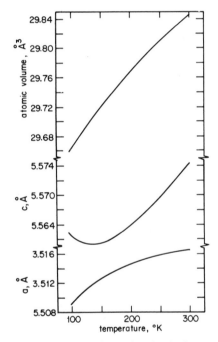

Fig. 71-2.   The thermal expansion of lutecium in the low temperature range (after Singh et al., 1971).

SUMMARY OF THE LANTHANIDES

The lattice constants, atomic volumes, and average nearest neighbor distances of the 15 lanthanides at room temperature and atmospheric pressure are summarized in Table 71-2, and the various allotropes observed at other temperature and pressures are summarized in Table 71-3.

Table 71-2. Summary of the Lanthanides at Room Temperature and 1 atm

Metal	Struc-ture	Lattice Constants, Å		Atomic Volume, Å^3	Average Nearest Neighbor Distance, Å
		a	c		
Lanthanum	hcp-4	3.760	12.143	37.17	3.746
Lanthanum	ccp	5.3045	–	37.30	3.751
Cerium	ccp	5.1608	–	34.36	3.649
Cerium	hcp-4	3.677	11.862	34.72	3.662
Praseodymium	hcp-4	3.684	11.937	35.08	3.674
Praseodymium	ccp	5.136	–	33.87	3.632
Neodymium	hcp-4	3.6570	11.7995	34.17	3.642
Promethium	hcp-4	3.65	11.65	33.60	3.62
Samarium	hcp-9	3.626	26.222	33.17	3.607
Samarium	hcp-4	3.618	11.66	33.05	3.602
Europium	bcc-2	4.5821	–	48.10	3.968
Gadolinium	hcp	3.6333	5.7794	33.04	3.602
Terbium	hcp	3.6041	5.6961	32.04	3.565
Dysprosium	hcp	3.5918	5.6518	31.57	3.548
Holmium	hcp	3.5769	5.6169	31.12	3.531
Erbium	hcp	3.5592	5.5885	30.65	3.513
Thulium	hcp	3.5376	5.5543	30.10	3.492
Ytterbium	ccp	5.4843	–	41.24	3.878
Ytterbium	hcp	3.8799	6.3859	41.63	3.890
Lutecium	hcp	3.5044	5.5504	29.52	3.469

Table 71-3. Summary of Allotropy in the Lanthanides[a]

	Room Temperature 1 atm	Low Temperature 1 atm	High Temperature 1 atm	Room Temperature High Pressure	High Temperature High Pressure
Lanthanum	hcp-4	–	ccp, bcc-2	ccp	ccp, bcc-2
Cerium	ccp	ccp', hcp-4	bcc-2	ccp'	ccp, ccp'
Praseodymium	hcp-4	–	bcc-2	ccp'	ccp
Neodymium	hcp-4	–	bcc-2	ccp'	ccp
Promethium	hcp-4	–	–	–	–
Samarium	hcp-9	–	bcc-2	hcp-4	bcc-2
Europium	bcc-2	bcc-2	bcc-2	–	–
Gadolinium	hcp	hcp	bcc-2	hcp-9	bcc-2
Terbium	hcp	orthorhombic	bcc-2	hcp-9	bcc-2
Dysprosium	hcp	orthorhombic	–	hcp-9	–
Holmium	hcp	–	hcp	hcp-9	–
Erbium	hcp	hcp	hcp	hcp	hcp
Thulium	hcp	–	hcp	–	–
Ytterbium	ccp	hcp	bcc-2	–	–
Lutecium	hcp	–	hcp	–	–

[a]Dash signifies not yet investigated in that region; ccp' refers to the collapsed form.

ACTINIDES

ACTINIUM

The actinium isotope of mass number 227 is cubic close packed with lattice constant 5.311 Å at room temperature (Farr, Giorgi, Bowman, and Money, 1961). Each actinium atom thus has 12 closest neighbors at 3.755 Å, and the atomic volume is 37.45 $Å^3$.

THORIUM

Thorium was found to be cubic close packed first by Bohlin (1920), and independently by Hull (1921b). These and subsequent determinations of the lattice constants are presented in Table 90-1. Evans and Raynor (1959) showed that abnormally high values of a are obtained if the samples are contaminated with nitrogen; such values were omitted in taking the average in Table 90-1.

A transition in thorium at about 1400°C was discovered by Chiotti (1950), who subsequently showed (1954) that the high temperature form is body centered cubic with two atoms per unit cell. Pertinent values of the lattice constants are included in Table 90-1. Within the accuracy of the measurements, the volume change on transition is zero.

In the cubic close packed form each atom has 12 closest neighbors at 3.596 Å at room temperature and at 3.663 Å at 1450°C; the values for the atomic volume are 32.87 and 34.75 $Å^3$, respectively; in the bcc(2) form at 1450°C, each atom has eight nearest neighbors at 3.56 Å, and the atomic volume is 34.7 $Å^3$.

PROTACTINIUM

The structure of protactinium was first determined by Zachariasen (1952a). Thirty lines on powder photographs were interpreted as being due to the presence of Pa, PaO, and PaO_2 in the samples under study. The free metal was said to be tetragonal, with two atoms in a body centered unit cell having a = 3.925 ± 0.005 Å, c = 3.238 ± 0.007 Å, c/a = 0.825, space group I4/mmm-D_{4h}^{17}). This structure is shown in Fig. 91-1.

As pointed out by Zachariasen, this structure can be regarded as being derived from the bcc(2) structure by compressing the latter along one of the three fourfold axes so that the axial ratio is decreased from the value unity to

Table 90-1.   Lattice Constants of Thorium

a, Å	Reference
Cubic close packed form at room temperature	
5.13[a]	Bohlin, 1920
5.05[a]	Hull, 1921b
5.0842[b]	Burgers & van Liempt, 1930
5.101[a]	Thompson, 1933
5.0874[c]	Wilhelm, Carlson, & Lunt, 1953
5.089[c]	Chiotti, 1954
5.0843	James & Straumanis, 1956
5.0843	Evans & Raynor, 1959
5.0854[b]	McMasters & Larsen, 1961
5.0850[b]	Evans & Raynor, 1961
5.0851[b]	Johnson & Honeycombe, 1961
5.0858[b]	Harris & Raynor, 1964
5.0863	Thompson, 1964
5.0856[b]	Norman, Harris, & Raynor, 1966
av. 5.0851 ± 0.0007	
Cubic close packed form at 1450°C	
5.180[d]	Chiotti, 1954
bcc(2) form at 1450°C	
4.11	Chiotti, 1954

[a]From kX, omitted from the average.   [c]Omitted from the average.
[b]From kX.   [d]Extrapolated value.

0.825. If this ratio had the value $\sqrt{2/3}$ = 0.816, each metal atom would have 10 neighbors at exactly the same distance. The true structure was stated to be distorted from this ideal structure such that each atom has eight neighbors at $\sqrt{a^2/2 + c^2/4}$ = 3.213 Å and two at c = 3.238 Å, average, 3.218 Å.

The ideal structure may also be described as stacking of hexagonal close packed layers, the second layer lying over the first such that each atom in it touches two atoms in the first layer, rather than three, as in cubic or hexagonal close packing. A different way of distorting the ideal structure was later suggested by Donohue (1959b), who pointed out that an orthorhombic structure based on space group Fmmm (D_{2h}^{23}), a = 5.566, b = 5.603, c = 3.214 Å, four atoms per cell, gives almost as good agreement with the powder data. In this structure each atom has six neighbors at $\frac{1}{2}\sqrt{a^2 + c^2}$ = c = 3.214 Å and four at $\frac{1}{2}\sqrt{b^2 + c^2}$ = 3.230 Å, average 3.220 Å. This structure, however, was rejected by Zachariasen (1959) on the basis of revised lattice constants of a = 3.932 ± 0.003 Å and c = 3.238 ± 0.003 Å, as obtained from remeasurement of the original photographs.

This question was later settled by Marples (1965), who obtained powder data of much better resolution, especially at higher temperatures. His results showed that the tetragonal structure is the correct one, with lattice constants $a = 3.945 \pm 0.002$ Å, $c = 3.240 \pm 0.001$ Å, $c/a = 0.8254$, at 18°C. The corresponding interatomic distances are eight at 3.214 Å and two at 3.240 Å, average, 3.219 Å.

Marples also measured the lattice constants from 18 to 1055°C. The results are presented in Fig. 91-2. He tentatively suggested that in the neighborhood of 1200°C, the a- and c-axes become equal in length, a change equivalent to a transition to the bcc(2) structure. On the basis of one observed powder line, the lattice constant at 1170°C is 3.81 Å.

Asprey, Fowler, Lindsay, White, and Cunningham (1971) found that very pure protactinium which had been arc-melted gave powder patterns (12 lines) corresponding to the cubic close packed structure, with $a = 5.019 \pm 0.003$ Å. This presumably is a quenched high temperature form of the metal. Flattening the sample in a press resulted in transformation to the tetragonal form.

The atomic volume of the cubic close packed form is 3.161 Å, or 26.6% larger than that of the tetragonal allotrope, and each protactinium atom has 12 nearest neighbors at 3.549 Å.

Direct measurements at high temperatures are needed to settle the question of the high temperature allotropy of protactinium.

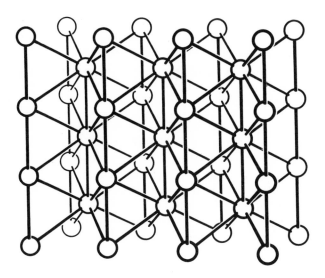

Fig. 91-1.   The structure of the room temperature form of protactinium.

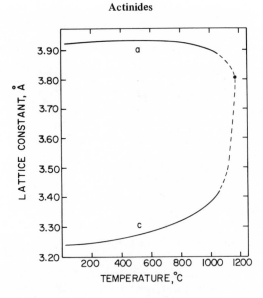

Fig. 91-2. The variation of the lattice constants of tetragonal protactinium with temperature; if the lattice constants become equal, at ca. 1179°C, the structure is then body centered cubic (from the data of Marples, 1965).

URANIUM

Uranium exists in three modifications:

$$\alpha \xrightarrow{\quad 662°C \quad} \beta \xrightarrow{\quad 772°C \quad} \gamma$$

The α form is stable down to at least 4.2°K (Barrett, Mueller, and Hitterman, 1963), while the γ form is stable from 772°C up to the melting point of 1132°C. (These transitions pertain to atmospheric pressure.)

α-Uranium

The first study of uranium at room temperature is that of McLennan and McKay (1930a), who reported that it is body centered cubic, two atoms per unit cell, $a = 3.44$ Å (from kX). In light of later work (see below) it now appears that what they examined was a sample of γ-uranium stabilized by an unknown amount of an unknown impurity.

A powder pattern for α-uranium was obtained by Wilson (1933a,b), who indexed it on the basis of a monoclinic unit cell having $a = 2.835$, $b = 4.897$, $c = 3.315$ Å (from kX), $\beta = 63°26'$, two atoms per unit cell. The space group was said to be C2/m–C_{2h}^3; the two atoms therefore lie in the special positions (000) and ($\frac{1}{2}\frac{1}{2}$0).

A quite different structure was deduced from more extensive and better re-solved powder data by Jacob and Warren (1937). They reported that α-uran-ium is orthorhombic, with a = 2.856, b = 5.877, c = 4.955 Å (from kX), four atoms per unit cell, space group Cmca–D_{2h}^{17}. The correctness of this lattice was later confirmed by single crystal work by Lukesh (1949), who also pointed out that there was a relation between the monoclinic cell of Wilson and the ortho-rhombic cell of Jacob and Warren. A primitive cell corresponding to the end-centered orthorhombic cell has lattice constants a = 2.856, b = 4.955, c = 3.268 Å, β = 64° 10′; these differ from Wilson's values (see above) by an aver-age of only 1%. These results illustrate the danger of accepting a structure, even one as simple as this, on the basis of powder data only.

Numerous determinations of the lattice constants followed that of Jacob and Warren. These are presented in Table 92-1.

The four atoms in the orthorhombic unit cell lie in the following positions: $\pm(0y\frac{1}{4})$, $(\frac{1}{2}\ \frac{1}{2}+y\ \frac{1}{4})$. There have also been numerous determinations of the y-parameter at room temperature. These are presented in Table 92-2.

The structure may be described in various ways. The nearest neighbors of each uranium atom are:

two at 2.753 Å
two at 2.854 Å
four at 3.263 Å
four at 3.343 Å

These 12 atoms are arranged at the vertices of a polyhedron which resembles that found in the hexagonal close packed structure but is quite distorted, as shown in Fig. 92-1.

On the other hand, if only the four shortest distances per atom are taken into consideration, then the structure may be thought of as consisting of pleated sheets of tightly bound atoms, with longer, subsidiary bonds between the sheets. Figures 92-2 and 92-3 are two views of this concept of the struc-ture. In this description, the four nearest neighbors lie very nearly at the ver-tices of trigonal bipyramid, with one of the equatorial positions unoccupied, as shown in Fig. 92-4.

The thermal behavior of α-uranium has been studied extensively. The vari-ation of the lattice constants at low temperatures is presented in Fig. 92-5, which is based on data of Barrett, Mueller, and Hitterman (1963). The ther-mal behavior of the lattice constants up to the transition temperature is pre-sented in Fig. 92-6. In this figure, smooth curves were drawn through the combined data of Bridge et al. (1956), Chiotti et al. (1959), and Barrett et al. (1963).

Table 92-1.   The Lattice Constants of α-Uranium at Room Temperature

a, Å	b, Å	c, Å	Reference
2.856[a,b]	5.877[a,b]	4.955[a,b]	Jacob & Warren, 1937
2.8540	5.8683	4.9576	Schramm, Gordon, & Kaufmann 1950
2,858[b]	5.877[b]	4.955[b]	Thewlis, 1951a
2.8541	5.8677	4.9563	Bridge, Schwartz, & Vaughan, 1956
2.8535	5.8648	4.9543	Chiotti, Klepfer, & White, 1959
2.854[b]	5.869[b]	4.955[b]	Sturcken & Post, 1960
2.848[b]	5.861[b]	4.957[b]	Cash, Hughes, & Murdock, 1961
2.85360[a]	5.86984[a]	4.95552[a]	Cooper, 1962
2.8537	5.8695	4.9548	Barrett, Mueller, & Hitterman, 1963
av. 2.8538	5.8680	4.9557	
±0.0004	±0.0020	±0.0013	

[a]From kX.
[b]Not included in the averaging.

Table 92-2.   The y-Parameter in α-Uranium at Room Temperature

Value	Error	Reference
0.105	±0.005	Jacob & Warren, 1937
0.107	0.003	Konobeevsky et al., 1958
0.1024	0.0005	Sturcken & Post, 1960, as revised by Mueller et al., 1962
0.105	—	Chebotarev, 1961
0.102	0.002	Cash, Hughes, & Murdock, 1961
0.1025[a]	0.0005	Mueller, Hitterman, & Knott, 1962; Barrett, Mueller, & Hitterman, 1963
0.1024[b]	0.0003	*Ibid.*
0.1021	0.0003	Eeles & Sutton, 1963
av.[c] 0.1023		
±0.0002		

[a]X-ray investigation.
[b]Neutron investigation.
[c]Only the four determinations having four significant figures were used in the averaging.

130

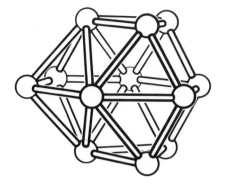

Fig. 92-1. The coordination polyhedron formed by the 12 nearest neighbors of a uranium atom in α-uranium. These nearest neighbors lie from 2.753 to 3.343 Å from the central atom, which is not shown.

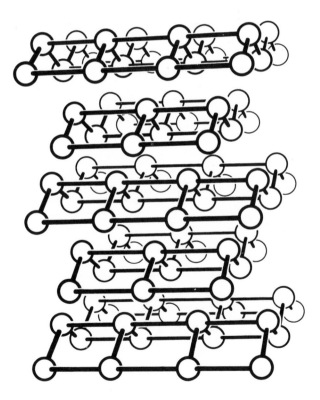

Fig. 92-2. The α-uranium structure, with only the four shortest U-U distances per atom depicted as bonds; b-axis vertical, c-axis horizontal.

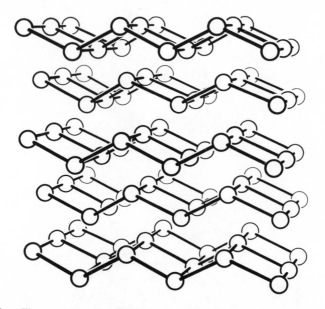

Fig. 92-3.   The α-uranium structure, with only the four shortest U-U distances per atom depicted as bonds; *b*-axis vertical, *c*-axis horizontal.

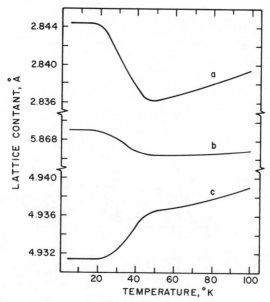

Fig. 92-5.   The variation with temperature of the lattice constants of α-uranium, in the low temperature range (drawn with the data of Barrett et al., 1963).

132

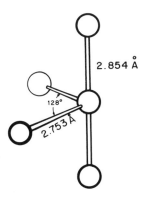

2.854 Å

128°

2.753 Å

Fig. 92-4.   The configuration of the four nearest neighbors of a uranium atom in α-uranium.

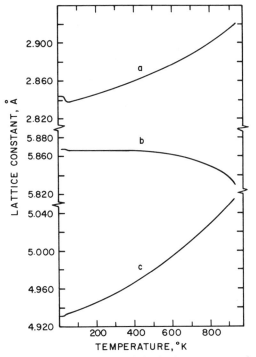

Fig. 92-6.   Variation with temperature of the lattice constants of α-uranium (combination of the data of Bridge et al., 1956, Chiotti et al., 1959, and Barrett et al., 1963).

The value of the y-parameter has also been determined at various temperatures. Barrett et al. (1963) carried out both x-ray and neutron diffraction studies in the range 4 to 298°K. Their results are summarized in Fig. 92-7. Various determinations of y at these and higher temperatures are presented in Fig. 92-8. At temperatures near the transition to the β structure (935°K) the agreement is not very good.

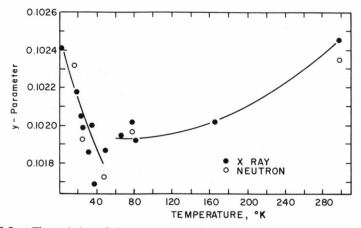

Fig. 92-7.   The variation of the y-parameter with temperature in α-uranium, in the low temperature range (from data of Barrett et al., 1963).

β-Uranium

The lengthy chronicle of the studies on the structure of β-uranium not only includes more than the usual number of contradictions and controversies, but the final chapter has yet to be written. Results to the present are summarized in Table 92-3. The letters used in the discussion below refer to the reference letters of Table 92-3.

The controversy has resolved itself as to which of the three possible space groups is the correct one. The groups $P\bar{4}n2\text{–}D_{2d}^8$, $P4nm\text{–}C_{4v}^4$, and $P4/mnm\text{–}D_{4h}^{14}$ are all consistent with the absences in the observed diffraction pattern. Of these, $P\bar{4}n2$, strangely, has not been seriously considered; the reason for this omission has never been explained. The acentric $P4nm$ was originally vigorously favored by Tucker, (C, H), but later dropped in favor of $P4/mnm$ by

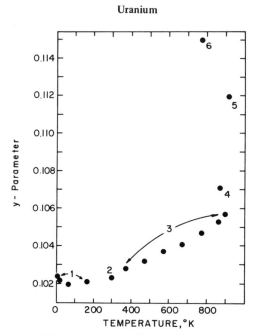

Fig. 92-8.   Variation of the y-parameter with temperature in α-uranium, low temperature up to the transition to the β structure. (1) Barrett et al., 1963; (2) Table 92-2;
(3) Mueller et al., 1962; (4) Eeles and Sutton, 1963; (5) Chebotarev, 1961;
(6) Konobeevski, 1958.

Tucker and Senio (L, N, Q). By analogy with the σ-phase, P4nm was first preferred by Dickins et al. (E), but P4/mnm was later preferred by them (P) for that
phase. Conversely, analogy with the σ-phase was early used to favor P4/mnm for
β-uranium (D, F, O).

A second point of confusion is whether the structure of pure β-uranium at
high temperature is the same as that of uranium-chromium alloys (0.3 to 4.0%
chromium) at room temperature. Unfortunately, the only examination of pure
uranium at high temperature was of powder (G, K), while the only examination of single crystals was of the alloy at room temperature (B, C, J, L). It
has been maintained that intensity differences among the powder patterns of
pure uranium at 720°C, the 1.4% chromium alloy at room temperature, and the
alloy at 720°C indicate that β-uranium and the alloy do not have identical
structures (G, K, Q). It has been suggested, on the other hand, that these differences may have arisen from preferred orientation in the powder samples,
and that β-uranium and the alloys are isostructural (J, L, Q); it therefore
follows that the single crystal studies of the alloys at room temperature (C,
H, J, L, N, Q) give a true picture of the structure of β-uranium itself.

A recent refinement (S) of the high temperature powder data of K was said

Table 92-3. Summary of Structural Studies on β-Uranium

Reference	Experimental	Results
A. Wilson & Rundle, 1948, 1949	Powder photographs of pure U at high temperature, and of quenched U-Cr and U-Mo alloys at room temperature	Unable to index the powder patterns
B. Tucker, 1950a	Powder photographs of pure U at 750°C, of U-Cr (2–4%) at room temperature, and single crystals of U-Cr (2%) at room temperature	High temperature powder pattern of pure β-U and room temperature pattern of quenched low Cr alloys essentially the same; single crystal data indicate an orthorhombic unit cell with $a=7.51$, $b=15.00$, $c=5.59$ Å, 30 atoms per cell
C. Tucker, 1950b, c	Single crystals of U-Cr (1.4%) quenched from 720°C, photographed at room temperature	Tetragonal, with $a=10.52$, $c=5.57$ Å, 30 atoms per cell; space group either P4/mmm (D_{4h}^{1}), P4nm(C_{4v}^{4}), or P4n2(D_{2d}^{8}); P4/mmm ruled out by intensity calculations, P4nm accounts adequately for the observed intensities $P\bar{4}n2$ not considered in detail
D. Bergman & Shoemaker, 1951	Single crystals of σ-FeCr, photographed at room temperature	σ-FeCr has space group P4/mmm; examination of the β-U data of Tucker (1950b) makes it appear likely that it has the same structure
E. Dickins, Douglas, & Taylor, 1951a, b	Single crystals and powders of σ-FeCr and σ-CoCr, photographed at room temperature	Excellent intensity agreement obtained if P4nm is assumed; extremely probable β-U and the σ-structure are the same

136

F.	Kasper, Decker, & Belanger, 1951	Single crystals and powders of σ-CoCr, photographed at room temperature	Space Group P4/mnm preferred; "seems little doubt that the sigma-phase structure is essentially that for β-U"
G.	Thewlis, 1951	Powder photographs of pure U at 720°C and of U-Cr (1.4%) at 720°C and room temperature	At 720°C β-U is tetragonal, with $a=10.759$, $c=5.656$Å, and at room temperature U-Cr (1.4%) has $a=10.590$, $c=5.634$Å, different from the values reported by Tucker (1950b); intensity differences show that β-U and U-Cr (1.4%) do not have identical structures
H.	Tucker, 1951	Same as Tucker (1950b, c)	Same as Tucker (1950b,c); P4/mnm "definitely excluded", P4nm accounts adequately for the intensities
I.	Tucker, 1952a	None	Corrects the date of a reference cited in Tucker (1951)
J.	Tucker, 1952b	Single crystals and powders of U-Cr (0.3–4%) photographed at room temperature	The lattice constants of Thewlis (1951) fit the data quite well; intensity considerations show that P4/mnm is definitely unacceptable; powder patterns of U-Cr (4.0, 1.8, 0.6, and 0.3% of Cr) show no significant intensity differences, suggesting the differences seen by Thewlis (1951) may be caused by preferred orientation
K.	Thewlis, 1952	Powder photographs of pure U and U-Cr (1.4%) at 720°C	Intensity differences suggest that refinement of the structure should be carried out on data from β-U itself, and not the alloy

Table 92-3 (Continued)

Reference	Experimental	Results
L. Tucker & Senio, 1953	Single crystals of U-Cr (1.4%) photographed at room temperature	P4/mnm now considered more probable, on the basis of detailed intensity considerations, R = 31%; intensity discrepancies observed by Thewlis (1951, 1952) do not occur in the single crystal data; β-U and the low Cr alloys therefore have identical structures; used lattice constants of Thewlis (1951)
M. Thewlis & Steeple, 1954	None: powder data of Thewlis (1952) used	The P4/mnm structure refines to R = 32%; the P4nm structure to R = 19%; the latter is therefore correct
N. Tucker, 1954	None: previously published results discussed	The R value for 74 planes in the Tucker and Senio (1953) data is 18%; the R for these same 74 planes, as observed in 54 lines, in the Thewlis and Steeple (1954) data is also 18%; intensity considerations indicate that the P4/mnm structure of Tucker and Senio is more probably correct
O. Bergman & Shoemaker, 1954	Single crystals and powders of σ-FeCr and σ-FeMo photographed at room temperature	Complete structure determinations; P4/mnm structure for both; β-U considered to have the same structure

P.	Dickins, Douglas, & Taylor, 1956	Single crystal photographs of σ-CoCr at room temperature	Space group P4/mnm now preferred for the σ-phase (cf. E above)
Q.	Tucker, Senio, Thewlis, & Steeple, 1956	None: previously published results discussed	No new conclusions: each pair of authors maintains its previous position
R.	Chiotti, Klepfer, & White, 1959	Powder diffractometry on pure U at 681–765°C	No structural studies; at 727°C the lattice constants are a=10.761, c=5.654 Å
S.	Steeple & Ashworth, 1966	None: powder data of Thewlis (1952) used	Additional refinements gave R=23% for the P4/mnm structure and R=16% for the P4nm structure; the latter "is the more likely"
T.	Donohue & Einspahr, 1971	None: powder data of Thewlis (1952) used, and comments made on the calculations of Steeple and Ashworth (1966)	Available data incapable of being used to fix the correct space group; new single crystal data should be collected

to favor the P4nm structure, but this conclusion was questioned (T): the powder data were stated in fact to be incapable of being used to reach a firm conclusion on which of the three space groups is the correct one. A similar analysis of the single crystal data of L has not been carried out, nor does this appear worthwhile because of the large absorption effects, as evidenced by serious inequalities between the observed intensities of symmetry-equivalent reflections. Surely these would cast serious doubts on any conclusions which might be reached concerning which space group is correct.

It seems fairly obvious that this situation will not be resolved to the satisfaction of all involved (and uninvolved) parties until single crystals of pure uranium are investigated in the temperature range 662 to 772°C.

Various determinations of the lattice constants are presented in Table 92-4. The equivalent positions for the three possible space groups are given in Table 92-5, and the values which have been reported for the various positional parameters are presented in Table 92-6.

Even though the space group of β-uranium has not yet been firmly established it does not appear unreasonable to follow the rule of parsimony here and assume, in the absence of compelling evidence to the contrary, that the structure is centric, space group P4/mnm, reserving the possibility that future evidence may show that distortions to either P4nm or P$\bar{4}$n2 occur. None of the three sets of positional parameters of Table 92-6 for P4/mnm is without criticism: those of L are based on a large set of single crystal intensity data, subject, however, to considerable absorption errors; those of S are based on a set of powder data corrected for absorption, but which, according to T, was, in part, incorrectly indexed; those of T, although corrected for the defects of S, still are based on a very small set of data as compared with that of L. As a compromise, therefore, the average of the parameters of L and T have been adopted, as shown in Table 92-6, plus the admittedly arbitrary decision to fix the value of z(III) at $\frac{3}{4}$, supported, however, by the fact that this value has been reported by D, F, O, and P for the corresponding atoms in the σ phase.

The structure has often been described in terms of "main layers" of atoms of types I, II, IV, and V parallel to the ab plane at $z = 0$ and $\frac{1}{2}$, interspersed with "subsidiary layers" of atoms of type III at $z = \frac{1}{4}$ and $\frac{3}{4}$. This is not, however, a particularly useful way of describing this structure, because the intra-"layer" bonds are not the shortest in the structure, which is really a complicated three-dimensional network. The nearest neighbors of each of the five types of atoms are listed in Table 92-7.

A projection of the structure down the z-axis is shown in Fig. 92-9. Atoms of types I and IV have coordination number 12, and their coordination polyhedra are distorted regular icosahedra, as shown in Figs. 92-10 and 92-11. Atoms of types III and V have coordination number 14. The coordination polyhedra in both cases closely approximate a hexagonal antiprism, with addi-

Table 92-4. The Lattice Constants of β-Uranium

a, Å	c, Å	Remarks	Reference
10.52[a]	5.57[a]	1.4% Cr, room temperature	Tucker, 1950c
10.590[b]	5.634[b]	1.4% Cr, room temperature	Thewlis, 1951
10.759	5.656	Pure U at 720°C	Thewlis, 1951
10.759[c]	5.654[c]	Pure U at 720°C	Chiotti, Klepfer, & White, 1959

[a]Low precision values.
[b]Accepted values at room temperature.
[c]Interpolated values.

Table 92-5. The Three Possible Descriptions of the Unit Cell of β-Uranium

Space group P$\bar{4}$n2[a]

   Two atoms in $2d$: $(00\frac{1}{2}),(\frac{1}{2}\frac{1}{2}0)$

   Four atoms in $4g$: $(xx0),(\bar{x}\bar{x}0),(\frac{1}{2}-x\ \frac{1}{2}+x\ \frac{1}{2}),(\frac{1}{2}+x\ \frac{1}{2}-x\ \frac{1}{2})$

   Twenty-four atoms in three sets of $8i$:

$$(xyz),(\bar{x}\bar{y}z),(\tfrac{1}{2}-x\ \tfrac{1}{2}+y\ \tfrac{1}{2}+z),(\tfrac{1}{2}+x\ \tfrac{1}{2}-y\ \tfrac{1}{2}+z)$$
$$(yx\bar{z}),(\bar{y}\bar{x}\bar{z}),(\tfrac{1}{2}-y\ \tfrac{1}{2}+x\ \tfrac{1}{2}-z),(\tfrac{1}{2}+y\ \tfrac{1}{2}-x\ \tfrac{1}{2}-z)$$

Space group P4nm

   Two atoms in $2a$: $(00\frac{1}{2}),(\frac{1}{2}\frac{1}{2}0)$[b]

   Twelve atoms in three sets of $4c$:

$$(xxz),(\bar{x}\bar{x}z),(\tfrac{1}{2}+x\ \tfrac{1}{2}-x\ \tfrac{1}{2}+z),(\tfrac{1}{2}-x\ \tfrac{1}{2}+x\ \tfrac{1}{2}+z)$$

   Sixteen atoms in two sets of $8d$:

$$(xyz),(\bar{x}\bar{y}z),(\tfrac{1}{2}+x\ \tfrac{1}{2}-y\ \tfrac{1}{2}+z),(\tfrac{1}{2}-x\ \tfrac{1}{2}+y\ \tfrac{1}{2}+z)$$
$$(yxz),(\bar{y}\bar{x}z),(\tfrac{1}{2}+y\ \tfrac{1}{2}-x\ \tfrac{1}{2}+z),(\tfrac{1}{2}-y\ \tfrac{1}{2}+x\ \tfrac{1}{2}+z)$$

Space group P4/mnm

   Two atoms in $2b$: $(00\frac{1}{2}),(\frac{1}{2}\frac{1}{2}0)$

   Four atoms in $4f$: $(xx0),(\bar{x}\bar{x}0),(\frac{1}{2}+x\ \frac{1}{2}-x\ \frac{1}{2}),(\frac{1}{2}-x\ \frac{1}{2}+x\ \frac{1}{2})$

   Eight atoms in $8j$: $(xxz),(\bar{x}\bar{x}z),(\frac{1}{2}+x\ \frac{1}{2}-x\ \frac{1}{2}+z),(\frac{1}{2}-x\ \frac{1}{2}+x\ \frac{1}{2}+z)$

         $(\bar{x}\bar{x}\bar{z}),(xx\bar{z}),(\frac{1}{2}-x\ \frac{1}{2}+x\ \frac{1}{2}-z),(\frac{1}{2}+x\ \frac{1}{2}-x\ \frac{1}{2}-z)$

   Sixteen atoms in two sets of $8i$:

$$(xy0),(\bar{x}\bar{y}0),(\tfrac{1}{2}+x\ \tfrac{1}{2}-y\ \tfrac{1}{2}),(\tfrac{1}{2}-x\ \tfrac{1}{2}+y\ \tfrac{1}{2})$$
$$(yx0),(\bar{y}\bar{x}0),(\tfrac{1}{2}+y\ \tfrac{1}{2}-x\ \tfrac{1}{2}),(\tfrac{1}{2}-y\ \tfrac{1}{2}+x\ \tfrac{1}{2})$$

[a]Origin moved to $(0\frac{1}{2}\frac{1}{4})$ from standard.
[b]z-coordinate arbitrarily chosen as $\frac{1}{2}$.

Table 92-6. Positional Parameters in β-Uranium

Atom Type	Parameter	P4nm^a				P4̄n2	P4/mmm			
Space Group:	Reference:^{b}	C, H	M	S^c	T	T	L^c	S^c	T	d
II	x	0.11	0.105	0.110	0.107	0.100	0.103	0.098	0.099	0.101
	z	0.07	0.04	0.017	-0.014	(0)^e	(0)	(0)	(0)	(0)
IIIa	x	0.32	0.290	0.301	0.303	0.323, 0.313^f	0.318	0.321	0.316	0.317
	z	0.84	0.82	0.815	0.769	(0.725)	0.730	0.723	0.720	0.750
IIIb	x	0.68	0.690	0.671	0.679	(0.677, 0.687)	(0.682)	(0.679)	(0.684)	(0.683)
	z	0.34	0.30	0.280	0.294	(0.275)	(0.270)	(0.277)	(0.280)	(0.250)
IV	x	0.56	0.547	0.563	0.558	0.557	0.561	0.561	0.556	0.559
	y	0.24	0.227	0.220	0.225	0.228	0.235	0.214	0.228	0.232
	z	0.09	0.09	0.075	0.052	0.043	(0)	(0)	(0)	(0)
V	x	0.38	0.367	0.374	0.372	0.368	0.367	0.370	0.368	0.368
	y	0.04	0.041	0.042	0.045	0.041	0.038	0.046	0.039	0.039
	z	0.04	0.00	-0.026	-0.023	-0.017	(0)	(0)	(0)	(0)

^{a}Origin moved to place atoms of type I at $z = \frac{1}{2}$ ($x = y = 0$).
^{b}As in Table 92-3.
^{c}Rounded to three decimal places.
^{d}Average of L and T, except that z(III) set at $\frac{3}{4}$.
^{e}Symmetry-fixed values of the parameters given in parentheses.
^{f}Value of y in position 8i; in the other two space groups $y = x$.

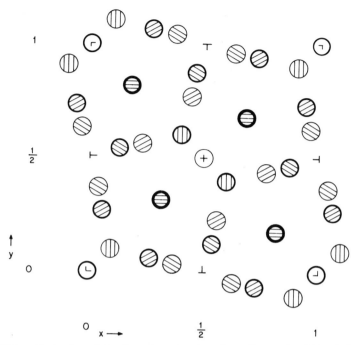

Fig. 92-9. Projection of the β-uranium structure down the c-axis. Thin circles, atoms at $z = 0$; medium circles, atoms at $z = \frac{1}{2}$; thick circles, atoms at $z = \frac{1}{4}$ and $\frac{3}{4}$. ⊕, type I; ⬦, type II; ⊖ type III; ⊘, type IV; ⬨, type V.

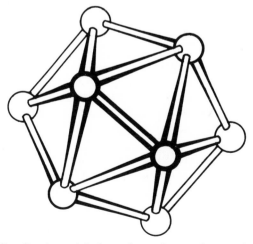

Fig. 92-10. Coordination polyhedron of type I atoms in β-uranium, viewed down the c-axis. The central atom (not shown) is at $x = y = 1/2, z = 0$ in Fig. 92-9.

143

Table 92-7.   Nearest Neighbors in β-Uranium (calculated with room
temperature lattice constants of Table 92-4 and positional
parameters of Table 92-6)

	Type	Distance	Number
Type I to:	IV	2.906	4
	III	3.080	4
	II	3.198	4
			12
Type II to:	V	2.902	2
	II	3.026	1
	I	3.198	2
	IV	3.355	4
	III	3.433	4
	III	3.528	2
			15
Type III to:	III	2.817	2
	IV	3.060	2
	I	3.080	1
	IV	3.117	2
	V	3.308	2
	V	3.369	2
	II	3.433	2

tional atoms located out from the centers of the two hexagonal faces. Views of these polyhedra are shown in Figs. 92-12 and 92-13. Comparison of Figs. 92-13a with Figs. 92-10 and 92-11 shows that the former has some features in common with the regular icosahedron. Type II atoms have coordination number 15; the coordination polyhedron is shown in Fig. 92-14. It, too, is related to the regular icosahedron, as is obvious from Fig. 92-14a.

Although future refinement of the parameters may change the individual interatomic distances in β-uranium by significant amounts the basic description of the structure as being based on the icosahedron and other polyhedra closely related thereto will remain.

Chiotti, Klepfer, and White (1959) determined lattice constants of pure β-uranium over the temperature range 662 to 772°C. The temperature dependence of the lattice constants, as calculated from their data, is $a = 10.596 + 22.6 \times 10^{-5} t$, and $c = 5.621 + 4.6 \times 10^{-5} t$. Although the constants in these equations are based on high temperature data, the extrapolated room temperature values of $a = 10.601$, $c = 5.622$ Å are surprisingly close to the room temperature values of the 1.4% Cr alloy (see Table 92-4).

Table 92-7. (Cont'd.)

	Type	Distance	Number
	II	3.527	1
			14
Type IV to:	V	2.876	1
	I	2.906	1
	V	2.972	1
	V	3.017	2
	III	3.060	2
	III	3.117	2
	IV	3.130	1
	II	3.355	2
			12
Type V to:	IV	2.876	1
	II	2.902	1
	V	2.915	1
	IV	2.972	1
	IV	3.017	2
	III	3.308	2
	III	3.369	2
	V	3.491	4
			14

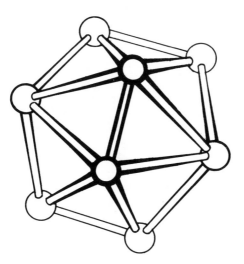

Fig. 92-11. The coordination polyhedron of type IV atoms in β-uranium, viewed down the c-axis. The central atom (not shown) is at $x = 0.559$, $y = 0.228$, $z = 0$ in Fig. 92-9.

145

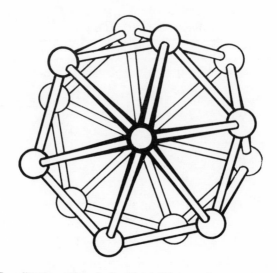

Fig. 92-12.   Coordination polyhedron of type III atoms in β-uranium viewed down the c-axis. The central atom (hidden) is at $x = y = 0.317$, $z = 1/4$ in Fig. 92-9.

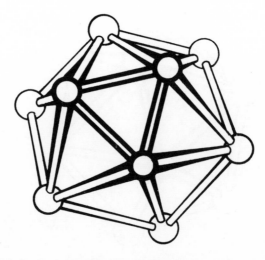

Fig. 92-13a.   Coordination polyhedron of type V atoms in β-uranium, viewed down the c-axis. The central atom (not shown) is at $x = 0.368$, $y = 0.039$, $z = 0$ in Fig. 92-9.

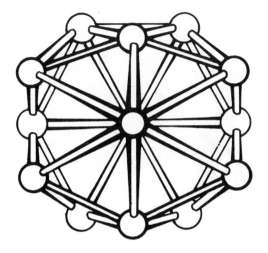

Fig. 92-13b.   Coordination polyhedron of type V atoms in β-uranium viewed in projection along a line through the atoms at the extreme right and left in Fig. 92-13a.

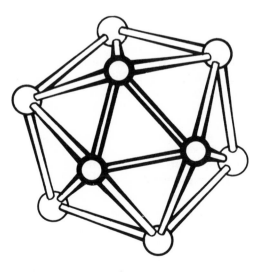

Fig. 92-14a.   Coordination polyhedron of the type II atoms in β-uranium. viewed down the c-axis. The central atom (not shown) is at $x = y = 0.101$, $z = 0$ in Fig. 92-9.

147

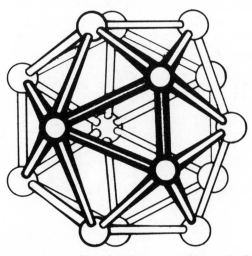

Fig. 92-14*b*. Same polyhedron as Fig. 92-14*a*, but viewed along the *b*-axis, *c*-axis vertical.

β-Uranium in Thin Film

Chatterjee (1958) reported that thin films of uranium obtained by vacuum evaporation had the β-structure (at room temperature). This result was questioned by Donohue (1961), who pointed out that although the spacings reported by Chatterjee may have matched those expected from β-uranium, the intensity discrepancies were so great as to exclude the possibility that that phase had actually been observed. This conclusion was confirmed by Kolomiels (1967), who found no trace of the β-uranium phase in thin films similarly obtained; the only diffraction lines observed were consistent with those expected from the room temperature of α form. The source of the pattern reported by Chatterjee remains unexplained.

γ-Uranium

The structure of the high temperature form of uranium was first determined by Wilson and Rundle (1949) from powder data obtained from pure uranium at 800°C. It was found to be body centered cubic, two atoms per unit cell, $a =$ 3.49 Å (from kX). Wilson and Rundle also found that this phase could be retained at room temperature in U-Mo alloys, and obtained the value $a =$ 3.474 Å (from kX) by extrapolation to zero Mo content. Thewlis (1951) reported $a = 3.524$ Å for pure body centered uranium at 805°C. Chiotti, Klepfer, and White (1959) studied pure uranium and uranium-zirconium alloys over the temperature range of 800 to 1060°C. From a figure given by them the equation $a = 3.472 + 77 \times 10^{-6}\,t$ may be deduced, where a is the lattice constant in angstroms and t the centigrade temperature. At 800°C the lattice constant is 3.534 Å, or somewhat higher than the two high temperature values cited above. The lengthy extrapolation to room temperature gives $a = 3.474$ Å, in exact agreement with Wilson and Rundle.

Uranium at High Pressure

The temperature-pressure behavior of uranium has been investigated by Klement, Jayaraman, and Kennedy (1963a). No new phases were discovered. Their phase diagram is shown in Fig. 92-15. It is noteworthy for the existence of a triple point among the three solid phases at about 31.5 kbar and 800°C.

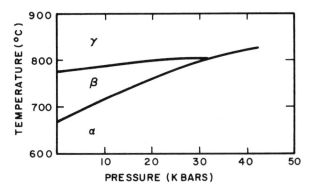

Fig. 92-15.   The phase diagram of uranium (after Klement et al., 1963).

Some Comparisons

Interatomic distances in the three uranium allotropes have all been determined at room temperature: directly determined in α, in the 1.4% Cr alloy in β, and by extrapolation in γ. The respective spectra of nearest neighbors are shown in Fig. 92-16.

The variation of the atomic volume with temperature is shown in Fig. 92-17. The low temperature behavior is remarkable for the minimum at about 50°K. The changes in volume at the transitions are $\alpha \rightarrow \beta$, +1.06% at 662°C, and $\beta \rightarrow \gamma$, +0.73% at 772°C.

? 3:01 ?

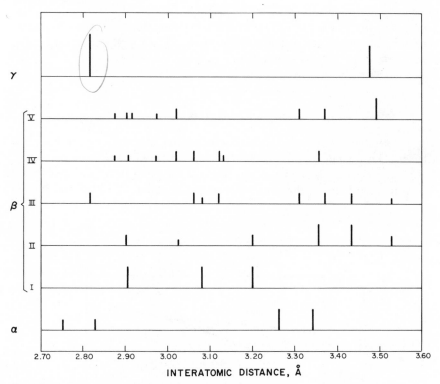

Fig. 92-16. The spectra of closest interatomic distances of a uranium atom in various uranium allotropes. Heights of lines proportional to number of neighbors, shortest line corresponding to one neighbor at that distance.

NEPTUNIUM

The following three modifications of neptunium are known:

$$\alpha \xrightarrow{\ 280°C\ } \beta \xrightarrow{\ 577°C\ } \gamma$$

No studies have been made below room temperature. The γ form is stable up to the melting point of 637°C.

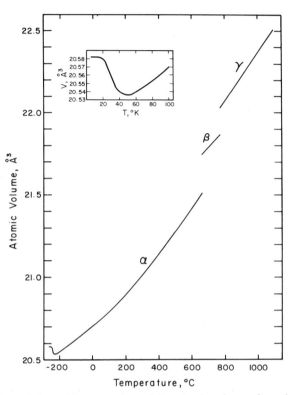

Fig. 92-17. The variation with temperature of the atomic volume of uranium.

α-Neptunium

The structure of α-neptunium was determined by Zachariasen (1952b), who observed 43 powder lines. He found that the substance is orthorhombic, $a = 4.723$, $b = 4.887$, $c = 6.663$ Å at room temperature, eight atoms per unit cell, space group Pmcn (a nonstandard setting of D_{2h}^{16}-Pnma). This unit cell was subsequently confirmed from powder data (24 lines) by Lee et al. (1959), who reported the values $a = 4.721$, $b = 4.888$, $c = 6.661$ Å. The eight atoms lie in two sets of fourfold positions, $\pm(\tfrac{1}{4}yz)$ $(\tfrac{1}{4} \ \tfrac{1}{2}-y \ \tfrac{1}{2}+z)$, with the following values for the parameters:

$$y_I = 0.208 \quad y_{II} = 0.842$$
$$z_I = 0.036 \quad z_{II} = 0.319$$

These were obtained by trial and error, and the uncertainty on each of these was stated to be ±0.006.

As discussed by Zachariasen, the atomic arrangement shows some similarity to the body centered cubic structure. A hypothetical body centered cubic form of neptunium with two atoms in a unit cell having a_c = 3.375 Å would have the same density as the orthorhombic form. This hypothetical bcc(2) form may also be described as having cell dimensions $a_c\sqrt{2}$, $a_c\sqrt{2}$, $2a_c$, with eight atoms in two sets of fourfold positions as above. The degree of distortion of the true structure from this hypothetical structure may be seen in the following comparison:

	True Structure	bcc(2) Structure
a	4.723 Å	4.773 Å
b	4.887 Å	4.773 Å
c	6.663 Å	6.750 Å
y_I	0.208	$\frac{1}{4}$
z_I	0.036	0
y_{II}	0.842	$\frac{3}{4}$
z_{II}	0.319	$\frac{1}{4}$

This distortion is considerable. In the hypothetical structure each neptunium atom would have eight nearest neighbors at 2.923 Å and six next-nearest neighbors at 3.375 Å. In the true structure, the nearest neighbors are:

Np_I–1Np_{II}	2.599 Å	Np_{II}–1Np_I	2.599 Å
1Np_{II}	2.632	1Np_I	2.632
2Np_{II}	2.634	2Np_I	2.634
1Np_{II}	3.052	1Np_I	3.052
2Np_I	3.153	2Np_I	3.351
2Np_{II}	3.351		
2Np_I	3.357		

All other interatomic distances are greater than 3.4 Å.

The structure as viewed in projection down the a-axis, is shown in Fig. 93-1. The four shortest bonds to each neptunium atom, which Zachariasen concluded are covalent, with strength about halfway between a single bond and a double bond, are approximately directed toward four of the five vertices of a trigonal bipyramid, but the unoccupied vertex is not the same for the two kinds of atoms. Details of this concept of the coordination are presented in Fig. 93-2. It is interesting that the coordination about the Np_I atom closely resembles that in α-uranium (cf. Fig. 92-4).

Lattice constants for α-neptunium in the range 20 to 275°C have been reported by Zachiarasen (1952c).

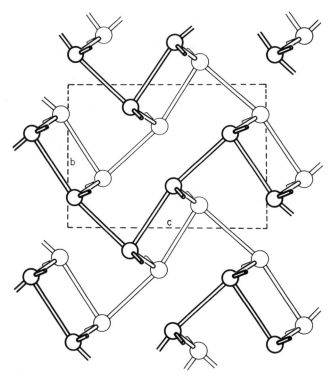

Fig. 93-1.   The structure of α-neptunium projected along the *a*-axis.

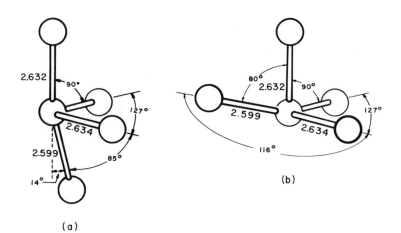

(a)

(b)

Fig. 93-2.   Configuration of the four nearest neighbors in α-neptunium. (*a*) At Np$_I$. (*b*) At Np$_{II}$.

β-Neptunium

The structure of β-neptunium was determined by Zachariasen (1952c) from powder data (35 lines). It is tetragonal, with a = 4.883, c = 3.389 Å at 282°C, four atoms per unit cell, space group P4/nmm.* Lee et al. (1959) reported that at 312°C the lattice constants are a = 4.895, c = 3.386 Å, values in excellent agreement with Zachariasen's values of a = 4.897, c = 3.388 Å at 313°C. The atomic positions are:

$$2Np_I \quad \text{at} \quad (000)\ (\tfrac{1}{2}\tfrac{1}{2}0)$$
$$2Np_{II} \quad \text{at} \quad (\tfrac{1}{2}0z)\ (0\tfrac{1}{2}\bar{z})$$

The value of z is 0.375 ± 0.015 (at 313°C).

The structure as viewed down the a-axis is shown in Fig. 93-3. Like that of α-neptunium, the structure may be discussed in terms of a distorted body centered cubic structure, but in this case the distortion is much simpler. A hypothetical body centered cubic form of neptunium with two atoms per unit cell having a = 3.431 Å would have the same density as the tetragonal form (at 282°C). An alternate description of this form is a face centered tetragonal unit cell with dimensions $a_c\sqrt{2}, a_c\sqrt{2}, a_c$. The degree of distortion may be seen as follows:

	True Structure	bcc(2) Structure
a	4.883 Å	4.852 Å
c	3.389 Å	3.431 Å
z	0.375	$\tfrac{1}{2}$

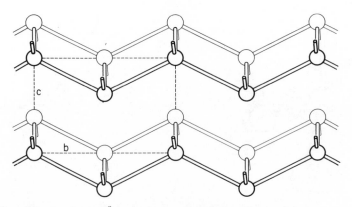

Fig. 93-3.   The structure of β-neptunium projected along the a-axis.

*Zachariasen, curiously, gave the space group as $P42_1$, but the structure described by him actually has space group P4/nmm.

Thus the coordination of each neptunium atom, instead of being eight nearest neighbors at 2.971 Å and six next-nearest neighbors at 3.431 Å, is:

Np_I–$4Np_{II}$	2.752	Np_{II}–$4Np_I$	2.752 Å
$4Np_{II}$	3.232	$4Np_I$	3.232
$2Np_I$	3.389	$2Np_{II}$	3.389
$4Np_I$	3.453	$4Np_{II}$	3.555

A portion of the structure is shown in Fig. 93-4, and the coordination of the two different kinds of neptunium atoms is shown in Figs. 93-5 and 93-6.

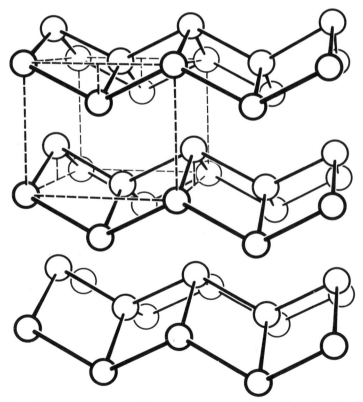

Fig. 93-4.   A view of a portion of the β-neptunium structure. The unit cell, c-axis vertical, is outlined with dashed lines.

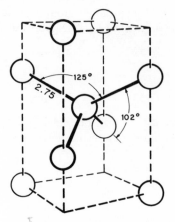

Fig. 93-5.   Configuration at the Np₁ atoms in β-neptunium.

γ-Neptunium

Zachariasen (1952c) stated that preliminary experiments strongly indicated
that at temperatures immediately below the melting point a third form of neptunium, body centered cubic with two atoms per unit cell, exists. All of the
lines save one on a powder pattern taken with the sample at 600°C corresponded to the lines of NpO. The one extra line, plus an observed enhanced intensity of *some* of the NpO lines, led Zachariasen to postulate the bcc(2) form of
neptunium, with a = 3.52 Å at 600°C. Partial confirmation of this result was
provided by Lee et al. (1959), who observed a transition near 577°C in thermal, dilatometric, and resistance experiments, and by Mardon and Pearce
(1959), who, on the basis of the apparently complete solid miscibility of
γ-Np and γ-U, tentatively suggested that the two were isostructural. A portion of the structure, showing its relationship to the structure of β-neptunium,
is presented in Fig. 93-7. Each atom has eight nearest neighbors at 3.05 Å
and six next-nearest neighbors at 3.52 Å.

The change in the atomic volume of neptunium with temperature is shown
in Fig. 93-8.

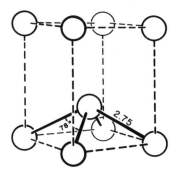

Fig. 93-6. Configuration at the Np$_{II}$ atoms in β-neptunium.

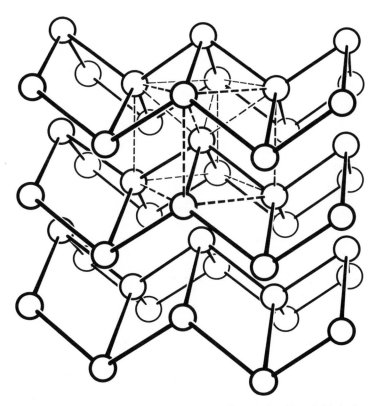

Fig. 93-7. The structure of γ-neptunium, showing the relationship of this body centered cubic structure to the structure of β-neptunium (cf. Fig. 93-4).

157

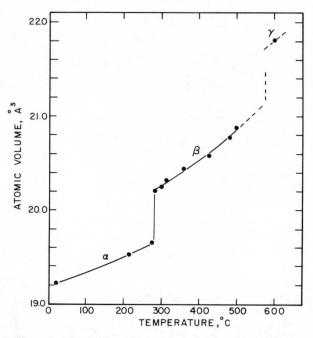

Fig. 93-8.   The variation of the atomic volume of the various allotropes of neptunium with temperature (data of Zachariasen, 1952b).

PLUTONIUM

Plutonium exists in the following six allotropic forms:

$$122°C \qquad 206°C \qquad 319°C \qquad 451°C \qquad 476°C$$
$$\alpha \longrightarrow \beta \longrightarrow \gamma \longrightarrow \delta \longrightarrow \delta' \longrightarrow \epsilon$$

This somewhat curious set of letter designations arises from the fact that the δ' form was overlooked in the earlier work. Furthermore, the designation η is assigned to the δ' phase by some authors. The α form is stable down to at least –196°C, and there is no evidence of an additional phase change between 476°C and the melting point of 640°C. The phase behavior at high pressure, shown in Fig. 94-1, is remarkable.

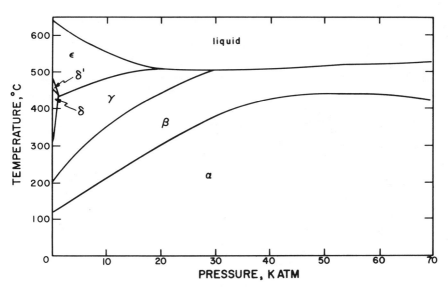

Fig. 94-1.   The phase diagram of plutonium (after Anonymous, 1969).

α-Plutonium

The structure of α-plutonium was briefly described by Zachariasen and Ellinger (1957), and their final refinement was published in 1963(b). The structure is based on about 165 powder lines observed with a diffractometer. Of these, 97 which were considered to have a reliably measured intensity were used in refining the structure.

The substance is monoclinic, with the following lattice constants (at 21°C): $a = 6.183$, $b = 4.822$, $c = 10.963$ Å, $\beta = 101.78°$, 16 atoms per unit cell. The space group is $P2_1/m\text{–}C_{2h}^2$, and the atoms lie on the mirror planes in eight different sets of positions $\pm(x\tfrac{1}{4}z)$. The values for the parameters are presented in Table 94-1.

The structure, as viewed along the b-axis, is shown in Fig. 94-2. It is a very irregular one. The atoms may be thought of as being arranged in highly distorted hexagonal close packed layers at $y = \tfrac{1}{4}$ and $\tfrac{3}{4}$, 2.41 Å apart, but the distortion is so great that the interatomic distances in these layers lie in the range 2.57 to 3.71 Å. The interatomic distances fall into three well-defined groups, 2.57 to 2.78, 3.19 to 3.71, and greater than 4 Å. The distribution of the first two groups at each of the eight crystallographically different plutonium atoms is shown in Fig. 94-3. Only the distances in the first, shorter, group are de-

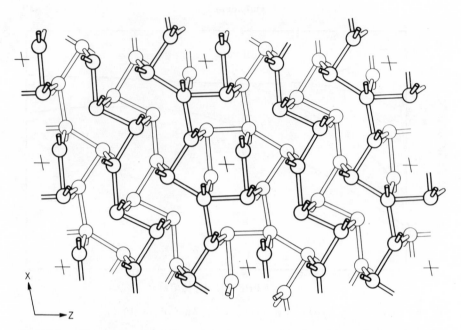

Fig. 94-2. Four unit cells of the α-plutonium structure, viewed in projection along the b-axis.

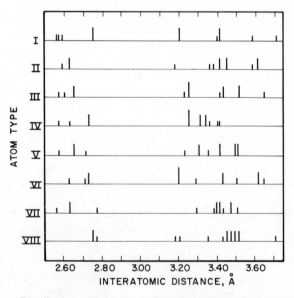

Fig. 94-3. Distribution of interatomic distances at the eight crystallographically different atoms in α-plutonium. Shortest lines denote one neighbor at that distance, double length two neighbors, etc.

160

picted as bonds in Fig. 94-2. The coordination of these short bonds is also very irregular. At Pu(I) the five are directed approximately toward the vertices of a trigonal bipyramid, while at Pu(VIII) the three are approximately planar trigonal. The four short bonds at Pu(II)-(VII) tend to lie in the same hemisphere; this is true also for the four short bonds in α-uranium (Fig. 92-4), α-neptunium (Fig. 93-2), and β-neptunium (Fig. 93-5).

The variation of the lattice constants with temperature has been reported by Zachariasen and Ellinger (1963b) and by Chebotarev and Beznosikova (1960). Their results are presented in Fig. 94-4.

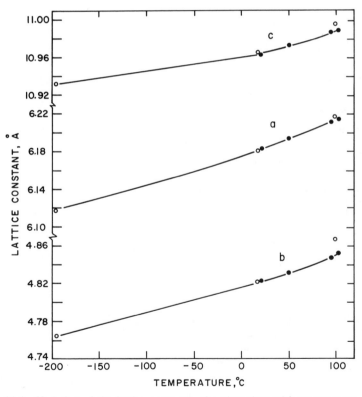

Fig. 94-4. Variation of the lattice constants of α-plutonium with temperature. Open circles, Chebotarev and Beznosikova, 1960; closed circles, Zachariasen and Ellinger, 1963b.

Table 94-1.   Positional Parameters in α-Plutonium[a]

Atom Type	x	z
I	0.345	0.162
II	0.767	0.168
III	0.128	0.340
IV	0.657	0.457
V	0.025	0.618
VI	0.473	0.653
VII	0.328	0.926
VIII	0.869	0.894

[a]The undertainties are 0.004–0.005 on x and 0.002–0.003 on z.

β-Plutonium

The unit cell of β-plutonium was deduced by Zachariasen and Ellinger (1959), and the refined structure was published by them in 1963(a). It is based on 70 uniquely indexed powder lines observed with a diffractometer. About 20 additional unresolved lines were also observed but not used in the structure determination.

β-Plutonium is also monoclinic, with the following lattice constants at 190°C: $a = 9.284$, $b = 10.463$, $c = 7.859$ Å, $\beta = 92.13°$, 34 atoms per unit cell. The space group is $I2/m-C_{2h}^3$. The atoms lie in the following positions:

Atom Type	Number	Coordinates, all with (000), $(\frac{1}{2}\frac{1}{2}\frac{1}{2})+$:
I	2	000
II	4	$\pm(x0z)$
III	4	$\pm(x0z)$
IV	4	$\pm(x0z)$
V	4	$\pm(\frac{1}{2}y0)$
VI	8	$\pm(xyz)\,(x\bar{y}z)$
VII	8	$\pm(xyz)\,(x\bar{y}z)$

The values for the parameters are presented in Table 94-2. These were obtained from diffraction data of a sample containing 2% of uranium, at temperatures ranging from 83 to 252°C. The presence of this amount of uranium was found to cause a shrinkage of about 0.05% of the unit cell dimensions, but it had no effect on the intensities, a result which apparently led to the tacit assumption that the change in temperature had no effect on the values of the positional parameters.

Table 94-2.   Positional Parameters for β-Plutonium[a]

Atom Type	x	y	z
I	0	0	0
II	0.146	0	0.387
III	0.337	0	0.082
IV	0.434	0	0.672
V	$\frac{1}{2}$	0.220	0
VI	0.145	0.268	0.108
VII	0.167	0.150	0.753

[a]The uncertainties are 0.003–0.004 on x, 0.002–0.003 on y, and 0.003–0.005 on z.

The structure, like that of α-plutonium, is also quite irregular. Zachariasen and Ellinger (1963a) divided the nearest neighbors of each of the seven crystallographically different plutonium atoms into two groups, those in the range 2.59 to 3.10 Å, called "short bonds", and those in the range 3.14 to 3.36 Å, called "long bonds", but, as may be seen in Fig. 94-5, this classification is much less convincing than in the case of α-plutonium.

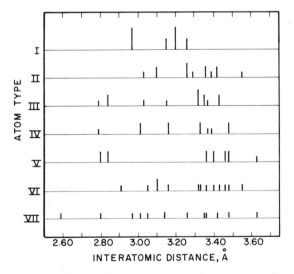

Fig. 94-5.   Distribution of interatomic distances at the seven crystallographically different atoms in β-plutonium. Shortest lines denote one neighbor at that distance, double length two neighbors, etc.

γ-Plutonium

The structure of γ-plutonium was determined by Zachariasen and Ellinger (1955) from powder data. The pattern consisted of 49 lines. It is ortho-rhombic, with $a = 3.1587$, $b = 5.7682$, $c = 10.162$ Å at 235°C, with eight atoms per unit cell. The space group is Fddd–D_{2h}^{24}, with the eight atoms lying in the special positions (000), $(\frac{1}{2}\frac{1}{2}0)$, $(\frac{1}{2}0\frac{1}{2})$, $(0\frac{1}{2}\frac{1}{2})$, $(\frac{1}{4}\frac{1}{4}\frac{1}{4})$, $(\frac{3}{4}\frac{3}{4}\frac{1}{4})$, $(\frac{3}{4}\frac{1}{4}\frac{3}{4})$, $(\frac{1}{4}\frac{3}{4}\frac{3}{4})$. In this structure, shown in Fig. 94-6, each plutonium atom has 10 nearest neighbors, four at 3.026 Å, two at 3.159 Å, and four at 3.288 Å (at 235°C). The structure may be described as consisting of slightly distorted hexagonal close packed layers lying perpendicular to the c-axis. (These layers would be precisely hexagonal close packed if the axial ratio b/a were $\sqrt{3}$ instead of 1.83.) The layers are stacked over one another in such a way that each atom in one layer is equidistant from two atoms in the layer above and two in the layer below. The bond lengths within the layers are 3.159 and 3.288 Å, and those between the layers, 3.026 Å.

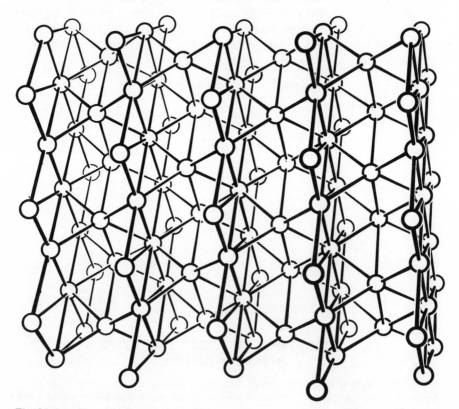

Fig. 94-6.   The structure of γ-plutonium.

Zachariasen and Ellinger also measured the lattice constants at numerous temperatures between 213 and 312°C. The lengths of b and c both increase with temperature, while the length of a decreases, as shown in Fig. 94-7.

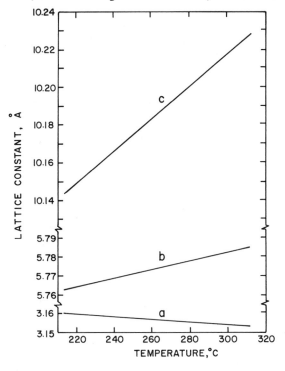

Fig. 94-7.   The variation of the lattice constants of γ-plutonium with temperature (smoothed from the data of Zachariasen and Ellinger, 1955).

δ-Plutonium

The structure of δ-plutonium was first reported by Jette (1955); the more detailed description by Ellinger (1956) followed soon afterward. The substance is cubic close packed, with the lattice constant given by the equation $a = 4.6379, - 4.01 \times 10^{-5}(t - 300)$, where t is the temperature in degrees centigrade. This equation was derived from data collected with the sample between 317 and 440°C. The calculated values of the lattice constant at the lower and upper transition temperatures of 319 and 451°C, respectively, are 4.6371 and 4.6319 Å. The corresponding values for the 12 shortest interatomic distances are 3.279 and 3.275 Å. Note that δ-plutonium contracts as the temperature increases.

The value of 4.631 Å for the lattice constant at 420°C given by Greenfield et al. (1961) is in satisfactory agreement with the equation of Ellinger.

Konobeevsky and Chebotarev (1961, 1963) proposed that the true space group of δ-plutonium is F23-T^2, with a structure derived from the cubic close packed one by small distortions from it. This structure would give rise to extra lines in the diffraction pattern, which, according to Marples and Lee (1961, 1962), should be detectable by diffractometer methods. Their failure to detect them casts considerable doubt on the validity of the distorted structure.

δ'-Plutonium

As in the case of δ-plutonium, the structure of δ'-plutonium was first reported by Jette (1955), closely followed by the details given by Ellinger (1956). δ'-Plutonium is body centered tetragonal, with two atoms per unit cell, at (000) and ($\frac{1}{2}\frac{1}{2}\frac{1}{2}$). The space group is I4/mmm-D$_{4h}^{17}$. The lattice constants determined by Ellinger at various temperatures are superseded by those of Eliot and Larson (1961), who investigated samples of both "ordinary" and "high" purity. For the latter they gave the following equations, based on observations from 452 to 480°C: $a = 3.3261 \pm 14.80 \times 10^{-4}(t - 450)$, $c = 4.4630 - 47.46 \times 10^{-4}(t - 450)$. The resulting values of the lattice constants and interatomic distances are given in Table 94-3.

A structure, like that of δ'-plutonium, which consists of atoms at the lattice points of a body centered tetragonal lattice, becomes face centered cubic (cubic close packed, like that of δ-plutonium) when the axial ratio c/a equals $\sqrt{2}$. It is interesting that the equations of Eliot and Larson predict 415°C for the temperature of this hypothetical transition. The calculated lattice constants are 4.630 Å for the δ' phase and 4.633 Å for the δ phase, the latter calculated from the equation of Ellinger given in the preceding section.

Table 94-3.   Lattice Constants and Interatomic Distances in δ'-Plutonium

Temperature, °C	a, Å	c, Å	Distance, Å (number)	Atomic Volume, Å^3
450	3.3261	4.4630	3.326(4), 3.242(8)	24.69
476	3.3646	4.3396	3.365(4), 3.220(8)	24.56

ε-Plutonium

The structure of ε-plutonium was also first published by Jette (1955). More details were subsequently given by Ellinger (1956). It is body centered cubic, with two atoms per unit cell. The lattice constant, as determined from data collected from 486 to 552°C, is given by the equation $a = 3.6348 + 13.25 \times 10^{-5}(t - 480)$. The value of 3.64 Å at 530°C given by Greenfield et al. (1961) is in excellent agreement with this equation. Calculated values for the lattice constant and derived quantities are given in Table 94-4.

Table 94-4.   Lattice Constants and Other Quantities in ε-Plutonium

Temperature °C	a, Å	Distance, Å (number)	Atomic Volume, Å³
476	3.6343	3.147(8), 3,634(6)	24.00
640	3.6560	3.166(8), 3.656(6)	24.43

The tetragonal body centered structure of δ′-plutonium is the body centered cubic structure of ε-plutonium when the axial ratio equals unity. This hypothetical transition, according to equations of Eliot and Larson given in the preceding section, would take place at 633°C, at which temperature the calculated lattice constants are 3.595 Å for the δ′ phase and 3.655 Å for the ε phase. The agreement here is not as good as is obtained for the hypothetical δ-δ′ transition, but, on the other hand, a longer extrapolation is involved. There are numerous other metals which crystallize with both the cubic close packed and body centered cubic structures, but plutonium is the only one where this transition is attended by the appearance of the intermediate body centered tetragonal structure.

The volume behavior of plutonium with temperature in the range –196 to 640°C is presented in Fig. 94-8.

AMERICIUM

α-Americium

Results of x-ray studies on americium at room temperature are conflicting. Graf et al. (1956) reported that it is double hexagonal close packed, with $a = 3.642$, $c = 11.76$ Å, $c/a = 2 \times 1.614$, atomic volume = 33.77 Å³. The sample had been obtained by reduction of AmF_3 with barium vapor, with subsequent annealing by reducing the temperature slowly from 800 to 25°C over a 10-hour period. The sample was stated to be of over 99% purity.

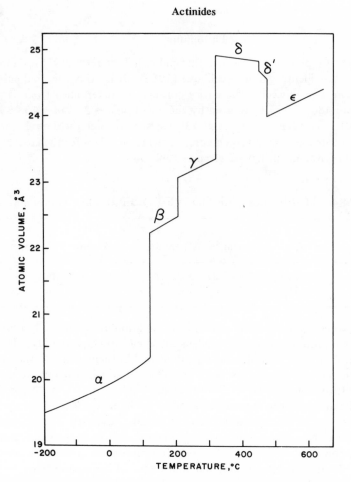

Fig. 94-8.   The variation of the atomic volume of the various allotropes of plutonium with temperature.

McWhan et al. (1960) examined samples prepared by two different methods. The first involved reduction of the trifluoride with barium vapor, followed by slow (10–16 hours) cooling. The resulting samples contained 1.6% Ba, 0.1% Ca, 0.1% Mg, and 0.2% Al. In the second method americium dioxide was reduced with lanthanum metal at 1200°C. These samples, after distillation, contained less than 0.5% La. Samples prepared by the first method were found to be double hexagonal close packed, with a = 3.474, c = 11.25 Å, c/a = 2 × 1.619, atomic volume 29.39 Å^3. There is no explanation of why these lattice constants are some 5% smaller than those found by Graf et al.

The samples prepared by the second method were found to be an intimate mixture of a cubic form (see below) and a close packed hexagonal form with stacking faults. The lattice constants of the hexagonal form were not given.

McWhan, Cunningham, and Wallman (1962) also found the smaller lattice constants for the double hexagonal close packed form, as obtained from samples prepared by reduction of the trifluoride with barium vapor. Impurities present in a representative sample were 1.45% Ba, 0.10% Ca, 0.16% Mg, and 0.45% Al. Unit cell constants are $a = 3.4681$, $c = 11.241$ Å, $c/a = 2 \times 1.621$, atomic volume 29.27 Å^3, at 20°C. These values supersede the earlier ones of McWhan et al. (1960).

This structure has space group $P6_3/mmc-D_{6h}^4$, with two americium atoms of type I at (000), (00$\frac{1}{2}$) and two of type II at ($\frac{1}{3}\frac{2}{3}\frac{1}{4}$), ($\frac{2}{3}\frac{1}{3}\frac{3}{4}$). Each atom of both types has six neighbors at 3.468 Å and six at 3.451 Å, average, 3.460 Å.

β-Americium

Americium which had been condensed onto tantalum wire or quartz (sic) fibers was found by McWhan et al. (1960) to be a mixture, as mentioned previously. The cubic phase, which was the predominant one, was found to be cubic close packed, with a lattice constant of 4.894 Å at room temperature. Storage at dry ice temperature for one week caused almost complete transition to the double hexagonal close packed α phase to take place. McWhan et al. (1962) reported additional studies of the β form, giving the lattice constant 4.894 Å at 22°C. In high temperature work they detected no indication of any change in structure of the α form, as prepared by the trifluoride reduction, up to 605°C. However, at 700 and 850°C powder lines of the cubic close packed form were detected. They stated that these observations imply a transition α→β between 600 and 700°C, but the evidence is marginal. A lattice constant of 4.893 Å was reported by Ellinger, Johnson, and Streubing (1966).

At room temperature the atomic volume of the cubic form is 29.30 Å^3, and each atom has 12 closest neighbors at 3.461 Å.

CURIUM

Cunningham and Wallman (1964) prepared metallic curium by reduction of the trifluoride with barium vapor. The substance is double hexagonal close packed with $a = 3.496$, $c = 11.331$ Å, $c/a = 2 \times 1.621$, atomic volume 29.98 Å^3 at 20°C. Each curium atom has six neighbors at 3.496 Å and six at 3.479 Å, average 3.488 Å.

Curium metal was also prepared by Smith, Hale, and Thompson (1969), who reduced curium(IV) oxide with molten Mg-Zn alloy. This preparation was found to be cubic close packed, with $a = 4.382$ Å. In this structure each curium atom has 12 closest neighbors at 3.099 Å, and the atomic volume is 21.04 Å^3. Smith et al. "explained" the corresponding high density on the grounds that the valence in the cubic close packed form is +4, as opposed to +3 in the hexagonal close packed form.

BERKELIUM

Peterson, Fahey, and Baybarz (1971) prepared berkelium metal by reduction of BkF$_3$ with Li in a tantalum crucible system at about 1000°C. In all but one of 19 samples two forms were identified, at room temperature, by x-ray diffraction methods: cubic close packed berkelium has a lattice constant of 4.997 ± 0.004 Å. Each atom thus has 12 closest neighbors at 3.533 Å, and the atomic volume is 31.19 Å^3. The second form is double hexagonal close packed with a = 3.416 ± 0.003 Å, c = 11.069 ± 0.007 Å, c/a = 2 × 1.620. Each atom has six neighbors at 3.416 Å and six at 3.398 Å, average 3.407 Å, and the atomic volume is 27.96 Å^3.

The relative atomic volumes of the two forms are thus inverted from those observed in the two forms of curium: in curium the double hexagonal close packed form is observed to be larger by 42.5%, whereas in berkelium the cubic close packed form is larger by 11.5%.

SUMMARY OF THE ACTINIDES

The allotropic behavior of the actinides with change in temperature is summarized in Fig. 97-1. For details, see the respective sections. Structures of
actinides heavier than berkelium have not been reported.

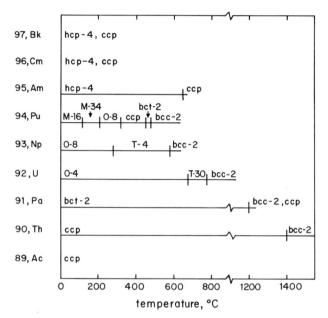

Fig. 97-1.   Allotropes observed in the actinides at various temperatures.

Chapter 6

Transition Metals

TITANIUM

At atmospheric pressure titanium changes, at 882°C, from the hexagonal close packed structure to the body centered cubic structure, two atoms per unit cell.

The structure of the room temperature form was first reported by Hull (1920) to be body centered cubic, but he later changed this result (1921b), finding it to be hexagonal close packed. His and subsequent determinations of the lattice constants are presented in Table 22-1. Some of the high values for the length of the c-axis in Table 22-1 are due to the presence of oxygen and/or nitrogen in the samples examined: Clark (1949a,b) found that 0.5% of oxygen had little effect on the a-axis but lengthened the c-axis by 0.008 Å, and 0.4% of nitrogen likewise had little effect on a but lengthened c by 0.009 Å.

The structure of the high temperature bcc(2) form was first determined by Burgers and Jacobs (1936), who reported a lattice constant of 3.33 Å (from kX) at about 900°C. Eppelsheimer and Penman (1950a) found $a = 3.3065$ Å at 900°C, while the data of Spreadborough and Christian (1959b) are reproduced within ±0.0001 Å by the equation $a = 3.2572 + 40 \times 10^{-6}\ t$, where a is the lattice constant in Ångstroms (from kX) and t is the temperature in degrees centigrade. A better value for the lattice constant at room temperature is probably estimated by extrapolating the values obtained from various alloys which have the bcc(2) structure at room temperature to 100% titanium content. The average of the values given by Levinger (1953) and Donohue (1963), both of whom used this procedure, is 3.284 ± 0.002 Å. The volume change for the hypothetical transition hcp→bcc(2) at room temperature is thus + 0.31 ± 0.19%.

172

Table 22-1.   Lattice Constants of Hexagonal Close Packed Titanium at 25°C

a, Å	c, Å	Reference
2.976[a]	4.732[a]	Hull, 1921b
2.957[a]	4.701[a]	Patterson, 1925a, b
2.959[a]	4.739[a]	Hägg, 1930b
2.951[a]	4.670[a]	Fast, 1939
2.9509[b]	4.6940[a]	Greiner & Ellis, 1949
2.9504	4.6833	Clark, 1949a, b
2.952[a]	4.705[a]	Gosner, 1949
2.9498[b]	4.6825[b]	Berry & Raynor, 1953
2.951[c]	4.685[c]	Worner, 1953
2.950[c]	4.686[c]	Swanson, Fuyat, & Ugrinic, 1954
2.9506	4.6788[c]	Szántó, 1955
2.95[c]	4.70[c]	Sofina, Azarkh, & Orlova, 1958
2.9489[b]	4.6838[b]	Spreadborough & Christian, 1959b
2.9510	4.6850	Makarov & Kuznetsov, 1960
2.9511	4.6843	Wood, 1962
2.9506[d]	4.6835[d]	Roberts, 1962
2.9497[d]	4.6822[d]	Willens, 1962

av.   2.9503   4.6836
       ±0.0007   ±0.0009
$c/a = 1.5875 \pm 0.0005$

[a]From kX, omitted from average.
[b]From kX, corrected to 25°C, using $\alpha_a = \alpha_c = 10 \times 10^{-6}$ deg^{-1}.
[c]Omitted from average.
[d]Corrected to 25°C.

At high pressure and room temperature the hexagonal close packed form undergoes a transition to a second hexagonal form (Jamieson, 1963c). The pressure at the transition was not determined, but it was found that this phase persists, on pressure release, at atmospheric pressure. Under these conditions, the lattice constants are $a = 4.625$, $c = 2.813$ Å, $c/a = 0.608$. The space group is P6/mmm-D_{6h}^1, and the unit cell contains one Ti$_I$ at (000) and two Ti$_{II}$ at $(\frac{2}{3}\frac{1}{3}\frac{1}{2})$, $(\frac{1}{3}\frac{2}{3}\frac{1}{2})$. The coordination number for Ti$_I$ is 14: each Ti$_I$ atom lies in the center of an hexagonal prism of Ti$_{II}$ atoms, distance 3.018 Å, plus two Ti$_I$ atoms above and below, at 2.813 Å. The coordination number for Ti$_{II}$ is 11: each Ti$_{II}$ atom lies in the center of a trigonal prism of Ti$_I$ atoms, distance 3.018 Å, plus Ti$_{II}$ atoms out from the center of each prism face, distance 2.670 Å, plus two Ti$_{II}$ atoms above and below at 2.813 Å. These unusual coordinations are seen in Figs. 22-1 and 22-2.

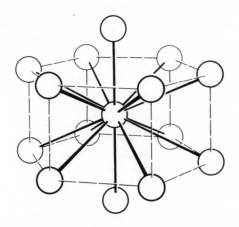

Fig. 22-1.   The coordination of the type I atoms in high pressure titanium.

The high pressure form is closely related to the body centered cubic form, as may be seen in Fig. 22-3.

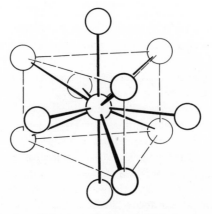

Fig. 22-2.   The coordination of the type II atoms in high pressure titanium.

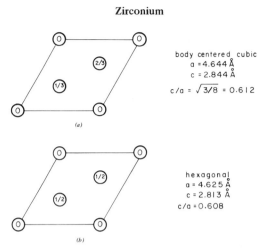

Fig. 22-3.   (a) The body centered cubic structure described in terms of a doubly centered hexagonal lattice. (b) The actual lattice of high pressure titanium.

ZIRCONIUM

At room temperature zirconium is hexagonal close packed, but changes to the
body centered cubic structure, two atoms per unit cell, at 865°C (at atmospheric pressure).

The structure of the room temperature form was first determined by Hull
(1921b). His and subsequent determinations of the lattice constants are presented in Table 40-1.

The structure of the high temperature form was first reported by Burgers
(1932a, b; also Burgers and Jacobs, 1936). At about 900°C, the lattice constant was found to be 3.62 Å (from kX). Later, more precise determinations
of the lattice constant are those of Skinner and Johnston (1953), 3.6162 Å
(from kX) at 979°C, and of Russell, (1954), 3.6090 Å at 862°C. The lattice
constant at room temperature, as obtained by extrapolation to 100% Zr of
alloys having the bcc(2) structure, is 3.587 Å (Donohue, 1963). The volume
change for the hypothetical transition hcp→bcc(2) at room temperature is thus
-0.9%.

A cubic close packed form of zirconium was observed by Chopra, Randlett,
and Duff (1967), who prepared thin films by sputtering onto glass, rocksalt,
or mica. They reported a lattice constant of 4.61 Å; this value corresponds
to an atomic volume 5.2% larger than that of the hexagonal close packed form.

As is the case of titanium, zirconium, at room temperature, undergoes a
transition to a second hexagonal form at high pressure (Jamieson, 1963c).
This form has lattice constants a = 5.036, c = 3.109 Å, c/a = 0.617 at

Table 40-1.   Lattice Constants of Hexagonal Close Packed Zirconium at 25°C

a, Å	c, Å	Reference
3.24[a]	5.15[a]	Hull, 1921
3.24[a]	5.15[a]	Noethling & Tolksdorf, 1925
3.230[a]	5.133[a]	van Arkel, 1927
3.236[a]	5.151[a]	Hägg, 1930b
3.25[a]	5.16[a]	Shinoda, 1934a, b
3.230[a]	5.133[a]	Burgers & Jacobs, 1936
3.234[a]	5.147[a]	Fitzwilliam, Kaufmann, & Squire, 1941
3.229[a]	5.139[a]	Kiessling, 1949a
3.231[a]	5.144[a]	Fast, 1952
3.232[b]	5.147[b]	Swanson & Fuyat, 1953
3.2321[c]	5.1475[c]	Treco, 1953
3.2323[d]	5.1477[d]	Treco, 1953
3.236[a]	5.153[a]	Domagala, & McPherson, 1953
3.2312	5.1477	Russell, 1953, 1954
3.2312[e]	5.1477[e]	Skinner & Johnston, 1953
3.2309[f]	5.1457[f]	Skinner & Johnston, 1953
3.23[b]	5.14[b]	Sofina, Azarkh, & Orlova, 1958
3.2316	5.1475	Lichter, 1959, 1960
3.2337[a]	5.1497[a]	Evans & Raynor, 1961

av.   3.2317   5.1476
       ±0.0005   ±0.0001
c/a = 1.5928 ± 0.0002

[a]From kX, omitted from the average.
[b]Omitted from the average.
[c]From kX, "iodide Zr."
[d]From kX, "Mg-reduced Zr," estimated values for zero oxygen content.
[e]Sample with 0.005% Hf content.
[f]Sample with 1.2% Hf content, omitted from average.

atmospheric pressure, under which condition it is metastable, although heating for 4 hours at 195°C caused partial transition to the normal hexagonal close packed form (heating at 110°C for 17 hours resulted in no conversion). The space group is $P6/mmm-D_{6h}^1$, and there are two kinds of zirconium atoms in the unit cell (see titanium, above). For Zr_I the 12 Zr_{II} neighbors at the vertices of the hexagonal prism lie at 3.297 Å, plus the two above and below at 3.109 Å; for Zr_{II}, the six Zr_I neighbors at the vertices of the trigonal prism lie at 3.297 Å, the three Zr_{II} out from the face centers at 2.908 Å, and the two Zr_{II} above and below at 3.109Å. (These coordinations are shown in Fig. 22-1 and 22-2.)

HAFNIUM

Hafnium was reported by Noethling and Tolksdorf (1925) to be hexagonal close packed, but the lattice constants they gave are in such great disagreement with those found later that it is doubtful that hafnium was the substance they examined. Theirs and subsequent determinations of the lattice constants are presented in Table 72-1. The presence of even small amounts of zirconium increases both lattice constants significantly.

Table 72-1.   Lattice Constants of Hafnium at Room Temperature

a, Å	c, Å	Purity	Reference
3.33[a]	5.47[a]	96%	Noethling & Tolksdorf, 1925
3.206[a]	5.087[a]	?	van Arkel, 1927
3.1947[a]	5.0524[a]	99%	Litton, 1951
3.1952	5.0569	?	Duwez, 1951
3.1934[a]	5.0512[a]	100%	Fast, 1948, 1952
3.200	5.061	95%	Sidhu & McGuire, 1952
3.193	5.052	97%	Glaser, Moskowitz, & Post, 1953
3.1946	5.0510	100%	Russell, 1953
3.1992	5.0602	98%	Russell, 1953
3.1967	5.0578	98%	Swanson, Fuyat, & Ugrinic, 1954
3.198	5.055	?	Sidhu, 1954
av.[b] 3.1940	5.0511		
c/a = 1.5814			

[a]From kX.
[b]For 100% Hf samples only.

A transition occurs in hafnium at about 1995°C. Duwez (1951), on the basis of examination of hafnium-niobium alloys, deduced that the high temperature form is body centered cubic, two atoms per unit cell, a conclusion verified by Ross and Hume-Rothery (1963) who studied pure hafnium in the temperature range 20 to 2100°C. At about 2000°C the lattice constant is 3.610 Å. The transition from the hexagonal close packed form is accompanied by a volume decrease of 1.1%.

A high pressure transition in hafnium, such as occurs in titanium and zirconium, was sought by Jamieson (1963c) but not found.

A cubic close packed form of hafnium was observed by Chopra, Randlett, and Duff (1967), who prepared thin films deposited on glass, mica, or rocksalt by sputtering. The lattice constant of 5.02 Å corresponds to an atomic volume of 31.63 Å^3, or 42% larger than the normal hexagonal close packed form.

GROUP VB

VANADIUM

Vanadium was found by Hull (1922) to be body centered cubic, two atoms per unit cell. The various determinations of the lattice constant are presented in Table 23-1. Beatty (1952), Gurevic and Ormont (1957), and James and Straumanis (1961) all found that the presence of very small amounts of impurities, such as oxygen and nitrogen, increase the lattice constant significantly. This effect doubtless explains some of the abnormally large values found in Table 23-1.

Rostoker and Yamamoto (1955), on the basis of studies of the vanadium-oxygen system, suggested that below −30° vanadium is body centered tetragonal. No transition was observed, however, by Hren and Wayman (1960) or Burger and Taylor (1961). Westlake (1967) suggested that many of the observed anomalies might be due to the presence of a vanadium hydride. The

Table 23-1.   The Lattice Constant of Vanadium at 25°C

a, Å	Reference
3.05[a]	Hull, 1922
3.050[a]	Osawa & Oya, 1929
3.017[a]	Hägg, 1930b
3.030[a]	Osawa & Oya, 1930
3.030[a]	Mathewson, Spire & Samans, 1931
3.0399[a]	Neuberger, 1936c
3.031[a]	van Arkel, 1939
3.0282[b]	Beatty, 1952
3.0278[b]	Seybolt & Sumison, 1953
3.0258[b]	Seybolt & Sumison, 1953
3.0240	Gurevic & Ormont, 1957
3.0236[c]	Straumanis, 1959
3.0241	James & Straumanis, 1960
3.026[b]	Carlson & Owen, 1961
3.0240	James & Straumanis, 1961
3.0335[b]	Storms & McNeal, 1962
3.0231	Brauer & Schnell, 1964
av. 3.0238 ± 0.0004	

[a]From kX, not included in the average.
[b]Not included in the average
[c]From kX.

tetragonal structure was then revived by Finkel, Glamazda, and Kovtun (1970), but Westlake, Ockers, Mueller, and Anderson (1972) observed no allotropic change between 80 and 300°K, in observations of both the electrical resistivity and x-ray diffraction maxima from single crystals.

NIOBIUM

Niobium was found by McLennan and Monkman (1929) to be body centered cubic, two atoms per cell. Lattice constants are presented in Table 41-1; the three precise, but high values were probably obtained from samples containing very small amounts of oxygen.

No transition between 18 and 2200°C was observed by Edwards, Speiser, and Johnston, (1951).

Hutchinson and Olsen (1967) reported that thin films of niobium deposited

Table 41-1.   The Lattice Constant of Niobium at 25°C

a, Å	Reference	a, Å	Reference
3.298^b	McLennon & Monkman, 1929	3.299^b	Eremenko, 1954
3.32^b	Meisel, 1930	3.3001^f	Seybolt, 1954
3.310^b	Neuberger, 1931a	3.3067^e	Nadler & Kempter, 1959a
3.306^b	Quill, 1932b	3.3063^b	Goldschmidt, 1960
3.300^b	Hidnert & Krider, 1933	3.300^g	Terao, 1963
3.301^b	Burgers & Basart, 1934	3.3059^b	Catterall & Barker, 1964
3.3009^c	Neuberger, 1936a	3.301^g	Giessen, Koch, & Grant, 1964
3.299^b	Bückle, 1946	3.3001^h	Gebhardt, Dürrschnabel, & Hörz, 1966
3.309^b	Horn & Ziegler, 1947	3.2986^g	Taylor & Doyle, 1967
3.3007^d	Edwards, Speiser, & Johnston, 1951	3.3002	Barns, 1968
		3.3004	Straumanis & Zyszczynski, 1970
av. 3.3007 ± 0.0003			

[a]Correction to 25°C made with $\alpha = 7.6 \times 10^{-6}$ when appropriate.
[b]From kX, not included in the average.
[c]From kX and 20°C.
[d]From kX and 18°C.
[e]Average value from two samples, not included in the average.
[f]Extrapolated value.
[g]Not included in the average.
[h]From 20°C.

by evaporation on single crystals of magnesium oxide were cubic close packed, with $a = 4.38 \pm 0.02$ Å. This form was said to be stabilized by a small amount of oxygen. The atomic volume is 21.01 Å^3, or 17% larger than that of the normal body centered cubic form. This form was also observed by Klechkovskaya (1970), who reported $a = 4.39$ Å for these films of niobium deposited *in vacuo*.

TANTALUM

Tantalum was found by Hull (1920) to be body centered cubic, two atoms per unit cell. Various determinations of the lattice constant are presented in Table 73-1. Abnormally high values obtained in apparently precise determinations may be due to the presence of small amounts of impurities such as oxygen, nitrogen, and carbon; no explanation is offered for the abnormally low value of Summers-Smith (1952).

No change in structure was observed from 18 to 2222°C by Edwards, Speiser, and Johnston (1951), nor was one found from 20 to 2700°C by

Table 73-1.   The Lattice Constant of Tantalum at 25°C

a, Å	Reference	a, Å	Reference
3.279[a]	Hull, 1920, 1921a	3.3030[d]	Edwards, Speiser, &
3.33[a]	Becker & Ebert, 1923		Johnston, 1951
3.298[a]	McLennan & Monkman, 1929	3.3041[b]	Geach & Summers-Smith, 1951
3.305[a]	Hägg, 1930b	3.2978[a]	Summers-Smith, 1952
3.288[a]	Agte & Becker, 1930	3.3058[c]	Swanson & Tatge, 1953
3.3181[a]	Owen & Iball, 1932	3.311[c]	Schönberg, 1954
3.305[a]	Quill, 1932b	3.302[c]	Waite, Wallace, & Craig, 1956
3.303[a]	Burgers & Basart, 1934		
3.3027[b]	Neuberger, 1936b	3.3035	Smirnova & Orment, 1956
3.303[a]	Bückle, 1946	3.3030	McMasters & Larsen, 1961
3.304[a]	Horn & Ziegler, 1947	3.3021	Vasyutinski, Kartmazov, & Finkel, 1962
3.303[a]	Kiessling, 1949b		
3.3078[c]	Schramm, Gordon, & Kaufmann, 1950	3.3030	Amonenko, Vasyutinski, Kartmazov, Smirnov, & Finkel, 1963

av. 3.3031 ± 0.0006

[a]From kX, omitted from the average.
[b]From kX, corrected to 25°C.
[c]Omitted from the average.
[d]Corrected to 25°C.

Vasyutinski, Kartmazov, and Finkel (1962).

At 25°C each tantalum atom has eight nearest neighbors at 2.8606 Å.

Other Forms of Tantalum

Tetragonal ("β") Tantalum

Read and Altman (1965) obtained a new form of tantalum, in the form of partially oriented films from 100 to 20,000 Å thick, by sputtering. They indexed the x-ray diffraction pattern of 30 maxima on a tetragonal cell having $a = 5.34$, $c = 9.94$ Å, $c/a = 1.861$. If this cell contains 16 atoms, the atomic volume is 17.72 Å^3, as compared with 18.02 Å^3 for the body centered cubic form. No structure, nor even a space group, was suggested by Read and Altman. This form of tantalum has also been observed by Sosniak, Polito, and Rozgonyi (1967), who found that its formation was independent of gaseous impurities, and by Marcus and Quigley (1968); both sets of investigators used sputtering techniques, but neither reported any values for the lattice constant. Westwood and Livermore (1970) expressed the opinion that β-tantalum is an "impurity phase which is formed to accommodate impurities at levels higher than the solubility limit of the impurity in body centered tantalum". They reported spacings for twelve powder lines from β-tantalum. (The nine resolved lines of this pattern give lattice constants of $a = 5.33$, $c = 9.88$ Å.) Baker (1970) also observed the beta form in thin films. The spacings he reported for (200) and (400) give the value 5.34 Å for a.

Das (1972), on the other hand, gave an entirely different picture based on electron diffraction data obtained from thin films of sputtered tantalum. He interpreted these data as coming from a tetragonal superlattice with $a = 10.29$ and $c = 9.20$ Å, and stated that this structure corresponds to the β-tantalum reported by Read and Altman (1965). He further stated that this tetregonally distorted superlattice is a result of interstitial ordering of nitrogen and oxygen atoms and of biaxial stresses in the thin films.

Schauer and Roschy (1972), in sputtering experiments, obtained only the body centered form from very pure tantalum in a pure argon atmosphere. The beta phase was observed on only slightly preheated substrates, or with small amounts of added oxygen or nitrogen. It would thus appear that, in the strict sense, β-tantalum is not an allotrope of the pure element.

Cubic Close Packed Tantalum

Denbigh and Marcus (1966) found that while growing thin films of tantalum a cubic close packed phase developed in the early stages of growth. A lattice constant of 4.42 Å was reported for films ca. 100 Å thick; the corresponding atomic volume is 21.59 Å^3, or 19.8% larger than that of the normal body centered cubic variety. Chopra, Randlett, and Duff (1967) also observed cubic close packed tantalum, with $a = 4.39$ Å; the corresponding atomic volume is 17.4% larger than normal. These discrepancies are so large that they might be thought to cast doubt on the validity of cubic close packed tantalum, but Marcus and Quigley (1968) showed that in thin films of bcc(2) tantalum the lattice constant increased as the thickness decreased. Their results are presented in Fig. 73-1. If, as they suggested, the transition ccp→bcc(2) takes place with zero volume change, then the thickness of their ccp films was about 700 Å. They also found that with increasing film thickness the ccp form converts to the bcc(2) form, and by ca. 200 Å thickness all of the material is in the latter form. A hypothetical cubic close packed bulk sample of tantalum would then have a lattice constant of 4.162 Å, and each atom would have 12 nearest neighbors at 2.943 Å. An even larger lattice constant of 4.48 Å for cubic close packed tantalum in sputtered films was observed by Schrey, Mathis, Payne, and Murr (1970). In this case, the atomic volume is

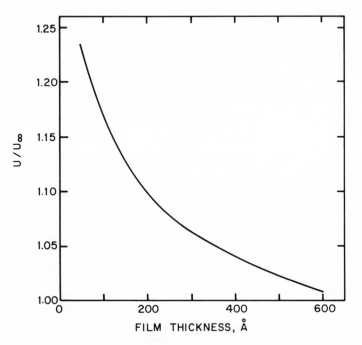

Fig. 73-1.   The decrease of the atomic volume, U, of tantalum, as a function of film thickness, U_∞ being the value for bulk tantalum (calculated from the data of Marcus & Quigley, 1968).

24.8% larger than normal. This form was observed with a substrate temperature of 325°C. When the substrate temperature was in the range 13 to 200°C, the bcc(2) form, with a lattice constant of 3.33 Å, was observed, but when the temperature was 270°C, a *smaller* value of 3.25 Å was found.

GROUP VIB

CHROMIUM

Chromium was found by Hull (1919) to be body centered cubic, two atoms per unit cell. The various determinations of the lattice constant are presented in Table 24-1. The atomic volume is 12.002 Å^3, and each chromium atom has eight nearest neighbors at 2.498 Å.

The existence of allotropic forms of chromium has been the subject of considerable confusion and controversy. What was termed (close packed) *hexagonal chromium* has been reported by Bradley and Olland (1926), Sasaki and Sakito (1930), Wright, Hirst, and Riley (1935), and Yoshida (1943). The lattice constants reported are in the range a = 2.719 to 2.722, c = 4.419 to 4.427 Å (from kX), c/a = 1.626. These samples all were obtained by electrolysis. The larger atomic volume of the hexagonal variety, 14.18 Å, as opposed to 12.00 Å^3 for body centered chromium, was explained (Hume-Rothery, 1936; Pauling, 1947) on the basis of different electronic structures of the atoms in the two forms. However, later work by Snavely (1947), Nemnonov (1948), and Snavely and Vaughan (1949) established that the so-called "hexagonal chromium" is, in fact a hydride. Thus the explanations referred to above become unnecessary.

A second cubic form of chromium, obtained by electrolysis of chromic sulfate solutions, was reported by Sasaki and Sekito (1930). The lattice constant was stated to be 8.735 Å (from kX), and it was suggested that this form was isostructural with α-manganese, in which case the atomic volume would be 11.49 Å^3. However, in more recent work Nemnonov (1948) considered this variety of "chromium" an intermediate between hexagonal chromium hydride (see above) and body centered cubic chromium.

A third cubic form, *cubic close packed chromium,* said to be stable above ~1830°C, was reported by Bloom and Grant (1951), who examined various chromium-nickel alloys at high temperatures. Bloom, Putnam and Grant (1952) reported a transition at 1840°C on the basis of melting curves. Confirmation of a high temperature modification was offered by Stein and Grant (1955), who also examined various chromium-nickel alloys. They stated that in pure chromium a β phase, probably cubic close packed, is stable above

Table 24-1.　　The Lattice Constant of Chromium at $25°C$

a,Å	Reference	a, Å	Reference
2.91[a]	Hull, 1919	2.885[a]	Yoshida, 1943
2.901[a]	Hull, 1921a	2.8849[b]	Carlile, Christian, & Hume-
2.881[a]	Phebus & Blake, 1925		Rothery, 1949
2.878[a]	Patterson, 1925a, b	2.8849[b]	Zwicker, 1949
2.878[a]	Sillers, 1927	2.8849	Fine, Greiner, & Ellis, 1951
2.878[a]	Wever & Hashimoto,	2.8843[c]	Taylor & Floyd, 1952
	1929	2.8846[c]	Pearson & Hume-Rothery,
2.883[a]	Sasaki & Sekito, 1930		1953
2.8844[b]	Preston, 1932	2.8850	Sully, Brands, & Mitchell,
2.8845[b]	Jette, Nordstrom,		1953
	Queneau, & Foote,	2.8839	Swanson, Gilfrick, &
	1934		Ugrinic, 1955
2.8839[b]	Wright, Hirst, & Riley,	2.8850[c]	Straumanis & Weng, 1955
	1935	2.8850[c]	Straumanis & Weng, 1956
2.8855[c]	Wood, 1937	2.8850[c]	James, Straumanis, & Rao,
2.8856[b]	van Arkel, 1939		1961
2.852[a]	Söchtig, 1940	2.8843[d]	Smirnov & Finkel, 1963
2.884[a]	Fricke, 1941		
2.885[a]	Kubaschewski &		
	Schneider, 1942		

av. 2.8847 ± 0.0005

[a]From kX, omitted from the average.
[b]From kX.
[c]From kX, corrected to $25°C$.
[d]From kX, read from a figure.

$1840°C$. Abrahamson and Grant (1956) presented the results of their examination of this system; they gave 3.68 Å for the lattice constant of quenched
cubic close packed chromium at room temperature; the corresponding atomic
volume is 12.46 $Å^3$. On the basis of cooling curves, Grigoryev et al. (1959,
1960), also proposed a transition, at $1830°C$; they stated, on the other hand,
the the form stable between $1830°C$ and the melting point (ca. $1900°C$) is
body centered cubic, and not cubic close packed. The situation was further
complicated by Grigoryev et al. in their report of the existence of no less
than four forms of chromium at high temperature! They proposed the
following set of transitions:

$$\beta \xrightleftharpoons{1300°C} \gamma \xrightleftharpoons{1650°C} \delta \xrightleftharpoons{1830°C} \epsilon$$

The β form was stated to be face centered cubic, the γ and ε forms body centered cubic, and the δ form hexagonal.

Other investigators have found other results: Vasyutinski, Kartmazov, and Finkel (1961), using x-ray diffraction methods, found no transitions in pure chromium in the range 700 to 1700°C; Svechnikov (1963), using dilatometry, could not detect any phase transformations, other than freezing, when chromium was cooled from 2040°C to room temperature; Wyder and Hoch (1963), on the basis of x-ray diffraction of chromium in the range 1005 to 1400°C, concluded that pure chromium exists only in the body centered cubic form at high temperature; the same result was obtained by Ross and Hume-Rothery (1963), who photographed the x-ray diffraction pattern of chromium up to 1800°C.

Yet another form of chromium was proposed by Muldawer, Hoffman, and Riseman (1958). They examined thin films which had been deposited *in vacuo* on glass and annealed in air at 250°C. They found that the material is face centered cubic, with a = 5.6 Å, and stated that the results obtained could not be explained in terms of oxide or nitride formation.

No transition in chromium up to 55 kbar (at 28°C) was detected by Evenson and Hall (1965).

As if all this were not enough, a *fifth* cubic form of chromium was said to exist by Kimoto and Nishida (1967), who obtained fine particles of chromium by evaporation at low temperature. The substance was found to be cubic, with a = 4.588 ± 0.001 Å. With eight atoms per unit cell, the calculated density of 7.151 g cm^{-3} is close to that of 7.193 g cm^{-3} observed in normal bcc(2) chromium. The structure described by Kimoto and Nishida is based on space group Pm3-T_h^1, with two atoms of type I in position 2(a) at (000), ($\frac{1}{2}\frac{1}{2}\frac{1}{2}$), and eight atoms of type II in position 24(ℓ) (disordered, with an occupancy factor of $\frac{1}{4}$ at each position), at $\pm(xyz)\Omega$, $\pm(\bar{x}yz)\Omega$, $\pm(x\bar{y}z)\Omega$, $\pm(xy\bar{z})\Omega$, where $x = \frac{1}{4} - u$, $y = v$, $z = \frac{1}{2} + w$, with u = 0.002, v = 0.04, w = 0.01. This represents a distortion from the β-tungsten structure, which is described in detail later, (see section on tungsten); that structure is obtained when $y = v = w = 0$, the space group then becoming Pm3n. Kimoto and Nishida remarked that some weak reflections eliminated that space group for this form of chromium. The coordination of the chromium atoms is highly irregular and will not be discussed in detail here, but the *average* interatomic distances are: from Cr(I), 12 Cr(II) at 2.565 Å; and from Cr(II), two Cr(II) at 2.294Å, four Cr(I) at 2.565 Å, and eight Cr(II) at 2.810 Å. The reasons for the distortions from the more symmetrical β-tungsten structure are not known.

The same form of chromium was also reported by Forssell and Persson (1969) who observed it simultaneously with the bcc(2) form in thin films deposited on rocksalt, but gave no lattice constants.

Additional work is needed to resolve some of the inconsistencies contained in the preceding paragraphs.

MOLYBDENUM

The structure of molybdenum was first determined by Hull (1921a). It is body centered cubic, two atoms per unit cell. The various determinations of the lattice constant are presented in Table 42-1. Small amounts of carbon cause an increase in the lattice constant; Speiser et al. (1952) give the equation $a = a_0 + 0.05705P$, where P is the weight percent of carbon present. There is no change in structure up to about $1800°C$ (Edwards, Speiser, and Johnston, 1951; Casselton and Hume-Rothery, 1964).

According to Aggarwal and Goswami (1957) deposition *in vacuo* of molybdenum on rocksalt or glass often gives a cubic close packed form, with lattice constant 4.16 Å. This corresponds to an increase in atomic volume of 16%. Similar results in the same kind of experiments were reported by Chopra, Randlett, and Duff (1967), who found $a = 4.19$ Å for thin films of cubic close packed molybdenum.

Table 42-1.   The Lattice Constant of Molybdenum at $25°C$

a, Å	Reference	a, Å	Reference
3.149[a]	Hull, 1921a	3.1479[a]	Kieffer & Cerwenka, 1952
3.09[a]	Stoll, 1922	3.1472	Swanson & Tatge, 1953
3.142[a]	Davey, 1924	3.146[a]	Eremenko, 1954
3.148[a]	Davey, 1925, 1926	3.1468[c]	Geach & Summers-Smith,
3.146[a]	van Arkel, 1926		1954
3.1464[a]	van Arkel, 1928	3.141[e]	Pipitz & Kieffer, 1955
3.145[a]	Hägg, 1930a	3.146[a]	Goldschmidt & Brand, 1961
3.17[a]	Zeidenfeld, 1931	3.1470	Taylor, Doyle, & Kagle, 1962
3.1467[c]	Owen & Iball, 1932	3.1471[d]	Ross & Hume-Rothery, 1963
3.1474[b]	Jette & Foote, 1935	3.1470	Niemiec, 1963
3.1467[c]	Lu & Chang, 1941	3.1468[b]	Catterall & Barker, 1964
3.147[a]	Kubaschewski & Schneider, 1942	3.1466	Casselton & Hume-Rothery, 1964
3.146[a]	Bückle, 1946	3.1470	Taylor & Doyle, 1965
3.1475[d]	Edwards, Speiser, & Johnston, 1951	3.14700	Straumanis & Shodhan, 1968
3.1470[d]	Geach & Summers-Smith, 1951	3.14702	Woodard & Straumanis, 1971
av. 3.1470 ± 0.0003			

[a] From kX, omitted from the average.
[b] From kX.
[c] From kX, corrected to $25°C$.
[d] Corrected to $25°C$.
[e] Omitted from the average.

TUNGSTEN

The structure of tungsten was first determined by Debye (1917). It is body centered cubic, two atoms per unit cell. Various determinations of the lattice constant are presented in Table 74-1. The structure is unchanged up to at least 3100°C (Ross and Hume-Rothery, 1963).

There is some controversy regarding a second form of tungsten. This form, which was first discovered by Hartmann, Ebert, and Bretschneider (1931), is usually termed β-tungsten, the ordinary variety being α-tungsten, but care must be exercised because some authors interchange the two Greek letter designations. The weight of the evidence at present appears to favor the opinion that it is a true allotrope of tungsten, although it is possible that the presence of minute amounts of oxygen are necessary for its formation. It has been pre-

Table 74-1.   The Lattice Constant of bcc(2) Tungsten at 25°C

a, Å	Reference	a, Å	Reference
3.19[a]	Debye, 1917	3.164[a]	Bückle, 1946
3.161[a]	Davey, 1924, 1925, 1926	3.1647[b]	Jaffee & Nielsen, 1948
3.161[a]	Davey & Wilson, 1926	3.1651	Schramm, Gordon, &
3.16[a]	Becker, 1926		Kaufmann, 1950
3.160[a]	van Arkel, 1926	3.1648	Swanson & Tatge, 1953
3.1647[b]	van Arkel, 1928	3.1653[b]	Pines & Kaluzhinova, 1954
3.148[a]	Agte & Becker, 1930	3.1653	Umanski, Kheiker, & Zevin,
3.162[a]	Hägg, 1930a		1960
3.17[a]	Zeidenfeld, 1931	3.1652	Umanski, Zubenko, &
3.1657[c]	Owen & Iball, 1932		Zolina, 1960
3.1647[b]	Neuberger, 1933a	3.1652	Parrish, 1960
3.1654[c]	Neuberger, 1934	3.1649	Dutta & Dayal, 1963b
3.1647[b]	Jette & Foote, 1935	3.1652	Delf, 1963
3.1651[c]	Straumanis & Ievinš, 1936	3.1654[b]	Taylor, Mack, & Parrish,
3.1647[b]	Cohen, 1936		1964
3.1652[b]	Moeller, 1937	3.1650	Beu, 1964
3.1650[c]	Lu & Chang, 1941	3.1652	Taylor & Doyle, 1965
3.159[a]	Kubaschewski & Schneider,	3.1650[d]	Gerdes, Chapman, & Clark,
	1942		1970

av. 3.1651 ± 0.0003

[a]From kX, not included in the average.
[b]From kX.
[c]From kX, corrected to 25°C.

pared by the electrolysis of melts (Hartmann et al., 1931; Burgers and van
Liempt, 1931; Hägg and Schönberg, 1954; Millner et al., 1957), by sputtering
onto glass *in vacuo* (Petch, 1944; Rooksby, 1944; Moss and Woodward, 1959),
and by reduction of tungsten oxide with hydrogen (Charlton, 1952; Mannella
and Hougen, 1956; Millner et al., 1957). Largely on the basis of a low value
for the observed density, Hägg and Schönberg (1954) concluded that β-tung-
sten is a metallic oxide with probable ideal formula W_3O; this view was supported by Charlton (1954) and Charlton and Davis (1954). The contrary view
was expressed by Mannella and Hougen (1956): ". . .The [x-ray] spectrum
identified with β-W is that of a crystalline form of tungsten metal"; by
Millner et al. (1957), who examined samples "völlig frei von Sauerstoff"; and
by Moss and Woodward (1959), who stated that "only about 0.01% of the total
film deposited could be the low oxide, W_3O. Thus the occurrence of A15
[β] tungsten . . . cannot be interpreted on the basis that it is an oxide". The
low density observed by Hägg and Schönberg, 15.0 g cm^{-3}, as opposed to the
calculated value of 19.0 g cm^{-3}, remains unexplained. (A still lower value for
the density of 12.3 g cm^{-3} was observed by Hartmann et al., who attributed it
to the extremely small size of the crystals.)

Proceeding on the basis that this substance is tungsten, the crystallographic
description is: simple cubic, space group $Pm3n-O_h^3$, eight atoms per cell, with
two W(I) at (000), $(\frac{1}{2}\frac{1}{2}\frac{1}{2})$ and six W(II))at $(\frac{1}{4}0\frac{1}{2})$, $(\frac{1}{2}\frac{1}{4}0)$, $(0\frac{1}{2}\frac{1}{4})$, $(\frac{3}{4}0\frac{1}{2})$, $(\frac{1}{2}\frac{3}{4}0)$,
$(0\frac{1}{2}\frac{3}{4})$; the lattice constant is 5.048 Å (see Table 74-2). A projection of the
structure down one of the cubic axes is shown in Figs. 74-1 and 74-2. Each
W(I) atom is surrounded by 12 W(II) atoms at 2.822 Å; these lie at the ver-

Table 74-2.   The Lattice Constant of β-Tungsten at
Room Temperature

a, Å	Reference
5.05[a]	Hartmann, Ebert, & Bretschneider, 1931
5.048[b]	Neuberger, 1933a
5.051[b]	Petch, 1944
5.051$_2$[b]	Rooksby, 1944
5.05[a]	Charlton, 1952
5.046[b]	Hägg & Schönberg, 1954
5.047[b]	Millner, Hegedüs, Sasvari, & Neugebauer, 1957
5.046	Moss & Woodward, 1959
av. 5.048 ± 0.002	

[a]From kX, omitted from the average.
[b]From kX.

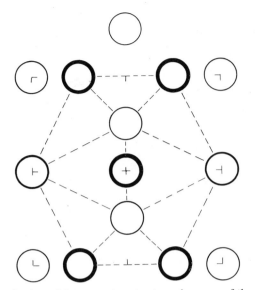

Fig. 74-1. The projection of the β-tungsten structure along one of the cubic axes. Thinnest circles, atoms at height 0 and 1; intermediate circles, atoms at height ¼ and 3/4; thickest circles, atoms at height ½. Dashed lines connect the atoms of the coordination polyhedron of type I atoms shown in Fig. 74-3.

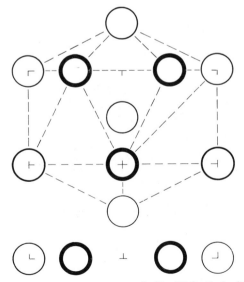

Fig. 74-2. Atoms of the β-tungsten structure as in Fig. 74-2. Dashed lines connect the atoms of the coordination polyhedra of type II atoms shown in Fig. 74-4.

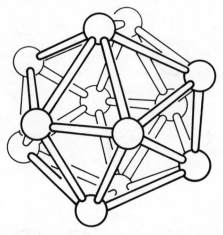

Fig. 74-3.   The coordination polyhedron of the type I atoms in β-tungsten. The central atom is not shown. (See Fig. 74-1.)

tices of a slightly distorted regular icosahedron.   Each W(II) has 14 near neighbors, two W(II) at 2.524 Å, four W(I) at 2.822 Å, and eight W(II) at 3.091 Å.   The polyhedron of W(II) is a distorted hexagonal antiprism, with added atoms above and below the hexagonal faces.   The coordination polyhedra are shown in Figs. 74-3 and 74-4.

Another form of tungsten at room temperature was reported by Horn

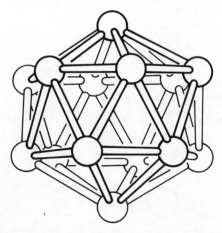

Fig. 74-4.   The coordination in polyhedron of the type II atoms in β-tungsten. The central atom is not shown. (See Fig. 74-2.)

(1950), who found a cubic close packed variety in thin films which had been evaporated on to a cold surface. The lattice constant was determined to be 4.17 Å. This value corresponds to an atomic volume of 18.13 Å, much larger than that in bcc(2) tungsten (15.854 ± 0.004 Å^3) or in β-tungsten (16.08 ± 0.02 Å^3). In similar experiments Chopra, Randlett, and Duff (1967) found cubic close packed tungsten with $a = 4.13$ Å; the corresponding atomic volume is 17.61 Å^3.

GROUP VIIB

MANGANESE

There are four allotropic forms of manganese:

$$\alpha \xrightarrow{727°C} \beta \xrightarrow{1095°C} \gamma \xrightarrow{1133°C} \delta \xrightarrow{1244°C} \text{liquid}$$

Low temperature studies apparently have not been carried out.

α-Manganese

The powder pattern of manganese at room temperature was reported first by Young (1923), who was unable to index it. At about the same time Becker and Ebert (1923) had the same difficulty and remarked that the pattern was not cubic. Westgren and Phragmen (1925a) were the first to index the powder pattern successfully as cubic, with $a = 8.912$ Å (from kX); they gave 56 as the number of atoms per unit cell. Various determinations of the lattice constant are presented in Table 25-1.

The structure was first determined by Bradley and Thewlis (1927) from powder data. It is body centered cubic, space group $I\bar{4}3m$-T_d^3, with 58 atoms in the unit cell, two of type I in position $2a$ at $(000, \frac{1}{2}\frac{1}{2}\frac{1}{2})$, eight of type II in position $8c$ at $(000, \frac{1}{2}\frac{1}{2}\frac{1}{2})$ + $(xxx, x\bar{x}\bar{x}, \bar{x}x\bar{x}, \bar{x}\bar{x}x)$, 24 of type III in position $24g$ at $(000, \frac{1}{2}\frac{1}{2}\frac{1}{2})$ + $(xxz\Omega, x\bar{x}\bar{z}\Omega, \bar{x}x\bar{z}\Omega, \bar{x}\bar{x}z\Omega)$ plus 24 of type IV in a second set of position $24g$. The five positional parameters were determined by Bradley and Thewlis, and the structure was described by them in terms of the shortest interatomic distances. The parameters were also determined at about the same time by Preston (1928a) from single crystal data, and have subsequently been re-fined by Gazzara, Middleton, Weiss, and Hall (1967), who used powder data, and by Kunitomi, Yamada, Nakai, and Fujii (1969)[*] and Oberteuffer and Ibers (1970), who used single crystal data. These two

---

[*]These values were subsequently modified slightly Yamada and Fujii (1970).

Table 25-1.   The Lattice Constant of α-Manganese at Room Temperature

a, Å	Reference	a, Å	Reference
8.912[a]	Westgran & Phragmen, 1925	8.9130[b]	Dean, Potter, & Huber, 1948
8.921[a]	Bradley & Thewlis, 1927	8.908[a]	Zwicker, Jahn, & Schubert, 1949
8.912[a]	Preston, 1928a	8.9120[b]	Carlile, Christian, & Hume-Rothery, 1949
8.92[a]	Sekito, 1929a		
8.912[a]	Öhman, 1930a, b	8.9125	Gazzara, Middleton, Weiss, & Hall, 1967
8.9135[b]	Johannsen & Nitka, 1938		
8.9139[b]	Carapella & Hultgren, 1942	8.9129	Swanson, McMurdie, Morris, & Evans, 1969
8.9128[b]	Potter & Huber, 1945	8.911[c]	Oberteuffer & Ibers, 1970
av. 8.9129 ± 0.0006			

[a]From kX, not included in the average.
[b]From kX.
[c]Not included in the average.

Table 25-2.   Positional Parameters in α-Manganese

Atom	Parameter	Bradley & Thewlis (1927)	Preston (1928a)	Gazzara et al. (1967)	Kunitomi et al. (1969)	Oberteuffer & Ibers[a] (1970)	Yamada & Fujii[a] (1970)
II	x	0.317	0.319	0.316	0.3176	0.31787(10)	0.31765(12)
III	x	0.356	0.347	0.356	0.3569	0.35706 (6)	0.35711 (8)
III	z	0.042	0.056	0.034	0.0346	0.03457 (9)	0.03470(11)
IV	x	0.089	0.092	0.089	0.0898	0.08958 (6)	0.08968 (8)
IV	z	0.278	0.281	0.282	0.2820	0.28194 (9)	0.28211(11)

[a]Standard errors × 10^5 given in parentheses.

groups of investigators used the same method of obtaining single crystals of α-manganese that had been used over 40 years earlier by Preston. The results of the various parameter determinations are presented in Table 25-2. The interatomic distances in the following discussion were calculated with the positional paramaters of Oberteuffer and Ibers and the lattice constant of Table 25-1.

A projection of the structure is shown in Fig. 25-1. The coordinations of the four different types of manganese are all quite different. The nearest neighbors of each are listed in Table 25-3. The environment of the type I atoms is the easiest to describe: it consists of 12 atoms of type IV at 2.754 Å lying at the vertices of a truncated tetrahedron, shown in Fig. 25-2, plus four

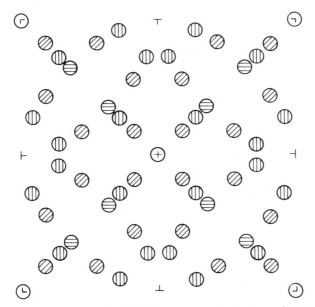

Fig. 25-1.   Projection of one unit cell of the α-manganese structure. Atom types as follows: ⊙ , I; ⊖ , II; ◎ , III; ⊘ , IV.

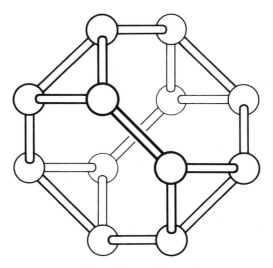

Fig. 25-2.   The 12 nearest neighbors of a type I atom in α-manganese (central atom not shown); these lie at the vertices of a truncated tetrahedron.

of type II at 2.812 Å lying out from the centers of the hexagons, as shown
in Fig. 25-3. The coordination of the atoms of type II consists of a triangle
of type III atoms at 2.573 Å on one side, and a hexagon of type IV atoms at
2.709 Å on the other side, plus an atom of type I at 2.812 Å out from the
center of the hexagon, as shown in Fig. 25-4. (Six more distant atoms, three

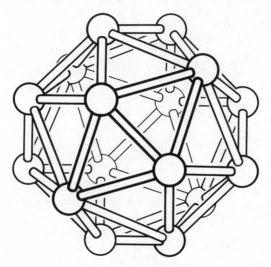

Fig. 25-3.   The complete coordination polyhedron of type I atoms in α-manganese, showing the atoms of Fig. 25-2 plus the four lying out from the centers of the hexagons.

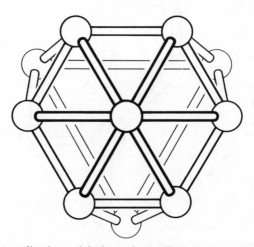

Fig. 25-4.   The coordination polyhedron of type II atoms in α-manganese. Central
atom hidden; see text for identification of other atoms.

of type IV at 2.895 Å and three of type III at 2.931 Å, found in Table 25-3, are not included in Fig. 25-4; it is doubtful whether they contribute much to the bonding.) The coordination of the atoms of type III is more complicated, as shown in Fig. 25-5, while that of type IV atoms approximates the icosahedron with one vertex missing, as seen in Fig. 25-6.

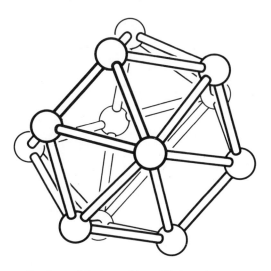

Fig. 25-5.   The coordination polyhedron of type III atoms in α-manganese.

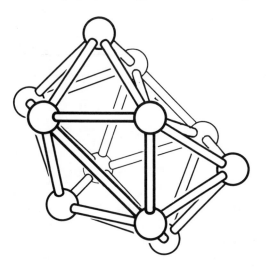

Fig. 25-6.   The coordination polyhedron of type IV atoms in α-manganese.

Table 25-3.   Shortest Interatomic Distances in α-Manganese

Atom Type	Site Symmetry	Neighbor Number	Neighbor Type	Distance, Å
I	$\overline{4}3m$–T_d	12	IV	2.754
		4	II	2.812
II	$3m$–C_{3v}	3	III	2.573
		6	IV	2.709
		1	I	2.812
		(3	IV	2.895)[a]
		(3	III	2.931)
III	m–C_s	1	IV	2.350
		2	IV	2.524
		1	II	2.573
		2	III	2.622
		4	III	2.661
		2	IV	2.683
		(1	II	2.931)
IV	m–C_s	1	IV	2.258
		1	III	2.350
		2	IV	2.425
		2	III	2.524
		2	III	2.683
		2	II	2.709
		1	I	2.755
		(1	II	2.895)

[a]Distances in parentheses are not included in the coordination polyhedra shown in Figure 25-2 to 25-6.

β-Manganese

Although the range of stability of β-manganese is 727 to 1095°C, it may be retained at room temperature by quenching. It was determined to be cubic, with 20 atoms per unit cell, by Olshausen (1925) and Westgren and Phragmen (1925). The latter, however, could not exclude the possibility of a cell with a doubled edge length containing 160 atoms. Various determinations of the lattice constant are presented in Table 25-4. All subsequent investigators except Wilson (1935) agree on the smaller unit cell.

A structure based on the smaller unit cell was published by Preston (1928b). This structure has space group $P4_332\text{-}O^5$ and the enantiomorphous $P4_132\text{-}O^7$, with eight atoms in position $8c$ and 12 in position $12d$. Preston gave values for the two positional parameters, the resulting interatomic distances, and co-ordinations of the two types of manganese atoms.

A structure based on the larger unit cell was published by Wilson (1935). This structure has space group $Fm3c\text{-}O_h^6$, with eight atoms in position $8a$, eight in position $8b$, 48 in position $48e$, and 96 in position $96i$. Wilson gave values for the three positional parameters, but no further information about the inter-atomic distances and coordination of the four different types of manganese atoms.

There does not appear to be much point in discussing the details of either of these structures until: (1) it is established which if them is correct, and (2) the correct alternative has been refined from a complete set of diffraction data. It should be pointed out, however, that the structure proposed by Preston, which was erected as Strukturbericht type $A13$, has been extensively cited and generally accepted, while the structure proposed by Wilson has been generally ignored (or overlooked).

Basinsky and Christian (1954) measured the lattice constant of β-manganese with samples in the temperature range 23 to $1095°C$. Their values, and the average value of Table 25-4, fit the following equation (smaller unit cell):

$$a = 6.3100 + 21.1 \times 10^{-5}t + 38.2 \times 10^{-9}t^2$$

where a is the lattice constant in Å (from kX) and t the temperature in degrees centigrade.

Table 25-4.   The Lattice Constant of β-Manganese at Room Temperature

a, Å	Reference	a, Å	Reference
6.303^a	Olshausen, 1925	6.303×2^a	Wilson, 1935
6.302^a	Westgren & Phragmen, 1925	6.3050^a	Johannsen & Nitka, 1938
6.30^a	Preston, 1928b	6.3145^b	Carlile, Christian, & Hume-Rothery, 1949
6.302^a	Sekito, 1929a	6.3159^c	Basinski & Christian,
6.313^a	Ohman, 1930a, b		1954
av. $6.3152^d \pm 0.0010$			

[a]From kX, omitted from the average.
[b]From kX.
[c]From kX, read from a graph.
[d]The true value may be twice this value; see text.

γ-Manganese

Early studies on the structure of γ-manganese agree that this allotrope is face centered tetragonal at room temperature, with c/a equal to somewhat less than unity. This conclusion was reached from examination of manganese obtained by electrolysis, manganese quenched from 800°C, and by extrapolation to zero Mn content of various alloys. These results are summarized in Table 25-5.

The first indications that the actual structure of γ-manganese in the high temperature range of its stability is not tetragonal were obtained by Zwicker (1951), who calculated that at 1100°C γ-manganese is face centered cubic with a = 3.84 Å (from kX), on the basis of high temperature observations of manganese-copper alloys. Basinski and Christian (1952) then found that although some of the copper-manganese alloys were tetragonal at room temperature, this structure resulted from a transition from a face centered cubic struc-

Table 25-5.   The Lattice Constants of "γ-Manganese" at Room Temperature (all from kX)

a, Å	c, Å	c/a	Reference	Remarks
3.782	3.540	0.936	Westgren & Phragmen, 1925a	a
3.772	3.563	0.945	Bradley, 1925	a
3.755	3.609	0.961	Sekito, 1929a	b
3.784	3.557	0.940	Sekito, 1929a	c
3.784	3.532	0.933	Persson & Öhmann, 1929	c
3.782	3.533	0.934	Persson, 1930	c
3.782	3.532	0.934	Köster & Rauscher, 1939	d
3.771	3.527	0.935	Grube, Oestreicher, & Winkler, 1939	e
3.778	3.551	0.940	Grube & Winkler, 1939	f
3.782	3.532	0.934	Ellsworth & Blake, 1944	c
3.782	3.548	0.938	Dean, Potter, & Huber, 1948	c
av. 3.780	3.542	0.937		
±0.005	±0.012	±0.003		

[a] Electrolytic.
[b] Quenched, probably impure; values omitted in averaging.
[c] Extrapolation of Cu-Mn alloys to 100% Mn.
[d] Extrapolation of Ni-Mn alloys to 100% Mn.
[e] Extrapolation of Cu-Mn alloys to 100% Mn, read from a graph.
[f] Extrapolation of Pd-Mn alloys to 100% Mn, read from a graph.

ture at high temperature. On the other hand, Morgan (1953) maintained that γ-manganese is tetragonal at all temperatures, on the basis of its magnetic properties. However, shortly thereafter Basinski and Christian (1953) took diffraction photographs of manganese at $1100°C$ and found that it is cubic close packed, not tetragonal.

Additional measurements were then reported by Basinski and Christian (1954), who reported lattice constants they obtained for a cubic close packed phase in the temperature range 1099 to $1131°C$. Their data are accurately reproduced by the equation $a = 3.6763 + 17 \times 10^{-5}t$, where a is the lattice constant in Å (from kX) and t the temperature in degrees centigrade. At the $\gamma \rightarrow \delta$ transition temperature of $1133°C$ the lattice constant is 3.869 Å and each manganese atom (in the γ phase) has 12 nearest neighbors at 2.736 Å.

δ-Manganese

The transformation $\delta \rightarrow \gamma$ occurs very rapidly on cooling, and attempts to obtain metastable δ structures by quenching alloys have all been unsuccessful (Basinski and Christian, 1954). The structure was determined by Basinski and Christian, who took photographs with samples held at temperatures of 1135 to $1239°C$. It is body centered cubic, with two atoms per unit cell. Their data may be fitted by the following equation: $a = 2.9384 + 12.5 \times 10^{-5}t$, where a is the lattice constant in Å (from kX) and t the temperature in degrees centigrade. At the $\gamma \rightarrow \delta$ transition temperature of $1133°C$ the lattice constant is 3.080 Å and each manganese atom in the δ phase has eight nearest neighbors at 2.667 Å. The atomic volume is 14.61 Å.

Summary of the Four Manganese Allotropes

The directly observed atomic volumes of the various forms of manganese are presented in Fig. 25-7. Also included in the figure are some of the atomic volumes as obtained from the equations given above; some of the extrapolations used are probably of dubious validity, but the resulting comparisons are nonetheless interesting. At the three transitions the percent changes in atomic volume are

$$\alpha \xrightarrow{\ +3.56\%\ } \beta \xrightarrow{\ +0.82\%\ } \gamma \xrightarrow{\ +0.90\%\ } \delta$$

The atomic volumes as determined by various methods are also presented in Table 25-6.

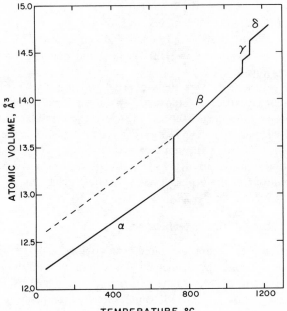

Fig. 25-7.   The variation of the atomic volume of manganese with temperature. For sources of the data used see text.

Table 25-6.   Atomic Volumes of Manganese Allotropes, Å^3

t, °C	Allotrope				
	α	β	γ	"γ"[a]	δ
25	12.207[b]	12.593[b]	12.47[c]	12.65	12.74[d]
727	13.160[b]	13.628[b]			
1095		14.289[b]	14.406[b]		
1133			14.478[b]		14.609[d]

[a]Hypothetical face centered tetragonal form as deduced from various alloys.
[b]Directly observed values.
[c]Extrapolated from data taken at 1099 to 1133°C.
[d]Extrapolated from data taken at 1135 to 1239°C.

TECHNETIUM

The structure of technetium was first determined by Mooney (1947); it is hexagonal close packed. Various determinations of the lattice constants are presented in Table 43-1. No studies at low or high temperature or high pressure have been reported.

Table 43-1.   The Lattice Constants of Technetium at Room Temperature

a, Å	c, Å	Reference
2.735	4.391	Mooney, 1947
2.738	4.393	Mooney, 1948a
2.735	4.388	Mooney, 1948b
2.743	4.400	Lam, Darby, Downey, & Norton, 1961
2.741	4.400	Niemiec, 1963
av. 2.738	4.394	
±0.003	±0.005	
c/a = 1.605 ± 0.002		

RHENIUM

The structure of rhenium was first determined by Goldschmidt (1929a); it is hexagonal close packed. There is no change in structure up to at least 1288°C (Wasilewski, 1961a). Various determinations of the lattice constants are presented in Table 75-1.

Table 75-1.   The Lattice Constants of Rhenium at Room Temperature

a, Å	c, Å	Reference
2.756[a]	4.457[a]	Goldschmidt, 1929a, b
2.761[a]	4.459[a]	Moeller, 1931
2.771[a]	4.479[a]	Agte et al., 1931
2.7609[b]	4.4583[b]	Stenzel & Weerts, 1932
2.7603[b]	4.4572[b]	Trzebiatowski, 1937
2.760[a]	4.452[a]	Winkler, 1943
2.760	4.458	Swanson & Fuyat, 1953
2.762[b]	4.459[b]	Trzebiatowski & Berak, 1954
2.760	4.458	Sims, Craighead, & Jaffee, 1955
2.7609	4.4576	Wasilewski, 1961
2.7612	4.4612[c]	Rudy, Kieffer, & Fröhlich, 1962
2.761[b]	4.503[a]	Savicki, Tylkina, & Polyakova, 1962
2.761	4.458	Niemiec, 1963
av. 2.7608	4.4580	
±0.0006	±0.0005	
c/a = 1.6147 ± 0.0003		

[a]From kX, omitted from the average.   [c]Omitted from the average.
[b]From kX.

Bond et al. (1965) found that the x-ray pattern of rhenium at $7°K$ corresponded to a cubic close packed structure with $a = 3.80$ Å. However, because the samples were unstable at room temperature, they suggested that they were not pure rhenium but possibly a subnitride.

Chopra, Randlett, and Duff (1967) also observed cubic close packed rhenium, but at room temperature, in thin films prepared by sputtering. Their reported lattice constant of 4.04 Å corresponds to an atomic volume of 16.48 $Å^3$, or 12% larger than that of the normal hexagonal close packed form.

GROUP VIII

IRON

The structure or iron at room temperature was first determined by Hull (1917a). It is body centered cubic, two atoms per unit cell. Various determinations of

$$\alpha \rightarrow \gamma \quad \text{at} \quad 910°C, \quad \triangle V = -1.0\%$$
$$\gamma \rightarrow \delta \quad \text{at} \quad 1390°C, \quad \triangle V = +0.3\%$$

the lattice constant are presented in Table 26-1.

The following structural transitions occur in iron at atmospheric pressure:

$$\alpha \xrightarrow{\ 910°C\ } \gamma \xrightarrow{\ 1390°C\ } \delta$$

The structure of γ-Fe was found by Westgren and Lindh (1921) to be cubic close packed, while Westgren and Phragmen (1922a, b, c) first showed that δ-Fe, like α-Fe, is body centered cubic, two atoms per unit cell. This form is stable up to the melting point of $1535°C$. These results were confirmed by Bach (1929), Esser and Mueller (1933), and Esser, Eilender, and Bunghardt (1938). Precision measurements of the lattice constants at various temperatures have been reported by Owen and Williams (1954a) (-194 to $18°C$), Basinski, Hume-Rothery, and Sutton (1955) (20 to $1502°C$), and Goldschmidt (1962) (20 to $1255°C$). The variation in atomic volume with temperature, as calculated from the data of the foregoing, is presented in Fig. 26-1. The changes in volume at the two transitions are:

Cubic close packed iron was observed *at room temperature* by Jesser and Matthews (1967), who deposited thin films of iron, up to 20 Å thick, on oriented single crystals of copper. They stated that the iron structure was strained to match the structure of the substrate copper "exactly". However, if this is the case, the lattice constant of cubic close packed iron is 3.615 Å

Table 26-1.

Table 26-1.   The Lattice Constant of Iron at 20°C[a] and Atmospheric Pressure

a, Å	Reference	a, Å	Reference
2.87[b]	Hull, 1917a, 1919	2.86646[f]	Jette & Foote, 1936
2.84[b]	Westgren, 1921a, b, c	2.8664[e]	Owen & Yates, 1937
2.84[b]	Westgren & Lindh, 192	2.8663[b]	Bradley, Jay, & Taylor,
2.87[b]	Westgren & Phragmen,		1937
	1922a, b, c	2.86653[e]	van Bergen, 1941
2.875[b]	Owen & Preston, 1923a	2.8663[e]	Lu & Chang, 1941
2.878[b]	McKeehan, 1923b	2.86646[f]	Troiano & Williams, 1943
2.87[b]	Wever, 1924b	2.8661[b]	Rovinski & Tagunova,
2.870[b]	Heindlhofer, 1924		1947
2.867[b]	Davey, 1924	2.86645	Thomas, 1948
2.865[b]	Wyckoff & Crittenden,	2.86609[e]	van Horn, 1949
	1925	2.86629[g]	Zhmudski, 1949
2.8661[b]	Blake, 1925	2.8666[c]	Kochanovska, 1949
2.861[b]	Davey, 1925	2.8662[d]	Owen & Williams, 1954a
2.853[b]	Osawa & Ogawa, 1928	2.8664[h]	van Batchelder &
2.86682[c]	Mayer, 1929		Raeuchle, 1954
2.8758[b]	Bach, 1929	2.8646[b]	Lihl, 1954
2.863[b]	Schmidt, 1929	2.8663	Grønvold, Haraldsen, &
2.864[b]	Roberts & Davey, 1930		Vihovde, 1954
2.8672[b]	van Arkel & Burgers,	2.8663	Swanson, Fuyat, &
	1931		Ugrinic, 1955
2.8665[d]	Phragmen, 1931	2.8662[e]	Basinski, Hume-Rothery,
2.8663[b]	Bradley & Jay, 1932		& Sutton, 1955
2.8670[b]	Nishiyama, 1932	2.86621[e]	Sutton & Hume-Rothery,
2.8642[b]	Iwase & Nasu, 1932		1955
2.8658[b]	Preston, 1932	2.8663[h]	Taylor & Jones, 1958a, b
2.86655[d]	Owen & Yates, 1933a	2.86648[h]	Gale, 1959
2.8669[e]	Esser & Mueller, 1933	2.8665[i]	Cotta & Gazzara, 1961
2.86618[f]	Jette & Foote, 1935	2.8660[e]	Lihl & Ebel, 1961
2.8663[c]	Straumanis & Ievinš,	2.8662[e]	Goldschmidt, 1962
	1936		

av. 2.86638 ± 0.00019

[a]Values determined at other temperatures corrected with a thermal expansion
coefficient of 11.7×10^{-6} deg^{-1}.
[b]From kX, not included in the average.
[c]From kX and 22°C.
[d]From kX and 18°C.
[e]From kX.
[f]From kX and 25°C.
[g]From kX, 19 and 27°C.
[h]From 25°C.

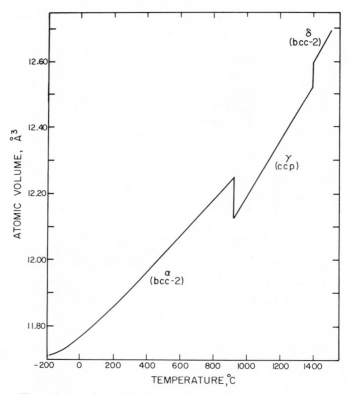

Fig. 26-1.   The variation of atomic volume with temperature of iron at atmospheric pressure (calculated from the data of Basinski et al., 1955).

(see Table 29-1); the corresponding atomic volume is 11.81 $\mathring{A}^3$, or almost exactly that of the normal bcc(2) form (11.78$\mathring{A}^3$) at room temperature. There is, accordingly, very little strain.

A phase transformation at 130 kbar and room temperature was discovered by Balchan and Drickamer (1961). One diffraction line of this new phase was observed by Jamieson and Lawson (1962), who postulated that the high pressure form is hexagonal close packed. This conjecture was soon verified by Takahashi and Bassett (1964), who observed the first six lines corresponding to the hexagonal close packed structure. On the basis of their and other measurements, they constructed the phase diagram for iron, Fig. 26-2.

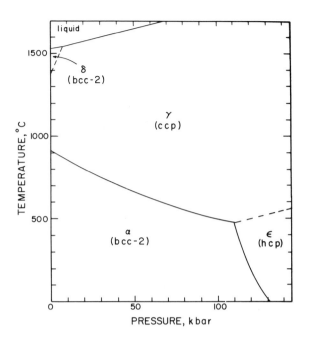

Fig. 26-2.   The phase diagram of iron (after Takahashi and Bassett, 1964).

Table 26-2.   Lattice Constants, Interatomic Distances, and Atomic Volumes in Iron

	a, Å	c, Å	Nearest Neighbors, Å (number)	Atomic Volume, Å³
At 20°C and atmospheric pressure				
α, bcc(2)	2.8664	—	2.482 (8)	11.78
ε, hcp	2.705	4.37	2.705(6), 2.686(6)	13.85
At 910°C and atmospheric pressure				
α, bcc(2)	2.9044	—	2.515 (8)	12.25
γ, ccp	3.6467	—	2.579(12)	12.12
At 1390°C and atmospheric pressure				
γ, ccp	3.6869	—	2.607(12)	12.53
δ, bcc(2)	2.9315	—	2.5388(8)	12.60
At 23°C and 130 kbar				
α, bcc(2)	2.805	—	2.429 (8)	10.43
ε, hcp	2.468	3.956	2.468(6), 2.408(6)	11.03

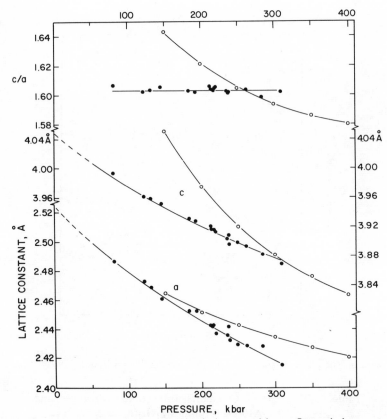

Fig. 26-3. Lattice constants and axial ratios in hexagonal iron. Open circles, Clendenan & Drickamer, 1964; closed circles, Mao, Bassett & Takahashi, 1966.

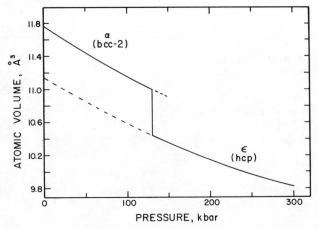

Fig. 26-4. The variation of the atomic volume of iron with pressure at room temperature (calculated with the data of Mao et al., 1966).

206

Hexagonal close packed iron *at atmospheric pressure* was observed by Bykov, Zdorovtseva, Troyan, and Khaimovich (1972), who examined thin films which had been deposited on KCl and then irradiated with helium ions, both processes having been carried out *in vacuo*. They reported lattice constants $a = 2.705$, $c = 4.37$ Å, $c/a = 1.616$. These are much larger than those predicted by the extrapolated values in Fig. 26-3, and the corresponding atomic volume of 13.85 Å^3 is similarly much larger than the extrapolated value in Fig. 26-4.

Additional determinations of the lattice constants of iron at high pressure were later reported by Clendenan and Drickamer (1964a) and by Mao, Bassett, and Takahashi (1966). The two sets of results are discordant, as may be seen in Fig. 26-3; these differences cannot be explained at present. The variation at room temperature of the atomic volume with pressure is presented in Fig. 26-4. Lattice constants and interatomic distances under various conditions are presented in Table 26-2.

COBALT

Cobalt exists in two forms, hexagonal close packed below 388°C and cubic close packed above that temperature. The transformations are sluggish, however, and the forms coexist both at room temperature and as high as 450°C. The structures were both first determined by Hull (1919). His and subsequent determinations of the lattice constants are presented in Table 27-1. At room temperature the ratio of the atomic volume of the cubic form to that of the hexagonal form is 1.0051. In the cubic form each cobalt atom has 12 nearest neighbors at 2.506 Å, while in the hexagonal form there are six at 2.507 Å and six at 2.497 Å, average 2.502 Å.

The pressure-temperature dependence of the phase change has been studied by Kennedy and Newton (1963). Their results are presented in Fig. 27-1. Owen and Jones (1954) concluded that the stable structure of cobalt at room temperature depends on the grain size: between room temperature and about 450°C, when the grain size is very small, as in cobalt sponge, the stable structure is cubic close packed, but in solid rods, with larger grain size, the stable structure is hexagonal close packed. The transformation temperature of 388°C pertains to samples with large grain size (Troiano and Tokich, 1948).

Owen and Jones observed what they considered a metastable form of cubic close packed cobalt, obtained by annealing at 600 to 840°C. This form, which had a lattice constant of 3.5612 Å, or 0.5% larger than normal, reverted to the other form with $a = 3.5441$ Å on annealing at higher temperatures,

Table 27-1.   The Lattice Constants of Cobalt at Room Temperature

a, Å	c, Å	Reference
Hexagonal Close Packed		
2.53[a]	—	Hull, 1919
2.519[a]	4.114[a]	Hull, 1921a
2.503[a]	4.060[a]	Sekito, 1927
2.862[a]	4.665[a]	Osawa, 1930
2.5068[b]	4.0711[b]	Hofer & Peebles, 1947
2.5074	4.0699	Taylor & Floyd, 1950
2.5067[b]	4.0698[b]	Taylor, 1950
2.501[c]	4.066[c]	Drain, Bridelle, & Michel, 1954
2.5054[a]	4.0892[a]	Owen & Jones, 1954
2.5071	4.0686	Anatharaman, 1958
av. 2.5070	4.0698	
±0.0003	±0.0009	
c/a 1.6233 ± 0.0004		
Cubic Close Packed		
3.58[a]		Hull, 1919
3.561[a]		Hull, 1921a
3.540[a]		Vegard & Dale, 1927
3.565[a]		Sekito, 1927
3.5442		Taylor & Floyd, 1950
3.5441[b]		Taylor, 1950
3.548[c]		Drain et al., 1954
3.5441[b]		Owen & Jones, 1954
3.5452		Luo & Duwez, 1963
3.5447		Swanson, Morris, & Evans, 1966
av. 3.5445 ± 0.0004		

[a]From kX, omitted from the average.
[b]From kX.
[c]Omitted from the average.

so contamination was ruled out as the explanation for this phenomenon for which no explanation was advanced.

A high temperature transformation in cobalt was claimed by Umino (1927), who observed a sharp drop in the specific heat at about 1150°C. Newkirk and Geisler (1953), on the other hand, observed no change in structure in diffraction experiments on cobalt wires at temperatures up to 1223°C. Evidence for and against the high temperature allotrope is summarized by Metcalf (1960).

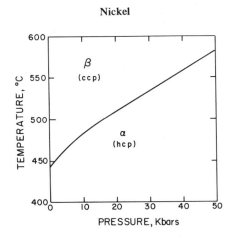

Fig. 27-1.   The phase behavior of cobalt (after Kennedy and Newton, 1963).

NICKEL

The structure of nickel was first determined by Hull (1917c). It is cubic close packed. Various determinations of the lattice constant are presented in Table 28-1. No change in structure between room temperature and 1200°C was observed by Jesse (1934). Reports of other forms of nickel are discussed below.

Hexagonal Nickel

The existence of an hexagonal allotrope of nickel has been the subject of controversy ever since the first report of it by Bredig and Allolio in 1927. References to papers in which the preparation of hexagonal nickel was claimed are listed in Table 28-2; some of these results have appeared later as pertaining to elemental nickel in standard compilations, for example, Neuberger (1933a), Barrett (1943), Wyckoff (1948), and Cullity (1956).

Jack (1950) considered that a preparation previously thought to be hexagonal nickel (that of reference d of Table 28-2) was in fact a nitride, Ni_3N; Trillat, Terao and Tertian (1961) suggested that the hexagonal nickel of reference m was the carbide reported by Nagakura (1957); Reddy (1964) concluded that the hexagonal substance reported in nickel electrodeposits (in references h, ℓ, and m) had yet to be unequivocally identified as hexagonal close packed nickel, largely, it appears, on the basis that the lattice constants of reference h are quite close to those of hexagonal *cobalt*, and that "cobalt is invariably present as an impurity in nickel salts and anodes".

Table 28-1.   The Lattice Constant of Nickel at 20°C[a]

a, Å	Reference	a, Å	Reference
3.53[b]	Hull, 1917c	3.5294[b]	Iwase & Nasu, 1932
3.55[b]	Hull, 1919	3.5233[b]	Bradley & Jay, 1932
3.54[b]	Bohlin, 1920	3.5241[b]	Bradley & Jay, 1932
3.547[b]	Hull, 1921a	3.5252[d]	Owen & Iball, 1932
3.517[b]	McKeehan, 1923b	3.5247[c]	Owen & Pickup, 1934
3.507[b]	Davey, 1924, 1925, 1926	3.535[b]	Jesse, 1934
3.525[b]	Wever, 1924a	3.5237[e]	Jette & Foote, 1935, 1936
3.55[b]	Levi & Tacchini, 1925	3.5244[c]	Owen & Yates, 1936
3.543[b]	Clark, Asbury, & Wick, 1925	3.5244[f]	LeClerc & Michael, 1939
		3.5244[f]	Lu & Chang, 1941
3.528[b]	Phebus & Blake, 1925	3.5237[e]	Fricke, 1941
3.525[b]	Lange, 1925	3.5240[d]	Esch & Schneider, 1944
3.526[b]	Holgersson, 1926	3.522[g]	Yang, 1950
3.523[b]	Bredig & Allolio, 1927	3.5240[d]	Taylor, 1950
3.524[b]	Hägg & Funke, 1929	3.5241[c]	Pearson & Hume-Rothery, 1952
3.522[b]	Greenwood, 1929		
3.52[b]	Valentiner & Becker, 1929	3.5236[h]	Swanson & Tatge, 1953
		3.5236[h]	Batchelder & Raeuchle, 1954
3.521[b]	Mazza & Nosini, 1929	3.5254[i]	Diament, 1956
3.515[b]	Osawa, 1930	3.5241[d]	Rovinski, Samilov, & Rovenski, 1959
3.525[b]	Burgers & Basart, 1930		
3.53[b]	Bredig & von Bergkampf, 1931	3.5225[i]	Kogan & Bulatov, 1962
		3.5238[g]	Casselton & Hume-Rothery, 1964
3.5252[c]	Phragmen, 1931		
3.524[b]	Kaya & Kussmann, 1931	3.524[j]	Fukano & Kimoto, 1967

av. 3.5241 ± 0.0007

[a]Values determined at other temperatures corrected to 20°C with a thermal expansion coefficient of 12.5×10^{-6} deg^{-1}.
[b]From kX, not included in the average.
[c]From kX and 18°C.
[d]From kX.
[e]From kX and 25°C.
[f]From kX and 21°C.
[g]Not included in the average.
[h]From 25°C.
[i]From kX and 27°C.
[j]Read from a graph; not included in the average.

Table 28-2. Lattice Constants and Derived Quantities Reported for Hexagonal Nickel

a, Å	c, Å	c/a	Atomic Volume, Å^3	Method of Preparation	Reference
2.689*	4.291*	1.596	13.44	Atomized in hydrogen	a
2.479*	4.07*	1.64	10.83	Atomized in air or argon	b
2.66*	4.33*	1.63	13.27	Heating cubic form to 170°C in carbon monoxide	c
Not reported				"Electric discharge conditions"	d
2.51*	4.03*	1.61	10.92	Sputtering *in vacuo*	e, f
Not reported				Electrodeposition	g
2.495	4.078	1.634	10.99	Electrodeposition	h
2.49	4.08	1.64	10.95	Electrodeposition	i
2.64	4.33	1.64	13.07	Vacuum evaporation	j
2.66	4.32	1.63	13.24	Vacuum evaporation	k
Not reported				Electrodeposition	ℓ, m
2.62	4.36	1.66	12.96	Neutron irradiation of thin films of Ni on glass or quartz (silica?) *in vacuo*	n
2.63	4.32	1.64	12.94	Vacuum evaporation	o
Not reported				Electrodeposition	p
2.622	4.320	1.648	12.86	Vacuum evaporation	q, r
2.50	3.98	1.59	10.77	Electrodeposition	s
Not reported				Vacuum evaporation	t
2.62	4.31	1.64	12.81	Recrystallization of thin films *in vacuo*	u

*From kX.
aBredig & Allolio, 1927.
bThomson, 1929.
cLeClerc & Michel, 1939.
dRooksby, 1942.
eColombani & Wyart, 1942.
fColombani, 1944.
gFinch, Wilman, & Yang, 1947.
hYang, 1950.
iEvans & Hopkins, 1952.
jBublik & Pines, 1952.

kTrillat, Tertian, & Terao, 1956.
ℓBanerjee & Goswani, 1959a.
mBanerjee & Goswani, 1959b.
nTeodorescu & Glodeanu, 1960.
oBonnelle & Jacquot, 1961.
pGraham, Lindsay, & Read, 1962.
qWeik & Hemenger, 1965.
rHemenger & Weik, 1965.
sWright & Goddard, 1965.
tBonnelle & Vergand, 1966.
uVergand, 1967.

Lattice constants of some selected substances are presented in Table 28-3. Comparison of these with the data in Table 28-2 strongly suggests that the hexagonal "nickels" of references a, c, j, k, n, o, q, and u are really carbides, nitrides, hydrides, or hydronitrides of uncertain composition; these all have volumes per nickel atom in the range 12.81 to 13.44 $Å^3$. This is far larger than the value 10.94 $Å^3$ observed in cubic close packed nickel. The remaining five preparations, those of references b, e, f, h, i, and s, all have volumes per nickel atom in the range 10.77 to 10.99 $Å^3$, or very close to that of the cubic close packed form. It is accordingly quite probable that these samples are indeed hexagonal close packed nickel. It is interesting that two of these were obtained by vacuum evaporation and three by electrodeposition. Furthermore, the electrodeposited samples of reference s contained the greatest metallic impurity, six parts per million of iron, a result which rules out the possibility that the substance was cobalt. Indicative information to the contrary being absent, the simple averages to five sets of lattice constants will be accepted as those of hexagonal nickel. Vacuum deposited nickel which was subsequently irradiated with helium ions was examined by Bykov et al. (1972), who reported hexagonal close packed lattice constants $a = 2.62$, $c = 4.30$ Å; the corresponding atomic volume is 12.78 $Å^3$. The experimental conditions described make it difficult to understand why these values are so large.

Table 28-3.   Some Lattice Constants

a, Å	c, Å	c/a	Volume Per Ni Atom/$Å^3$	Substance	Reference
2.654*	4.347*	1.638	13.26	Ni_3C	a
2.650*	4.321*	1.631	13.14	NiH	b
2.667*	4.269*	1.601	13.15	Ni(H,N)	b
2.6677	4.3122	1.616	13.29	Ni_3N	c
2.633	4.315	1.639	12.95	Ni_3C	d
2.507	4.070	1.623	(11.08)	Co	e

*From kX.
[a]Jacobsen & Westgren, 1933.
[b]Büssem & Gross, 1934.
[c]Jack, 1950.
[d]Nagakura, 1957, 1958.
[e]see above, p. 208.

Body Centered Cubic Nickel

A body centered cubic form of nickel, two atoms per unit cell, a = 2.77 Å (from kX) was first reported by Hull (1917c). It was stated to have been prepared by filing a sample which had been previously melted. This result was unanimously ignored by subsequent workers.

In 1955 Finch, Sinha, and Goswami, unaware of the earlier result of Hull, reported that they had prepared body centered cubic nickel by thermal deposition in high vacuum onto various substrates at 300 to 340°C. The lattice constant was found to be 2.78 Å, in excellent agreement with the one reported in 1917.

Lattice constants and other quantities for the various forms of nickel are presented in Table 28-4.

Table 28-4.   Lattice Constants and Various Other Quantities for Nickel
(at room temperature)

	Cubic Close Packed	Hexagonal Close Packed	Body Centered Cubic
a, Å	3.5241 ± 0.0007	2.495 ± 0.012	2.775 ± 0.014
c, Å	—	4.048 ± 0.043	—
c/a	—	1.622 ± 0.020	—
Atomic Volume, Å^3	10.94 ± 0.01	10.91 ± 0.16	10.68 ± 0.16
Nearest neighbor distances	2.492 (12)	2.495 (6) 2.484 (6)	2.403 (8)

RUTHENIUM

The hexagonal close packed structure for ruthenium was first established by Hull (1920). Various determinations of the lattice constant are presented in Table 44-1.

The existence of three high temperature allotropes of ruthenium was proposed by Jaeger and Rosenbohm (1931b, 1941) on the basis of measurements of the specific heat, electrical resistivity, and thermal behavior with respect to platinum; the transitions were stated to occur at 1035, 1190, and 1500°C. However, no evidence for such transitions was found in high temperature measurements of the lattice constants by Hall and Crangle (1957; 25 to 1400°C) or by Ross and Hume-Rothery (1963; 25 to 2180°C).

No transition at 25°C and pressures up to 400 kbar were detected by Clendenen and Drickamer (1964a).

At room temperature each ruthenium atom has six neighbors at 2.7053 Å and six at 2.6499 Å, average 2.6776 Å. The atomic volume is 13.568 $Å^3$.

Table 44-1.   The Lattice Constants of Ruthenium at Room Temperature

a, Å	c, Å	Reference
2.691[a]	4.279[a]	Hull, 1920, 1921a
2.700[a]	4.281[a]	Barth & Lunde, 1925, 1926
2.685[a]	4.270[a]	Levi & Haardt, 1926
2.7042[b]	4.2815[b]	Owen, Pickup, & Roberts, 1935
2.7039[b]	4.2817[b]	Owen & Roberts, 1937
2.7058[b]	4.2816[b]	Hellawell & Hume-Rothery, 1954a
2.7058	4.2819	Swanson, Fuyat, & Ugrinic, 1955
2.7056	4.2803	Hall & Crangle, 1957
2.7070[a]	4.2846[a]	Rudnicki & Polyakova, 1962
2.7058	4.2811	Černohorsky, 1960
2.7060	4.2837[c]	Rudy, Kieffer, & Fröhlich, 1962
2.7058	4.2816	Savicki, Tylkina, & Polyakova, 1962
2.700[c]	4.275[c]	Clendenen & Drickamer, 1964a
av. 2.7053	4.2814	
±0.0008	±0.0005	
c/a 1.5826 ± 0.0005		

[a]From kX, not included in the average.
[b]From kX.
[c]Omitted from the average.

RHODIUM

Rhodium is cubic close packed, first reported by Hull (1921a). Determinations of the lattice constant are presented in Table 45-1.

A second form of rhodium was reported by Jaeger and Zanstra (1931), who found a maximum in the specific heat curve at about 1200°C. Jaeger and Rosenbohm (1931a) reported that rhodium deposited electrolytically was a mixture of two cubic forms, one the normal form with lattice constant 3.80 Å (from kX), the second, also cubic, with lattice constant 9.230 Å (from kX). This second form has not been observed subsequently by anyone else. Furthermore, no transition was observed by Ross and Hume-Rothery (1963), who measured the lattice constant from room temperature up to 1950°C.

At 20°C each rhodium atom has 12 closest neighbors at 2.6893 Å. The atomic volume is 13.753 $Å^3$.

Table 45-1.   The Lattice Constant of Rhodium at 20°C[a]

a, Å	Reference	a, Å	Reference
3.828[b]	Hull, 1921a	3.8045[b]	Goldschmidt & Land, 1947
3.803[b]	Barth & Lunde, 1926	3.8029[e]	Swanson, Fuyat, & Ugrinic,
3.8021[b]	van Arkel, 1928		1954
3.80[b]	Jaeger & Rosenbohm,	3.8023[f]	Bale, 1958
	1931a	3.803[b]	Raub, Beeskow, & Menzel,
3.8032[c]	Owen & Iball, 1932		1959
3.8034[d]	Owen & Yates, 1933a, b	3.8036	Černohorsky, 1960
3.807[b]	Drier & Walker, 1933	3.8031[g]	Ross & Hume-Rothery, 1963
		3.8032	Singh, 1968
av. 3.8032 ± 0.0002			

[a]Values corrected to 20°C with a thermal expansion coefficient of 8.4 × 10^{-6} deg^{-1}.
[b]From kX, not included in the average.
[c]From 16.5° and kX.
[d]From 18° and kX.
[e]From 25°.
[f]Not included in the average.

PALLADIUM

Palladium is cubic close packed. There is no change in structure down to $4°K$ (Abrahams, 1963) or up to at least $878°C$ (Dutta and Dayal, 1963b). The structure was first determined by Hull (1920); his and later determinations of the lattice constant are presented in Table 46-1.

At $20°C$ each palladium atom has 12 closest neighbors at 2.7506 Å. The atomic volume is $14.716 Å^3$.

Table 46-1.   The Lattice Constant of Palladium at $20°C$[a]

a, Å	Reference	a, Å	Reference
3.93[b]	Hull, 1920	3.888[b]	Nowotny, Schubert, &
3.958[b]	Hull, 1921a		Dettinger, 1946
3.91[b]	McKeehan, 1922b	3.8897[f]	Swanson & Tatge, 1953
3.908[b]	McKeehan, 1923a	3.8908[b]	Hellawell & Hume-Rothery,
3.881[b]	Barth & Lunde, 1925		1954b
3.867[b]	Davey, 1925, 1926	3.891[b]	Raub & Wörwag, 1955
3.852[b]	Jaeger & Zanstra, 1931	3.8908[g]	Coles, 1956
3.8887[c]	Stenzel & Weerts, 1931	3.899[b]	Rudnicki & Polyakova,
3.893[b]	Owen & Iball, 1932		1959
3.8903[d]	Owen & Yates, 1933a,b	3.8899[f]	Abrahams, 1963
3.8900[e]	Owen & Jones, 1937	3.8898[f]	Dutta & Dayal, 1963b
3.888[b]	Michel, Bénard, &	3.8908[g]	Catteral & Barker, 1964
	Chaudron, 1945	3.888[h]	Fukano & Kimoto, 1967
3.8895[b]	Kuznecov, 1946	3.892[i]	Pratt, Myles, Darby, &
av. 3.8900 ± 0.0007			Mueller, 1968

[a]Values corrected to $20°C$ with a thermal expansion coefficient of 11.75×10^{-6} deg^{-1}.
[b]From kX, not included in the average.
[c]From kX.
[d]From kX and $18°C$.
[e]From kX and $10°C$.
[f]From $25°C$.
[g]From kX and $22°C$.
[h]Read from a graph; not included in the average.
[i]Not included in the average.

OSMIUM

The structure of osmium, which is hexagonal close packed, was first determined by Hull (1921b). Various determinations of the lattice constant are presented in Table 76-1.

At 20°C each osmium atom has six neighbors at 2.7348 Å and six at 2.6753 Å, average 2.7051 Å. The atomic volume is 13.988 $Å^3$.

Table 76-1. The Lattice Constants of Osmium at 20°C

a, Å	c, Å	Reference
2.719[a]	4.324[a]	Hull, 1921b
2.730[a]	4.323[a]	Barth & Lunde, 1925
2.719[a]	4.325[a]	Levi & Haardt, 1926
2.721[a]	4.340[a]	Swjaginzeff & Brunowsk, 1932
2.7359[b]	4.3186[b]	Owen, Pickup, & Roberts, 1935
2.7353[b]	4.3191[b]	Owen & Roberts, 1937
2.731[a]	4.323[a]	Winkler, 1943
2.7341	4.3197	Swanson, Fuyat, & Ugrinic, 1955
2.7341	4.3199	Taylor, Doyle, & Kagle, 1962
av. 2.7348	4.3193	
±0.0009	±0.0006	

c/a 1.5794 ± 0.0005

[a]From kX, not used in the averaging.
[b]From kX.

IRIDIUM

The structure of iridium was first determined by Hull (1920). It is cubic close packed. There is no change in structure down to $4.2°K$ (Schaake, 1968). Determinations of the lattice constant are presented in Table 77-1.

At $20°C$ each iridium atom has 12 closest neighbors at 2.7147 Å. The atomic volume is 14.146 $Å^3$.

Table 77-1.   The Lattice Constant of Iridium at $20°C$

a, Å	Reference
3.813[a]	Hull, 1920, 1921a
3.831[a]	Wyckoff, 1923
3.831[a]	Barth & Lunde, 1926
3.8391[b]	Owen & Iball, 1932
3.8389[b]	Owen & Yates, 1933a
3.8394	Swanson, Fuyat, & Ugrinic, 1955
3.8390	Schäfer & Heitland, 1960
3.8388	Singh, 1968
3.8396	Schaake, 1968
av. 3.8391 ± 0.0003	

[a]From kX, omitted from the average.
[b]From kX.

PLATINUM

Platinum is cubic close packed, as first determined by Hull (1921a). This structure is retained at least up to 1732°C (Edwards, Speiser, and Johnston, 1951). Determinations of the lattice constant are presented in Table 78-1; the early values exhibit a curiously large amount of scatter.

At room temperature each platinum atom has 12 closest neighbors at 2.7742 Å. The atomic volume is 15.097 Å^3.

Table 78-1. The Lattice Constant of Platinum at Room Temperature

a, Å	Reference	a, Å	Reference
3.938[a]	Hull, 1921a	3.9237[b]	Owen & Yates, 1933a, b
4.03[a]	Kahler, 1921	3.9240[b]	Owen & Yates, 1934
3.88[a]	Uspenski & Konobejewski, 1923	3.9224[b]	Moeller, 1937
3.921[a]	Davey, 1924	3.923[a]	Esser, Eilender, & Bungardt, 1938
3.920[a]	Davey, 1925		
3.911[a]	Barth & Lunde, 1926	3.924[a]	Popov, Simanov, Skutarov, &Suzdalceva, 1943
3.952[a]	Bredig & Allolio, 1927	3.9237[b]	Esch & Schneider, 1944
3.9221[b]	van Arkel, 1928	3.9240[b]	Goldschmidt & Land, 1947
3.92[a]	Thomson, 1930	3.9236[b]	Grube, Schneider, & Esch, 1951
3.922[a]	Johansson & Linde, 1930b		
3.920[a]	Greenwood, 1931	3.924[a]	Edwards, Speiser, & Johnston, 1951
3.9230[b]	Stenzel & Weertz, 1931	3.9231	Swanson & Tatge, 1953
3.94[a]	Bannister & Hey, 1932	3.924[a]	Raub & Wörwag, 1955
av. 3.9233 ± 0.0007			

[a]From kX, not included in the average.
[b]From kX.

Transition Metals

GROUP IB

COPPER

The structure of copper was first determined by Bragg (1914). He studied a single crystal, an etched mineralogical specimen, with PdK_α radiation, and concluded that it is cubic close packed with a = 3.61 Å (from kX). Subsequent determinations of the lattice constant are presented in Table 29-1. There is no change in structure down to 8°K (Simmons and Balluffi, 1957) or up to 1000°C (Esser, Eilender, and Bungardt, 1937).

Table 29-1.　The Lattice Constant of Copper at 20°C[a]

a, Å	Reference	a, Å	Reference
3.60[b]	Bragg, 1914	3.616[b]	von Göler & Sachs, 1929
3.622[b]	Kirchner, 1922	3.612[b]	Weinbaum, 1929
3.641[b]	Gerlach, 1922	3.609[b]	Sekito, 1929b
3.615[b]	Owen & Preston, 1923a	3.609[b]	Katoh, 1929
3.635[b]	Owen & Preston, 1923b	3.616[b]	Aborn & Davidson, 1930
3.61[b]	Young, 1923	3.615[b]	Ageew, Hansen, &
3.61[b]	Patterson, 1924		Sachs, 1930
3.618[b]	Jette, Phragmen, &	3.615[b]	Burgers & Basart, 1930
	Westgren, 1924	3.614[b]	Persson, 1930
3.605[b]	Davey, 1924	3.615[b]	Arrhenius & Westgren,
3.617[b]	Westgren & Phragmen,		1931
	1925b	3.615[b]	Katoh, 1930
3.614[b]	Lange, 1925	3.6151[c]	Owen & Iball, 1932
3.604[b]	Davey, 1925, 1926	3.6163[b]	Megaw, 1932
3.62[b]	Sacklowski, 1925	3.619[b]	LeBlanc & Wehner, 1932
3.627[b]	Holgersson, 1926	3.615[b]	Linde, 1932
3.627[b]	Jung, 1926a	3.6153[d]	Obinata & Wasserman, 1933
3.612[b]	Davey & Wilson, 1926	3.6152[d]	Owen & Pickup, 1933a
3.62[b]	Erdal, 1926	3.6152[d]	Owen & Yates, 1933a
3.62[b]	Vegard & Dale, 1927	3.6154[e]	Vegard & Kloster, 1934
3.615[b]	Bredig & Allolio, 1927	3.615[b]	Weibke, 1934
3.610[b]	Terrey & Wright, 1928	3.6151[d]	Owen & Pickup, 1934
3.615[b]	Westgren & Almin, 1928	3.6148[b]	Straumanis & Mellis, 1935
3.610[b]	Smith, 1928	3.6152[d]	Owen & Rogers, 1935
3.613[b]	van Arkel, 1928	3.6143[f]	Hume-Rothery, Lewin &
3.617[b]	Frohlich, Davidson, &		Reynolds, 1936
	Fenske, 1929		

Hexagonal Copper

The occurrence of a hexagonal close packed form of copper has been reported by Takahashi (1952, 1953), who prepared thin films electrolytically. The lattice constants are $a = 2.56$, $c = 4.17$ Å (from kX and c/a), $c/a = 1.63$. The atomic volume is 11.8 Å^3, a value in excellent agreement with that of the cubic close packed form, 11.808 ± 0.003 Å^3.

The hexagonal close packed form of copper, having $a = 2.72$, $c = 4.40$ Å, $c/a = 1.58$, atomic volume 14.1 Å^3, reported by Couderc et al. (1959), is probably a hydride.

Table 29-1. The Lattice Constant of Copper at 20°C[a] (Continued)

3.6148[d]	Esser, Eilender, & Bungardt, 1937	3.6145[g]	Anderson & Kingsbury, 1943
3.6148[e]	van Bergen, 1937, 1938	3.6145[i]	Rose, 1946
3.6148[d]	Owen & Roberts, 1939	3.6144[b]	Rovinski & Tagunova, 1947
3.6148[g]	Foote & Jette, 1940	3.6144[e]	Crussard & Aubertin, 1949
3.6147[e]	Felipe, 1940	3.61[b]	Eppelsheimer & Penman, 1950
3.6149[d]	Owen & Rowlands, 1940		
3.6143[g]	Carapella & Hultgren, 1941, 1942	3.6147[e]	Frohnmeyer & Glocker, 1953
		3.6147[j]	Swanson & Tatge, 1953
3.6144[h]	Lu & Chang, 1941	3.6143[k]	Mitra & Mitra, 1963
3.615[b]	Fricke, 1941	3.6146[j]	Straumanis & Yu, 1969
3.6148[d]	Hume-Rothery & Andrews, 1942	3.6146[k]	Krull & Newman, 1970

av. 3.6148 ± 0.0003

[a]Values corrected to 20°C with a thermal expansion coefficient of 16.6×10^{-6} deg^{-1} where appropriate.
[b]From kX, not included in the average.
[c]From kX and 16.5°C.
[d]From kX and 18°C.
[e]From kX.
[f]From kX and 23°C.
[g]From kX and 25°C.
[h]From kX and 19°C.
[i]From kX and 21°C.
[j]From 25°C.
[k]From 27°C.

SILVER

The structure of silver was first determined by Vegard (1916a). It is cubic close packed, and there is no change in structure up to 943°C (Hume-Rothery and Reynolds, 1938), nor down to -194°C (Owen and Williams, 1954).

Allotropes of Silver

Allard (1928) reported that a sample of silver obtained by precipitation from silver nitrate solution by copper is orthorhombic, $a = 4.24$, $b = 4.92$, $c = 5.71$ Å (from kX). This form has not been reported since.

Hexagonal close packed silver has been observed, in thin films deposited from the vapor, by Quarrell (1937), Andrushchenko, Tjapkina, and Dankov (1948), Bublik (1954), and König (1958). The averaged lattice constants (from kX) are $a = 2.90 \pm 0.02$, $c = 4.74 \pm 0.04$ Å (from kX), $c/a = 1.63$. These correspond to an atomic volume of 17.26 ± 0.28 Å^3, a value in satisfactory agreement with that of 17.051 ± 0.002 Å^3 for cubic close packed silver.

A metastable cubic close packed form of silver with $a = 4.385$ Å was reported by Leiga (1966), who stated that it is formed during the thermal decomposition of silver oxalate. The lattice constant is 7.3% larger, and the atomic volume 23.6% larger, than that of normal silver.

GOLD

The structure of gold is cubic close packed, as first determined by Vegard (1916b). The various determinations of the lattice constant are presented in Table 79-1. There is no change in structure up to 1050°C (Esser, Eilender, and Bungardt, 1938).

Conflicting results have been obtained concerning the structure of gold in thin films. Chatterjee (1957) found that it is still cubic close packed, but that the lattice constant varied with the thickness of the films, values from 3.984 to 4.187 Å being reported. These correspond to nearest neighbor distances of 2.817 to 2.961 Å, as compared with 2.884 Å in normal gold. Couderc et al. (1959), on the other hand, found that thin films of gold have the hexagonal close packed structure, with $a = 2.79$, $c = 4.40$ Å, $c/a = 1.58$. If this preparation is really gold, then the average nearest neighbor distance is 2.76 Å, and the atomic volume is 14.8 Å^3. The latter may be compared with the value 16.96 Å^3 in normal gold; it must be pointed out that a *decrease* (this case, 13%) in atomic volume in going from bulk material to thin film is most unusual.

Table 47-1. The Lattice Constant of Silver at $20°C$[a]

a, Å	Reference	a, Å	Reference
4.07[b]	Vegard, 1916a	4.086[b]	Weibke & Eggers, 1935
4.07[b]	Kahler, 1921	4.0854[b]	Owen & Rogers, 1935
4.09[b]	McKeehan, 1922b	4.0858[h]	Hume-Rothery, Lewin, &
4.086[b]	Wilsey, 1923		Reynolds, 1936
4.066[b]	Davey, 1924	4.0862[c]	Moeller, 1937
4.088[b]	Westgren & Phragmen,	4.0858[h]	Hume-Rothery & Reynolds,
	1925b		1938
4.087[b]	Davey, 1925, 1926	4.0858[e]	Owen & Roberts, 1939
4.07[b]	Sacklowski, 1925	4.0856[g]	Miller & DuMond, 1940
4.084[b]	Holgersson, 1926	4.0857[g]	Foote & Jette, 1940
4.086[b]	Barth & Lunde, 1926	4.0858[e]	Owen & Rowlands, 1940
4.05[b]	Erdal, 1926	4.0854[e]	Felipe, 1940
4.078[b]	Jung, 1926a	4.0857[g]	Foote & Jette, 1941b
4.0857[b]	van Arkel, 1928	4.0855[c]	Lipson, Petch, & Stockdate,
4.086[b]	Westgren & Almin, 1929		1941
4.077[b]	Osawa, 1929	4.085[b]	Hass, 1942
4.0859[b]	Sachs & Weerts, 1930	4.085[b]	Andrushchenko, Tjapkina, &
4.0859[b]	Ageew & Sachs, 1930		Dankov, 1948
4.085[b]	Nial, Almin, &	4.10[i]	Berry, 1949
	Westgren, 1931	4.080[i]	Karlsson, 1952
4.0854[c]	Stenzel & Weerts, 1931	4.0858	Swanson & Tatge, 1953
4.087[b]	Phelps & Davey, 1932	4.0855	Becherer & Ifland, 1954
4.0858[d]	Owen & Iball, 1932	4.0858[e]	Owen & Williams, 1954a
4.086[b]	Megaw, 1932	4.0858[j]	Smakula & Kalnajs, 1955
4.0847[b]	Jette & Gebert, 1933	4.0848[b]	Neff, 1956
4.085[b]	Stenbeck, 1933	4.0856[g]	Hill & Axon, 1956
4.0859[b]	Ageew & Shoyket, 1933	4.0858[c]	Spreadborough & Christian,
4.0856[e]	Owen & Yates, 1933a		1959a
4.0856[c]	Saini, 1933	4.08[i]	Vlach & Stehlik, 1960
4.0858[f]	Owen & Yates, 1934	4.0857[j]	King & Vassamillet, 1961
4.0852[b]	Straumanis & Mellis,	4.0856	Crockett & Davis, 1963
	1935	4.0859[j]	Straumanis & Riad, 1965
4.0857[g]	Jette & Foote, 1935		

av. 4.08570 ± 0.00018

[a]Values not determined at $20°C$ corrected with a thermal expansion coefficient of 18.9×10^{-6} deg^{-1} where appropriate.

[b]From kX, not included in the average.

[c]From kX.

[d]From kX and $16.5°C$.

[e]From kX and $18°C$.

[f]From kX and $15°C$.

[g]From kX and $25°C$.

[h]From kX and $19.6°C$.

[i]Not included in the average.

[j]From $25°C$.

Table 79-1. The Lattice Constant of Gold at 20°C[a]

a, Å	Reference	a, Å	Reference
4.10[b]	Vegard, 1916b	4.0784[c]	Jette & Foote, 1935
4.09[b]	Kahler, 1921	4.078[b]	Köster & Dannöhl, 1936
4.083[b]	McKeehan, 1922b	4.0783[c]	Straumanis & Ieviņš, 1936
4.075[b]	Kirchner, 1922	4.0783[c]	Owen & Rowlands, 1940
4.08[b]	Huber, 1924	4.078[b]	Popov, Simanov, Skuratov,
4.084[b]	Davey, 1924		& Suzdalceva, 1943
4.081[b]	Westgren & Phragmen,	4.0632[b]	Lu & Malmberg, 1943
	1925b	4.0782[c]	Owen & Roberts, 1945
4.078[b]	Lange, 1925	4.0782[b]	Jaffee, Smith, & Gonser, 1945
4.073[b]	Davey, 1925, 1926	4.0783[b]	Grube, Schneider, & Esch,
4.088[b]	Holgersson, 1926		1951
4.078[b]	Barth & Lunde, 1926	4.0792[b]	Kato, 1951
4.076[b]	Jung, 1926a	4.0783[e]	Swanson & Tatge, 1953
4.078[b]	van Arkel, 1928	4.0782[c]	Geach & Summers-Smith,
4.072[b]	Smith, 1928		1953
4.077[b]	Johansson & Linde,	4.0782[c]	Guntert & Faessler, 1956
	1930a	4.0780[c]	Weyerer, 1956b, c, d
4.0788[c]	Sachs & Weerts, 1930	4.0775[b]	Neff, 1956
4.0782[c]	Stenzel & Weerts, 1931	4.0780[c]	Weyerer, 1957
4.0795[b]	Owen & Iball, 1932	4.0793[b]	Warlimont, 1959
4.078[b]	LeBlanc & Wehner, 1932	4.0785[f]	Day, 1961
4.0787[b]	Weist, 1933	4.0777[e]	Dutta & Dayal, 1963a
4.0782[c]	Owen & Yates, 1933a,b	4.074[g]	Fukano & Kimoto, 1967
4.0783[d]	Vegard & Kloster, 1934		
4.0788[b]	Jette, Brunner & Foote,		
	1934		
av. 4.0782 ± 0.0002			

[a]Values corrected to 20°C with a thermal expansion coefficient of 13.9 × 10^{-6} deg^{-1} when necessary.
[b]From kX, not included in the average.
[c]From kX, corrected to 20°C.
[d]From kX.
[e]Corrected to 20°C.
[f]Not included in the average.
[g]Read from a graph; not included in the average.

GROUP IIB

ZINC

The structure of zinc was first determined by Hull (1920). It may be described as a distortion of hexagonal close packing, with the axial ratio increased from the ideal value of 1.633 to 1.856. Various determinations of the lattice constants are presented in Table 30-1.

There is no change in structure up to 415°C, or 5° below the melting point (Peirce, Anderson, and Van Dyck, 1925; Owen and Yates, 1934), but there is a steady increase in the axial ratio, as shown in Fig. 30-1. The structure is also unchanged down to liquid air temperature (McLennan and Monkman, 1929).

With increase in pressure (at room temperature) the atomic volume decreases monotonically (Lynch and Drickamer, 1965), as shown in Fig. 30-2, but the changes in the lengths of the axes (Fig. 30-3) are rather unexpected. In the region ~50 to 100 kbar the decrease in the compressibility parallel to the c-axis combines with the increase in the compressibility parallel to the a-axis to give the unusual behavior of the axial ratio shown in Fig. 30-4.

At room temperature and atmospheric pressure each zinc atom has six neighbors in the close packed layer at 2.644 Å, plus six more above and below at 2.912 Å; the atomic volume is 15.202 ± 0.003 Å^3.

Finch and Quarrell (1933) found that zinc deposited as thin films on platinum had the same value for a as bulk zinc, but the observed value for c was 4.7% larger, at 5.18 Å.

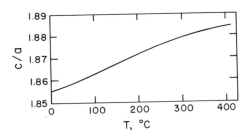

Fig. 30-1.   The variation of the axial ratio with temperature for zinc.

Table 30-1. The Lattice Constants of Zinc at 20°C[a]

a, Å	c, Å	Reference
2.675[b]	4.980[b]	Hull, 1920, 1921a
2.675[b]	4.976[b]	Hull & Davey, 1921a, b
2.68[b]	4.98[b]	Mark, Polanyi, & Schmid, 1922
2.68[b]	4.98[b]	Owen & Preston, 1923b
2.69[b]	5.00[b]	Simson, 1924
2.669[b]	4.938[b]	Phebus & Blake, 1925
2.662[b]	4.958[b]	Peirce, Anderson, & Van Dyck, 1925
2.654[b]	4.940[b]	Freeman, Sillers, & Brandt, 1926
2.670[b]	4.945[b]	Osawa & Ogawa, 1928
2.662[b]	4.958[b]	McLennan & Monkman, 1929
2.6638[c]	4.9453[c]	Boas, 1932
2.6644[c]	4.9451[c]	Stenzel & Weerts, 1932
2.664[b]	4.944[b]	Owen & Pickup, 1933b
2.6639[b]	4.9442[b]	Weigle, 1933
2.6635[b]	4.9441[b]	Jette & Gebert, 1933
2.664[b]	4.944[b]	Owen & Iball, 1933
2.63[c]	5.18[c]	Finch & Quarrell, 1933
2.6643[d]	4.9450[d]	Owen & Yates, 1934
2.6647[e]	4.9453[e]	Jette & Foote, 1935
2.6646[f]	4.9459[f]	Owen, Pickup, & Roberts, 1935
2.665[b]	4.945[b]	Wiedenach-Nostiz, 1946
2.665[g]	4.947[g]	Swanson & Tatge, 1953
2.6646[h]	4.9456[h]	Ancker, 1953
2.6645[i]	4.9454[i]	Brown, 1954
2.665[g]	4.947[g]	Lynch & Drickamer, 1965
av. 2.6644	4.9454	
±0.0003	±0.0003	

c/a 1.8561 ± 0.0002

[a]Values not reported for 20°C corrected when necessary with thermal expansion coefficients of $\alpha_\perp = 14 \times 10^{-6}$ deg^{-1} and $\alpha_\parallel = 61 \times 10^{-6}$ deg^{-1}.
[b]From kX, not included in the average.
[c]From kX, sample examined a thin film deposited on platinum
[d]From kX.
[e]From kX and 25°C.
[f]From kX and 18°C.
[g]Not included in the average.
[h]From 22°C.
[i]From 25°C.

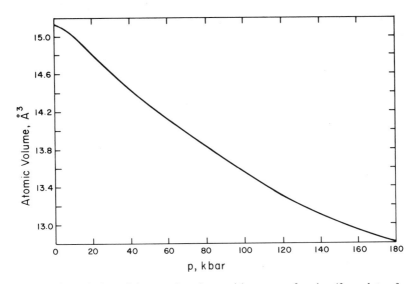

Fig. 30-2.   The variation of the atomic volume with pressure for zinc (from data of Lynch & Drickamer, 1965).

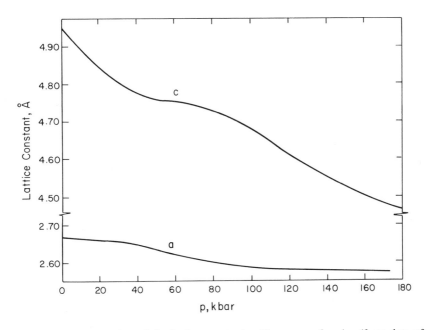

Fig. 30-3.   The variation of the lattice constants with pressure for zinc (from data of Lynch & Drickamer, 1965).

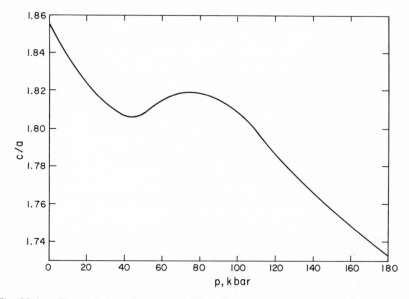

Fig. 30-4.   The variation of the axial ratio with pressure for zinc (from data of Lynch and Drickamer, 1965).

CADMIUM

The structure of cadmium was first determined by Hull (1920). Like that of zinc, it is a distortion of hexagonal close packing, with the axial ratio increased to 1.885. Determinations of the lattice constants are presented in Table 48-1.

There is no change in structure down to liquid air temperature (McLennan and Monkman, 1929) nor up to 310°C, or 12° below the melting point (Shinoda, 1934; Hume-Rothery and Raynor, 1940). The transitions reported at 64.9°C (Cohen and Halderman, 1914) and 112°C (Jänecke, 1915) apparently do not take place in pure cadmium.

The behavior under pressure, at room temperature (Lynch and Drickamer, 1965), is reminiscent of that of zinc, but somewhat less anomalous. The decrease in atomic volume with pressure is presented in Fig. 48-1, and the changes in the lengths of the axes and of the axial ratio in Figs. 48-2 and 48-3. The anomalous behavior is seen to occur in the region near 100 kbar.

At room temperature and atmospheric pressure each cadmium atom has six neighbors in the close packed layers at 2.979 Å plus six more above and below at 3.293 Å; the atomic volume is 21.579 ± 0.007 Å³.

Table 48-1.   The Lattice Constants of Cadmium at 20°C[a]

a, Å	c, Å	Reference
2.99[b]	5.65[b]	Hull, 1920, 1921a, b; Hull & Davey, 1921a, b
2.99[b]	5.65[b]	Simson, 1924
2.971[b]	5.609[b]	McLennan & Monkman, 1929
2.9784[c]	5.6158[c]	Jenkins & Preston, 1931
2.9796[d]	5.6171[d]	Stenzel & Weerts, 1932
2.969[b]	5.656[b]	Taylor, 1932
2.9773	5.6159	Jette & Gebert, 1933
2.9791[e]	5.6173[e]	Kossolapow & Trapesnikow, 1935
2.9786[f]	5.6168[f]	Jette & Foote, 1935
2.9785[g]	5.6158[g]	Owen & Roberts, 1936
2.9785[g]	5.6160[g]	Owen, Rogers, & Guthrie, 1939
2.9787[h]	5.6164[h]	Lu & Chang, 1941
2.9791[f]	5.6172[f]	Edwards, Wallace, & Craig, 1952
2.9790[i]	5.6167[i]	Swanson, Fuyat, & Ugrinic, 1954
2.976[j]	5.620[j]	Schneider & Heymer, 1956
2.979[j]	5.617[j]	Lynch & Drickamer, 1965

av.   2.9788   5.6164 c/a 1.8855 ± 0.0003
      ±0.0004   ±0.0006

[a]Values not determined at 20°C corrected when necessary with thermal expansion coefficients of $\alpha_\parallel = 50 \times 10^{-6}$ deg^{-1} and $\alpha_\perp = 18 \times 10^{-6}$ deg^{-1}.
[b]From kX, not included in the average.
[c]From kX and 19°C.   [g]From kX and 18°C.
[d]From kX.   [h]From kX and 21°C.
[e]From kX and 26°C.   [i]From 25°C.
[f]From kX and 25°C.   [j]Not included in the average.

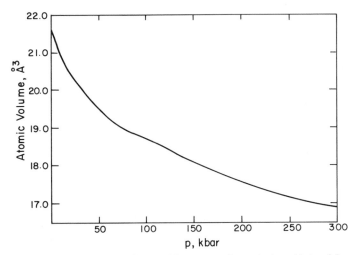

Fig. 48-1.   The change in atomic volume with pressure for cadmium (data of Lynch & Drickamer, 1965).

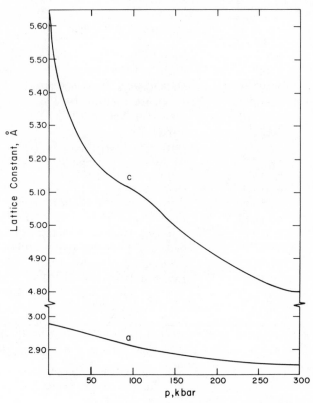

Fig. 48-2.   The change in the lattice constants with pressure for cadmium (data of Lynch & Drickamer, 1965).

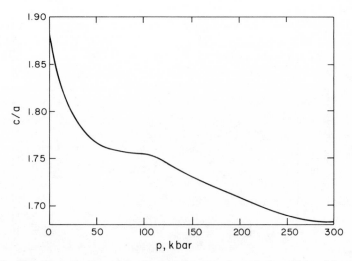

Fig. 48-3. The change in the axial ratio with pressure for cadmium (data of Lynch & Drickamer, 1965).

MERCURY

α-Mercury

A structure for mercury was first reported by Alsen and Aminoff (1922). On the basis of powder data they reported that at the temperature of dry ice it is hexagonal, with a = 3.85, c = 7.25 Å (from kX), c/a = 1.89, four mercury atoms per unit cell. These data correspond to a calculated density of 14.31 g cm^{-3}, as compared with an observed value of 14.2 g cm^{-3}. The mercury atoms were said to lie at $(\frac{1}{3}\ \frac{2}{3}\ z)$, $(\frac{1}{3}\ \frac{2}{3}\ \bar{z})$, $(\frac{2}{3}\ \frac{1}{3}+z)$, $(\frac{2}{3}\ \frac{1}{3}-z)$, but no attempt was made to determine the z-parameter.

At about the same time McKeehan and Cioffi (1922) published the results of their analysis of powder data obtained from a sample of mercury at -115°C. They stated that it is rhombohedral, with a = 3.031 Å (from kX), α = 70° 31.7', one mercury atom per unit cell. The density calculated from these data is 13.9 g cm^{-3}.

Ewald and Hermann (1931a), after commenting that an allotropic transformation between -78 and -115°C was possible, preferred the hexagonal structure of Alsen and Aminoff, terming it "wohl wahrscheinlich", and created Strukturbericht type A10 for it. Their preference was based on the calculated densities.

Terrey and Wright (1928) examined powders at liquid air temperature and confirmed the rhombohedral structure of McKeehan and Cioffi. They reported lattice constants a = 3.002 Å (from kX), α = 70°32'; these correspond to a density of 14.3 g cm^{-3}.

The rhombohedral structure was also found by Wolf (1928, 1929) at -80°C, by Mehl and Barrett (1930) at -46°C, and by Hermann and Ruhemann (1932) at ~-50°C. Further, Horovitz (1929) reported that the structure is the same at both liquid air and dry ice temperatures, and that statements of former investigators that a different form of mercury exists at -80°C are therefore erroneous.

Neuberger (1933b) then showed that the powder data of Alsen and Aminoff could be interpreted on the basis of a mixture of rhombohedral mercury and solid carbon dioxide. The original description of the Strukturbericht A10 type was then replaced by the new description (Ewald and Hermann, 1931b).

Determinations of the lattice constants obtained in the foregoing studies and in subsequent investigations are presented in Table 80-1 and in Fig. 80-1.

Each mercury atom is surrounded by six other mercury atoms which lie at the vertices of a trigonal antiprism, as shown in Fig. 80-2. At 78°K the Hg-Hg distance is 2.993 Å; there are six next nearest neighbors at 3.465 Å. This structure may be regarded as a distortion of the cubic close packed structure obtained by compression along a threefold axis which increases the rhombohedral angle from 60° to 70°44′, with a corresponding increase of the six equatorial interatomic distances. This is in sharp contrast to the behavior of the two congeners of mercury (see zinc and cadmium, above) in which the distortions *decrease* the six equatorial distances relative to the six above and below. The ratios of equatorial interatomic distance:interlayer interatomic distance for zinc, cadmium, and mercury are 0.908, 0.905, and 1.158, respectively. The axial ratio of a hypothetical distorted hexagonal close packed form of mercury having these distances is $c/a = 1.457$, as compared with 1.856 for zinc and 1.886 for cadmium, the ideal value being 1.633.

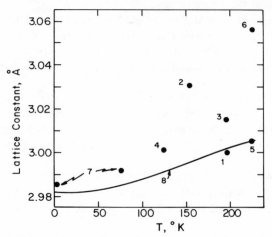

Fig. 80-1. The lattice constant of rhombohedral mercury. The numbers refer to the references in the order given in Table 80-1.

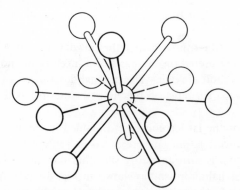

Fig. 80-2.   The configuration of the 12 nearest neighbors of a mercury atom in α-mercury. The interatomic distances are (at 78°K) six at 2.993 Å (shown as sticks) and six at 3.465 Å (shown as dashed lines).

Table 80-1.   The Lattice Constants of Rhombohedral Mercury

a, Å	α	T, °K	Reference
3.00[a]	–	195	Alsen & Aminoff, 1922; Neuberger, 1933b
3.031[b]	70° 31.7'	115	McKeehan & Cioffi, 1922
3.015[b]	70° 33'	193	Wolf, 1928, 1929
3.002[b]	70° 32'	123	Terrey & Wright, 1928
3.005[b]	70° 31.7'	227	Mehl & Barrett, 1930
3.056[b]	70° 32'	223	Hermann & Ruhemann, 1932
2.9863	70° 44.6'	5 ⎫	Barrett, 1957
2.9925	70° 44.6'	78 ⎭	
2.982[c]	–	0 ⎫	
2.987[c,d]	–	78 ⎬	Swenson, 1958
3.006[c]	–	234 ⎭	

[a]From kX; data of Alsen & Aminoff reinterpreted by Neuberger.
[b]From kX.
[c]Calculated from the molar volumes given by Swenson, assuming $\alpha = 70°\ 44.6'$
 (Barrett, 1957); for values at intermediate temperatures, see Fig. 80-1.
[d]Interpolated value.

β-Mercury

A high pressure transition in mercury described by Bridgman (1935) has been
investigated further by Swenson (1958), Schirber and Swenson (1962), and
Klement, Jayaraman, and Kennedy (1963b). The structure of the high pressure,
or β, form was first determined by Atoji, Schirber, and Swenson (1959). It is
body centered tetragonal, with two atoms per unit cell. At 77°K the lattice
constants are a = 3.995 ± 0.004, c = 2.825 ± 0.003 Å, c/a = 0.7071 ± 0.0011.
It is interesting that the axial ratio is $1/\sqrt{2}$, within experimental error. Each
mercury atom has 10 close neighbors, two at 2.825 Å and eight at 3.158 Å.
This structure is seen in Fig. 80-3. It may also be described in terms of distorted hexagonal close packed layers, parallel to (110), stacked such that each
atom is in contact with two atoms in adjacent layers, leading to an overall
coordination number of 10; it is thus reminiscent of the structure of the room
temperature form of protactinium. The transition $\alpha \rightarrow \beta$ is accompanied by a
volume decrease of 2.6% at 78°K. (In this calculation the lattice constants
of Barrett, 1957, were used.)
    A good deal of hysteresis is associated with the transitions, and, although
the β form is the stable one below 79°K, the $\alpha \rightarrow \beta$ transition does not take
place at temperatures below this except at high pressure. The so-called
"region of indifference" was studied in detail by Schirber and Swenson (1962);
the phase diagram for the low temperature-moderate pressure region is presented in Fig. 80-4. The phase behavior at high pressure has been investigated

by Klement et al. (1963). Their phase diagram is presented in Fig. 80-5. They found that the fusion curve and the phase boundary diverge with pressure, thus eliminating the possibility of a triple point among the three phases.

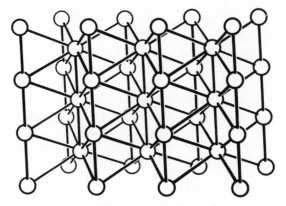

Fig. 80-3.   The structure of the high pressure (β) form of mercury.

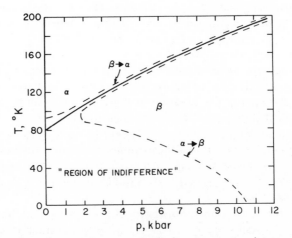

Fig. 80-4. The phase diagram of mercury in the low temperature-intermediate pressure region (after Schirber & Swenson, 1962).

γ-Mercury

A third form of mercury was reported by Abell and Crocker (1968), who found that partial transformation of α-mercury was induced by stress at 4°K. This new allotrope, termed γ-mercury by them, reverted to the α form at 50°K. γ-Mercury was also observed by Doidge and Eastham (1968), who made magnetic measurements on samples of mercury which had undergone partial transformation after tensile deformation at 4°K. Weaire (1968), on the basis of calculations of strain, suggested that γ-mercury has a simple rhombohedral structure, with the angle α about 50°; this corresponds to an opposite deformation to α-mercury from cubic closest packing, but an alternate, distorted hexagonal close packed structure with an axial ratio c/a of about 2.0 was also mentioned as being possible. Abell, Crocker, and King (1970) obtained γ-mercury by tensile strain at 20°K. The diffraction pattern of their preparations showed, in addition to the peaks expected from α-mercury, two peaks which are not part of the β-mercury pattern. These two peaks were said not to correspond to spacings of the structures proposed by Weaire. Although there is little doubt of the reality of this new allotrope of mercury, its structure remains unknown.

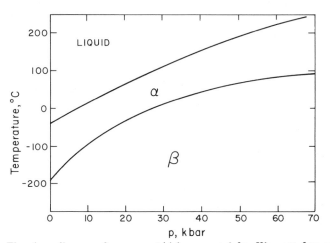

Fig. 80-5. The phase diagram of mercury at high pressure (after Klement, Jayaraman, & Kennedy, 1963).

GROUP IIIB

GALLIUM

Gallium at Atmospheric Pressure

The structure of gallium was first reported by Jaeger, Terpstra, and Westen-brink (1926, 1927), who stated that it is tetragonal. They described a structure based on eight atoms in position $8i$ of space group D_{4h}^{16}-P4/ncm, giving approximate values for the two positional parameters. Ewald and Hermann (1931c) later designated this structure as type $A11$.

It was shortly thereafter discovered by Laves (1932, 1933) that the symmetry of gallium is in fact only orthorhombic, with two of the axes very nearly equal in length. The space group is D_{2h}^{18}-Cmca,* with eight gallium atoms in position $8f$, $(000; \frac{1}{2}\frac{1}{2}0) \pm (0yz; \frac{1}{2} y \frac{1}{2} -z)$. Laves gave values of the positional parameters as determined from single-crystal data. This work was verified by Bradley (1935), who gave more precise values of the parameters and lattice constants. A revised Strukturbericht type $A11$ was then described by Gottfried and Schossberger (1937). Subsequent determinations of the lattice constants were made by Swanson and Fuyat (1953), Barrett (1961), and Barrett and Spooner (1965), and a redetermination of the positional parameters was carried out by Sharma and Donohue (1962). The above results are summarized in Tables 31-1 and 31-2. The calculated density is 5.91 g cm^{-3}.

Table 31-1.   The Lattice Constants of α-Gallium

a, Å	b, Å	c, Å	T	Reference
4.52[a]	7.66[a]	4.52[a]	room	Laves, 1932
4.515[a]	7.657[a]	4.515[a]	room	Laves, 1933
4.5198[b]	7.6602[b]	4.5258[b]	room	Bradley, 1935
4.523[c]	7.661[c]	4.524[c]	25°C	Swanson & Fuyat, 1953
4.5186	7.6570	4.5258	24°C	Barrett & Spooner, 1965
4.4904[c]	7.6328[c]	4.5156[c]	4.2°K	Barrett, 1961
av.[d] 4.5192	7.6586	4.5258		

[a]From kX, omitted from the average.   [c]Omitted from the average.
[b]From kX.   [d]Room temperature.

*In much of the work on gallium the nonstandard setting Abma has been used; when this was done the axial conversion $abc \rightarrow cab$ has been made in the present discussion so that all work is described here in terms of Cmca.

Table 31-2. Positional Parameters for α-Gallium

	y	z	Reference
	0.159	0.080	Laves, 1932
	0.153 ± 0.002	0.080 ± 0.001	Laves, 1933
	0.1525 ± 0.0005	0.0785 ± 0.0005	Bradley, 1935
	0.1549 ± 0.0008[a]	0.0810 ± 0.0006[a]	Sharma & Donohue, 1962
	0.1549 ± 0.0005[b]	0.0803 ± 0.0011[b]	Sharma & Donohue, 1962
Weighted average	0.1539 ± 0.0013	0.0798 ± 0.0011	

[a]Fourier values.
[b]Least squares values.

There has been some confusion regarding the thermal expansion of gallium. It was first reported (Barrett, 1961) that a reversal in the relative magnitudes of the a and c axes occurred between 4.2 and 297°K, the two being equal at about 255°K. This result, however, is in disagreement with the relative thermal expansion coefficients reported earlier by Powell (1949, 1950). The 1961 data were therefore reevaluated by Barrett and Spooner (1965), who concluded that the inequality $c>a$ holds throughout the temperature range. The revised lattice constants (see Table 31-1) are in excellent agreement with the thermal expansion coefficients reported by Powell.

The structure may be described as consisting of a stacking of distorted hexagonal close packed layers, with the bonds within the layers considerably weaker than the bonds between the layers. A gallium atom at $(0yz)$ has seven nearest neighbors:

    a. One at $(0\bar{y}\bar{z})$ 2.465 ± 0.020 Å
    b. Two at $(0\ \frac{1}{2}-y\ \pm\frac{1}{2}+z)$ 2.700 ± 0.010 Å
    c. Two at $(\pm\frac{1}{2}\ y\ \frac{1}{2}-z)$ 2.735 ± 0.005 Å
    d. Two at $(\pm\frac{1}{2}\ \frac{1}{2}-y\ \bar{z})$ 2.792 ± 0.011 Å

A view perpendicular to one of the layers, showing the bonds of type b, c, and d, is presented in Fig. 31-1. Figure 31-2 is a view of the bonding between the layers. This structure has also been discussed as consisting of singly bonded Ga_2 units, with each gallium atom also forming six weaker bonds of order approximately $\frac{1}{3}$.

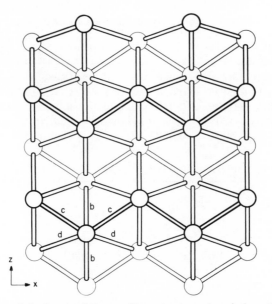

Fig. 31-1.   One of the layers of the α-gallium structure viewed along the y-axis. The letters identify the bond types in the text. The interlayer bonds, which are down from the lower atoms and up from the upper atoms, are not shown.

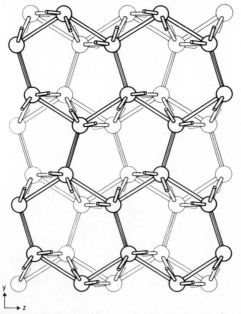

Fig. 31-2.   The α-gallium structure viewed along the x-axis, or edgewise to the layers. The strong interlayer bonds of type a are shown as triple lines.

238

High Pressure Polymorphs of Gallium

The phase behavior of gallium at high pressure has also been the subject of some confusion, because of the capricious appearance of metastable phases. This situation has been clarified by Jayaraman, Klement, Newton, and Kennedy (1963), whose phase diagram is shown in Fig. 31-3. (Some authors use Roman numerals to designate the forms of gallium, others use Greek letters; in the discussion here α, β, and γ refer to I, II, and III, respectively, α being the ordinary form.)

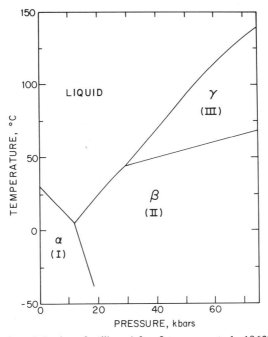

Fig. 31-3.   The phase behavior of gallium (after Jayaraman et al., 1963).

β-Gallium

Defrain and co-workers (Defrain, Epelboin, and Erny, 1959; Defrain and Epelboin, 1959; Defrain, 1960; Defrain and Epelboin, 1960) observed β-gallium in small droplets as a phase unstable at atmospheric pressure, with a melting point of about –16°C, while Bosio, Defrain, and Epelboin (1960) found that it was stable above about 12,000 bar (at room temperature). Curien, Rimsky, and Defrain (1961) found that this phase could be retained below –16°C and atmospheric pressure from 15 to 40 minutes, and proposed the following structure from powder data: orthorhombic, space group D_{2h}^{17}–Cmcm, $a = 2.90$, $b = 8.13$, $c = 3.17$ Å, four atoms in position $4c$, $(000; \frac{1}{2}\frac{1}{2}0) \pm (0y\frac{1}{4})$ with $y = 0.133$. The calculated density is 6.20 g cm^{-3}. These results were also reported by Bosio, Defrain, and Epelboin (1962).

This structure was later revised by Bosio, Defrain, Curien, and Rimsky (1969), who stated that they obtained better powder diagrams. The new structure is monoclinic, space group C_{2h}^{6}–C2/c, $a = 2.766$, $b = 8.053$, $c = 3.332$ Å, $\beta = 92° \; 2'$, four atoms in position $4e$, $(000, \frac{1}{2}\frac{1}{2}0) \pm (0y\frac{1}{4})$ with $y = 0.131$. The calculated density is 6.24 g cm^{-3}. In this structure each gallium atom has eight nearest neighbors, those for an atom at $(0y\frac{1}{4})$ being the following:

a. Two at $(0 \; \bar{y} \; \frac{\bar{1}}{4}, \frac{3}{4})$ 2.69 Å
b. Two at $(\pm 1 \; y \; \frac{1}{4})$ 2.77 Å
c. Two at $(\frac{1}{2} \; \frac{1}{2} - y \; \frac{\bar{1}}{4}, \frac{3}{4})$ 2.86 Å
d. Two at $(-\frac{1}{2} \; \frac{1}{2} - y \; \frac{\bar{1}}{4}, \frac{3}{4})$ 2.92 Å

The reliability index R was stated to be 8%, but no data – neither observed nor calculated spacings and intensities – were presented. This same form was observed at low temperature by Berty, David, and Lafourcade (1969), in thin films formed by evaporation on to an amorphous substrate.

An entirely different structure for β-gallium was described by Vereshchagin, Kabalkina, and Troitskaya (1965). On the basis of powder data obtained from a sample at 30 kbar they reported that it is body centered tetragonal, two atoms per unit cell, with $a = 2.80$, $c = 4.37$ Å, $c/a = 0.641$. (The alternate face centered cell has $a = 3.96$, $c = 4.37$ Å, $c/a = 0.906$.) In this structure each gallium atom has 12 nearest neighbors, four at 2.80 and eight at 2.95 Å. The calculated density is 6.76 g cm^{-3}.

This structure was verified by Weir, Piermarini, and Block (1971), who prepared single crystals of β-gallium at room temperature and a pressure specified merely as ">20 bar". The lattice constants reported by them are $a = 2.808 \pm 0.003$, $c = 4.458 \pm 0.003$ Å, $c/a = 0.6299 \pm 0.0008$. These give a calculated density of 6.59 g cm^{-3}, and for the nearest neighbors of a gallium atom four at 2.808 Å and eight at 2.985 Å.

It appears that additional work is necessary to establish the correct structure for β-gallium. It is possible that the metastable form observed at atmospheric pressure by Defrain and co-workers is not the same as the high pressure form observed by Vereshchagin et al. and by Weir et al.

γ-Gallium

A third form of solid gallium was sometimes observed by Bridgman (1935); its place in the phase diagram was established by Jayaraman et al. (1963)

(see Fig. 31-3), who suggested that γ-gallium has the body centered cubic structure which is common to many high temperature phases. Blanconnier (1964), on the other hand, reported that γ-gallium is orthorhombic, with a = 10.73, b = 13.66, c = 5.28 Å, space group D_{2h}^{17}-Cmcm, probably 40 atoms per unit cell. These data give a calculated density of 5.98 g cm^{-3}. Slightly different lattice constants of a = 10.60, b = 13.56, c = 5.19 Å at 237.5°K were given by Blanconnier, Bosio, Defrain, Rimsky, and Curien (1965), who stated that space groups C_{2v}^{12}-Cmc2 and C_{2v}^{16}-C2cm were also possible. The calculated density for this cell, with 40 atoms, is 6.21 g cm^{-3}. Blanconnier et al. prepared this phase by crystallization of small droplets which were said to melt at -35.6°C. No structural details were given, but in a later report Bosio, Curien, Dupont, and Rimsky (1972) described a structure based on some 300 diffraction maxima. At about 220°K the lattice constants are a = 10.593, b = 13.523, c = 5.203 Å, and the space group is Cmcm-D_{2h}^{17}, the other two possibilities being eliminated on the basis of the three-dimensional Patterson function. The calculated density is 6.21 g cm^{-3}. The 40 gallium atoms in the unit cell are distributed as follows: eight Ga(1) and Ga(6) in two sets of position 4c, at $(000\ \frac{1}{2}\frac{1}{2}0) \pm (0y\frac{1}{4})$; eight Ga(3) in one set of position 8f, at $(000\ \frac{1}{2}\frac{1}{2}0) \pm (0yz\ 0y\frac{1}{2}-z)$; and 24 Ga(2), Ga(4), and Ga(5) in three sets of position 8g, at $(000\ \frac{1}{2}\frac{1}{2}0) \pm (xy\frac{1}{4}\ \bar{x}y\frac{1}{4})$. The structure was refined by least squares, including anisotropic thermal parameters, to an R of 5.7%. The customary table of observed and calculated values for F_{hkl} was not presented. The final positional parameters are presented in Table 31-3.

All interatomic distances less than 3.10 Å are listed in Table 31-4. The coordination of the various kinds of gallium atom is quite irregular, as may be seen in Fig. 31-4. The structure viewed in projection along the c-axis is shown in Fig. 31-5.

Table 31-3. Positional Parameters in γ-Gallium

Atom	Position	x	y	z
(1)	4c	0	0.0009	$\frac{1}{4}$
(2)	8g	0.2794	0.0504	$\frac{1}{4}$
(3)	8f	0	0.3947	0.0000
(4)	8g	0.1256	0.2062	$\frac{1}{4}$
(5)	8g	0.2718	0.3612	$\frac{1}{4}$
(6)	4c	0	0.7853	$\frac{1}{4}$

Table 31-4.   Interatomic Distances in γ-Gallium

From Atom	To Atom	Distance, Å	Number
Ga(1)	Ga(1)	2.602	2
	Ga(6)	2.916	1
	Ga(2)	3.035	2
	Ga(5)	3.068	2
	Ga(4)	3.078	2
Ga(2)	Ga(5)	2.616	1
	Ga(4)	2.663	1
	Ga(3)	2.776	2
	Ga(5)	2.914	2
	Ga(2)	2.937	2
	Ga(1)	3.035	2
Ga(3)	Ga(3)	2.602	2
	Ga(6)	2.760	1
	Ga(2)	2.776	2
	Ga(3)	2.848	1
Ga(4)	Ga(5)	2.606	1
	Ga(4)	2.661	1
	Ga(2)	2.663	1
	Ga(6)	2.924	2
	Ga(5)	2.963	2
	Ga(1)	3.078	
Ga(5)	Ga(4)	2.606	1
	Ga(2)	2.616	1
	Ga(6)	2.626	1
	Ga(2)	2.914	2
	Ga(4)	2.963	2
	Ga(1)	3.068	1
Ga(6)	Ga(5)	2.626	2
	Ga(3)	2.760	2
	Ga(1)	2.916	1
	Ga(4)	2.924	2

β'-Gallium (metastable)

Bosio, Defrain, Epelboin, and Vidal (1968) observed a reversible transformation of β-gallium at higher pressure to another form which they termed β'-gallium. The phase boundaries they proposed are shown in Fig. 31-6. A structure was not proposed.

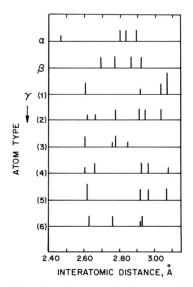

Fig. 31-4. The spectra of the closest interatomic distances of the atoms in the allotropes of gallium. Heights of lines are proportional to number of neighbors, with the shortest line corresponding to one neighbor. The distances shown for the β form are those reported by Bosio et al. (1969), those for the γ form by Bosio et al. (1972).

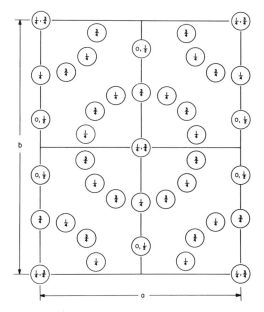

Fig. 31-5. The γ-gallium structure of Bosio et al. (1972), as viewed in projection down the c-axis. The numbers give the heights of the atoms above the ab plane.

243

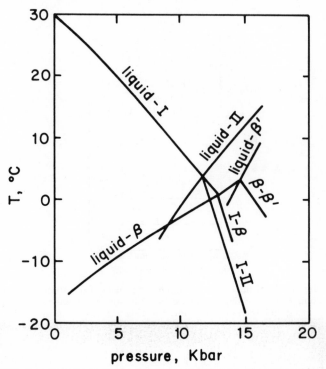

Fig. 31-6.   The phase boundaries in the gallium system showing the metastable β and β' forms (according to Bosio et al., 1968).

INDIUM

The structure of indium was first determined by Hull (1920). It is body centered tetragonal, two atoms per unit cell. The space group is thus D_{4h}^{17}-I4/mmm. The alternate face centered tetragonal unit cell, with $a' = \sqrt{2}a$, has often been used to describe this structure. All values in the present discussion have been converted to the smaller cell. Various determinations of the lattice constants are presented in Table 49-1. At room temperature each indium atom has 12 nearest neighbors, four at 3.252 and eight at 3.377 Å; the atomic volume is 26.16 Å^3.

There is no transformation down to 4.2°K and atmospheric pressure (Pearson, 1955; Swenson, 1955b) nor up to about 110 kbar at room temperature (Dudley and Hall, 1960; Vereshchagin, Kabalkina, and Troitskaya, (1965). No anomalies along the melting curve were found by Jayaraman, Klement,

Table 49-1.   The Lattice Constants of Indium

a, Å	c, Å	T, °C	Reference
3.25[a]	4.87[a]		Hull, 1920, 1921a; Hull & Davey, 1921a, b
3.251[a]	4.956[a]		Dwyer & Mellor, 1932
3.246[a]	4.943[a]		Shinoda, 1933
3.247[a]	4.946[a]		Zintl & Neumayr, 1933a
3.251[a]	4.948[a]	22	Frevel & Ott, 1935
3.284[a]	5.007[a]		Ageev & Ageeva, 1936
3.2513[b]	4.9457[b]	20	Betteridge, 1938
3.2517[b]	4.9471[b]	26.5	Fink, Jette, Katz, & Schnettler, 1945
3.2522	4.9478	24	Guttman, 1950
3.2517	4.9459	26	Swanson, Fuyat, & Ugrinic, 1954
3.263[c]	4.910[c]		Zorll, 1954b
3.2512[b]	4.9467[b]	20	Tyzack & Raynor, 1954b
3.2521[d]	4.9478[d]	20	Graham, Moore, & Raynor, 1955
3.251[a]	4.952[a]	18	Schneider & Heymer, 1956
3.2214[c]	4.9342[c]	-269	Barrett, 1962
3.2234[c]	4.9377[c]	-195	Barrett, 1962
3.2529	4.9490	25	Ievens & Livdinya, 1964
3.2530[c]	4.9455[c]	?	Smith & Schneider, 1964
3.242[c]	4.941[c]	?	Vereshchagin et al., 1965
3.2528	4.9463	27	Deshpande & Pawar, 1969
av. 3.2520	4.9470	23.6°	c/a 1.5212 ± 0.0003
±0.0006	±0.0011	±3.1°	

[a]From kX, omitted from the average. [c]Omitted from the average.
[b]From kX.   [d]Calculated value, see text.

Newton, and Kennedy (1963) up to about 420°C and 75 kbar.

The thermal expansion of indium is unusual in that the c-axis passes through a maximum near room temperature (Graham, Moore, and Raynor, 1955). Their data corresponds to the following equations, which give a and c, in Å, as functions of the absolute temperature, T:

$$a = 3.2182 + 8.47 \times 10^{-5}\ T + 3.61 \times 10^{-10}\ T^3$$
$$c = 4.9201 + 15.15 \times 10^{-5}\ T - 6.64 \times 10^{-10}\ T^3$$

These equations are valid between -183 and 135°C.

Values of the lattice constants at various temperatures are presented in Fig. 49-1. The axial ratio is seen to decrease with increasing temperature, as shown in Fig. 49-2. The point at which it would reach the value $\sqrt{2}$, where the structure would properly be described as cubic close packed, lies well above the melting point of 157°C.

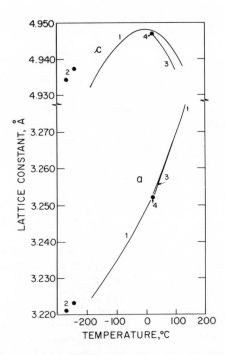

Fig. 49-1. The lattice constants of indium as a function of temperature. (1) Graham et al., 1955; (2) Barrett, 1962; (3) Deshpande & Pawar, 1969; (4) Table 49-1.

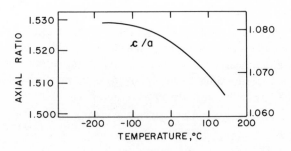

Fig. 49-2. The variation of the axial ratio of indium with temperature. Left ordinate, body centered cell; right ordinate, face centered cell. (Drawn with the corrected equations of Graham et al., 1955.)

THALLIUM

The first x-ray study of thallium is that of Nishikawa and Asahara (1920), who confirmed, by the use of Laue photography, that the transition at 230°C previously observed in thermal studies did, in fact, correspond to a change in structure. They did not, however, determine the structure of either form.

Becker and Ebert (1923), on the basis of powder patterns, reported the room temperature form of thallium to be face centered tetragonal, four atoms per unit cell, with $a = 4.76$, $c = 5.41$ Å (from kX), calculated density 11.07 g cm^{-3} (The observed density is 11.75 g cm^{-3}.) A different result was obtained by Levi (1924, 1925, 1926), who interpreted powder data on the basis of a hexagonal close packed structure with $a = 3.48$, $c = 5.53$ Å (from kX), $c/a = 1.59$, calculated density 11.70 g cm^{-3}. The situation was then further complicated by Terpstra (1926), who stated that thallium is tetragonal, but with a unit cell different from that of Becker and Ebert; his data yield $a = 5.20$, $c = 8.17$ Å (from kX), calculated density 12.29 g cm^{-3}. Becker (1927) next maintained that the powder diagrams did not correspond to a hexagonal close packed structure and that one could consider the Laue diagrams as tetragonal. Perlitz (1928) then concluded that, on the basis of the volume and resistance changes both at the transition and at the melting point, the low and high temperature forms are, respectively, hexagonal and cubic close packed. The low temperature form was soon thereafter reported by Persson and Westgren (1928) to be hexagonal close packed, a result confirmed by all subsequent investigators. The basis for the two earlier erroneous tetragonal structures is not known Various determinations of the lattice constants are presented in Table 81-1. The calculated density is 11.87 g cm^{-3}.

There is no change in structure down to 4.2°K, even after coldworking, (Swenson, 1955b; Barrett, 1958).

Thallium at High Temperature

The structure of thallium above the transition at 230°C has also been a subject of confusion. Sekito (1930), who examined specimens that had been quenched in ice water from above the melting point, stated that it is cubic close packed, with $a = 4.851$ Å (from kX). This structure was also reported by Schneider and Heymer (1956), who found that at 244°C, $a = 4.889$ Å.

Different results were obtained by Lipson and Stokes (1941) and by Ponyatovski and Zakharov (1962), who found the high temperature form to be body centered cubic, two atoms per unit cell. The former reported $a = 3.882$ Å (from kX) at 262°C, the latter, $a = 3.879$ Å (from kX) at 250°C.

Table 81-1.   The Lattice Constants of Hexagonal Close Packed
Thallium at Room Temperature

a, Å	c, Å	Reference
3.48[a]	5.53[a]	Levi, 1924, 1925, 1926
3.422[a]	5.557[a]	Asahara & Sasahara, 1926
3.456[b]	5.524[b]	Persson & Westgren, 1928
3.457[b]	5.531[b]	Sekito, 1930
3.4566[b]	5.5248[b]	Lipson & Stokes, 1941
3.456	5.530	Schneider & Heymer, 1956
3.456[c]	5.522[c]	Barrett, 1958
av. 3.4563	5.5263	
±0.0005	±0.0039	

c/a 1.5989 ± 0.0011

[a]From kX, not included in the average.
[b]From kX.
[c]Average values from two specimens.

Alexopoulos (1955) occasionally observed bcc(2) thallium on the surface of single crystals of hexagonal thallium at room temperature. The reported lattice constant of 3.77 Å is 3% smaller than those observed at ~250°C, a difference too large to be attributed to thermal contraction. It is thus possible, according to Alexopoulos, that impurities were present.

The structure of the high temperature form of thallium is thus not known with certainty, although the weight of evidence appears to favor the body centered cubic structure. It is interesting that the above lattice constants give the same atomic volume of 29.21 Å^3 for both the ccp and bcc(2) structures.

Thallium at High Pressure

A third form of thallium, stable at high pressure, was first proposed by Bridgman (1935) and later observed by Kennedy and La Mori (1961). The phase behavior, shown in Fig. 81-1, was later more extensively investigated by Jayaraman et al. (1963). They accept the body centered cubic structure for the high temperature form and state that Piermarini (1963) has tentatively identified the structure of the high pressure form as cubic close packed, but no lattice constants for it have been reported.

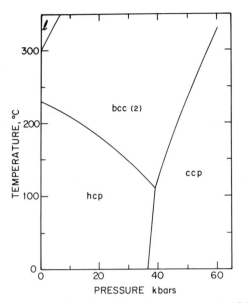

Fig. 81-1.   Phase behavior of thallium (after Jayaraman et al., 1963).

Chapter 7

Group IV

CARBON

There are seven crystallographically characterized forms of carbon: two kinds of diamond, cubic and hexagonal; two kinds of graphite, hexagonal and rhombohedral; a high pressure cubic form metastable at room temperature; and two hexagonal forms prepared under conditions of high temperature and low pressure. The existence of an eighth form was postulated on the basis of the phase behavior at high pressure and temperature. These eight allotropes are discussed below in the same order as they appear above.

Yet another cubic form of carbon, prepared by heating diamond to 1600°C, and having $a = 12.3$ Å (Seal, 1958), was later shown to be a silicon carbide which arose from an impurity in the sample (Seal, 1960; Raal, 1960).

Discussion of various amorphous forms of carbon, such as those found in coal, coke, carbon black, and pyrolitic products of numerous organic compounds, is not included here.

Before proceeding with the discussion of the individual allotropes, some nomenclatural problems must be dealt with. For the uncommon forms of unknown structure the same system used for boron is applied here: a capital letter for the crystal system, C for cubic and H for hexagonal, followed by a number denoting the number of atoms in the unit cell.

hexagonal diamond. Naturally occurring examples of hexagonal diamond (all, so far, from meteorites) have sometimes been termed "lonsdaleite". The geologically oriented purist would then term the laboratory-prepared samples of this same substance "synthetic lonsdaleite". This usage does not seem very appropriate.

hexagonal carbon, H-168(?). The appellation "chaoite" has been applied

250

to specimens of a hexagonal form of carbon first found in gneiss from a crater in Bavaria. It has since been synthesized in the laboratory, but the name "synthetic chaoite" for these preparations is unnecessarily arch. When the structure has been determined, the number of atoms in the unit cell can be stated with certainty. Meanwhile, the name H-168(?) carbon is unambiguous.

hexagonal carbon, H(?)-44(?). A second low pressure-high temperature hexagonal form of carbon was termed "carbon VI" by its first (and only) synthesizers. The rationale of the "VI" was based on five previously recognized forms of the two graphites, diamond, "lonsdaleite", and "chaoite". The true symmetry may be only trigonal, and the number of atoms per unit is not firmly established. Nevertheless, the terminology H(?)-44(?) carbon is certainly to be preferred over "carbon VI".

"carbon III". The phase diagram of carbon customarily contains the regions of (cubic) diamond, (hexagonal) graphite, plus a region at very high pressure which has been labelled "solid III". This rather unsatisfying name will have to be used for this phase until the structure of it has been determined.

Diamond

Cubic Diamond

The structure of diamond was first determined by W. H. and W. L. Bragg (1913a, b). It is the first element the structure of which was determined by the use of x-ray diffraction. It is cubic, space group Fd3m–O_h^7, eight atoms per unit cell, at (000, $0\frac{1}{2}\frac{1}{2}$, $\frac{1}{2}0\frac{1}{2}$, $\frac{1}{2}\frac{1}{2}0$) $\pm$ ($\frac{1}{8}\frac{1}{8}\frac{1}{8}$). Various determinations of the lattice constant are presented in Table 6-1. The average value for 25°C is 3.56688 ± 0.00015 Å, but this precision is probably too high: Lonsdale (1944) observed a spread of 0.00040 Å among a large number of individual diamonds. Furthermore, there is disagreement regarding the value of the thermal expansion coefficient at room temperature.

In this structure each carbon atom is tetrahedrally bonded to four others, at 1.54450 ± 0.00007 Å. The atoms associated with one unit cell are depicted in Fig. 6-1. A projection of the structure nearly along one of the cube body diagonals is shown in Fig. 6-2. In this view it is seen that the structure may be described as layers of fused cyclohexane-type rings bonded together in such a way that in all of the C_2 units, $-\overset{\diagup}{\underset{\diagdown}{C}}-\overset{\diagup}{\underset{\diagdown}{C}}-$, the conformation is staggered, with a repeat in the stacking sequence after three layers, $\cdots [ABC] \cdots$.

Two types of diamond, termed I and II, are known. They differ in certain physical properties, such as birefringence, ultraviolet and infrared absorption,

Table 6-1.   The Lattice Constant of Diamond at 25°C[a]

a, Å	Reference	a, Å	Reference
3.55[b]	Bragg & Bragg, 1913a, b	3.56688	Skinner, 1957
		3.56714	Vogel & Kempter, 1959
3.56672[c]	Ehrenburg, 1926	3.56686	Kaiser, Bond & Tannenbaum, 1959
3.56699[c]	Tu, 1932		
3.56669[d]	Renninger, 1937	3.56675[e]	Perdok, 1960
3.56678[d]	Trzebiatowski, 1937	3.56688	Straumanis & James, 1960
3.56696[c]	Lonsdale, 1944		
3.56692[c]	Riley, 1944	3.56703	Beu, 1960
3.56680[c]	Straumanis & Aka, 1951a	3.56719	Kempter, 1960
		3.56707[f]	Tournarie, 1960
3.56681[c]	Straumanis & Aka, 1951b	3.56696	Wilkens, 1960
		3.56670	Sokhor & Vitol, 1970
3.56681[c]	Straumanis, 1953		
3.56674	Swanson & Fuyat, 1953		

av. 3.56688 ± 0.00015

[a]Values determined at other temperatures converted to 25°C with $\alpha = 1.1 \times 10^{-6}$ deg^{-1}.
[b]From kX, not included in the average.
[c]From 18°C and kX.
[d]From 20°C and kX.
[e]Average of three values.
[f]Average of two values.

fluorescence, photoconductivity, and the occurrence of anomalous diffuse x-ray reflections. There is, on the other hand, no consistent variation of the lattice constant among individual crystals of the two types. The discussion of theoretical reasons for the other observed differences is still continuing: for a fairly recent treatment, see a review of 68 papers in *Structure Reports* (Anonymous, 1951).

Hexagonal Diamond

By analogy with the cubic and hexagonal forms of zinc sulfide, sphalerite or zinc blende, and wurtzite, respectively, Ergun and Alexander (1962) suggested the existence of a hexagonal polymorph of diamond. This would consist of cyclohexane-like layers, as in diamond, but with a repeat after two layers instead of three.

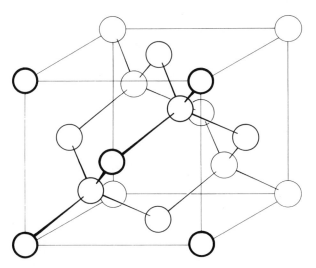

Fig. 6-1.   One unit cell of the diamond structure. (The origin of the description in the text lies halfway between the two atoms at the lower left.)

The crystallographic description of this structure is: space group P6$_3$/mmc–D$_{6h}^4$, four atoms per unit cell in position 4f, $\pm(\frac{1}{3} \frac{2}{3} z, \frac{2}{3} \frac{1}{3} \frac{1}{2}+z)$. If the carbon-carbon bond length is the same as in diamond, and if the bonding is also strictly tetrahedral, then the lattice constants would be $a_{hex} = a_{diam}/\sqrt{2}$, $c_{hex} = 2a_{diam}/\sqrt{3}$, $c/a = \sqrt{8/3}$, and the z-parameter exactly $\frac{1}{16}$. The resulting numerical values for the hexagonal lattice constants are $a = 2.522$, $c = 4.119$ Å, $c/a = 1.633$.

The occurrence of hexagonal diamond in the meteorites from Canyon Diablo, Arizona, and Goalpara, Assam, was reported in 1967 by Hanneman, Strong, and Bundy, who gave lattice constants $a = 2.52$, $c = 4.12$ Å, $c/a = 1.63$. Closely similar values were given very shortly thereafter by Frondel and Marvin (1967) for hexagonal diamond from the Canyon Diablo meteorite: $a = 2.51$, $c = 4.12$ Å, $c/a = 1.64$. Both sets of values are very close to the ideal values.

Bundy and Kasper (1967) reported the synthesis of hexagonal diamond by subjecting graphite to a pressure of 130 kbar at temperatures in excess of 1000°C. They reported lattice constants identical to those of Hanneman et al.

A view of this structure is shown in Fig. 6-3. The main difference between this structure and that of cubic diamond is that in one-quarter of the C$_2$ units the bonds are eclipsed. Hexagonal diamond would thus be expected to be unstable with respect to the cubic form. This energy difference has not been measured.

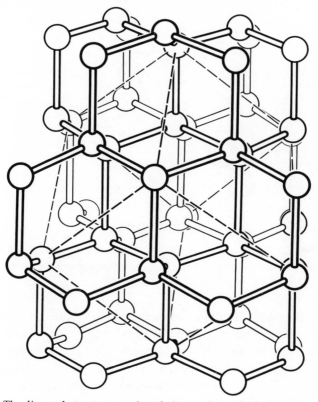

Fig. 6-2.   The diamond structure, as viewed almost along a body diagonal of the unit cell, which is shown with dashed lined.

Graphite

Hexagonal Graphite

The structure of hexagonal graphite, the normally occurring form, was first determined by Hull (1917c). Various determinations of the lattice constants are presented in Table 6-2. The space group is $P6_3mc$-C_{6v}^4, with four atoms in the unit cell, two in position $2a$ at $(000, 00\frac{1}{2})$ and two in position $2b$ at $(\frac{2}{3} \frac{1}{3} z, \frac{1}{3} \frac{2}{3} \frac{1}{2}+z)$. Hull's original determination of the z-parameter gave the value 0.071, but subsequent determinations pushed it down to at least 0.004, and it seems probable that the true value is exactly zero. If this is correct, then the proper space group is $P6_3/mmc$-D_{6h}^{14}, with two atoms in position $2b$ at $\pm(00\frac{1}{4})$, and two in position $2d$ at $\pm(\frac{2}{3}\frac{1}{3}\frac{1}{4})$. This gives a structure consisting

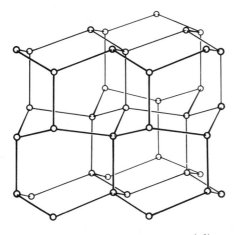

Fig. 6-3.   The structure of hexagonal diamond.

of flat layers stacked in the sequence $\cdots[AB]\cdots$, that is, a repeat after two layers. A projection of the structure down the c-axis is shown in Fig. 6-4, and a general view in Fig. 6-5. Half of the atoms in a layer are directly above atoms in the layer below, the other half being above holes.

The length of the c-axis depends on the condition of the specimen. Bacon (1950a) found that while the length of a remained constant with decreasing crystallite size, the length of c increased. Franklin (1951) discussed this effect in terms of disorder and gave the relation $d = 3.440 - 0.0861(1 - p^2)$ where d is interlayer spacing, $c/2$, in Ångstroms, and p the probability that a random disorientation occurs between any two adjacent layers. This effect probably accounts for some of the high values for c in Table 6-2.

There has been much discussion concerning the true symmetry of graphite. Lukesh (1950) found evidence for a superstructure, possibly with orthorhombic symmetry, in diffraction experiments with single crystals. He suggested that the true unit cell has $a = 4.263$, $b = 36.91$, $c = 6.709$ Å, or $\sqrt{3}a_{hex}$, $15a_{hex}$, and c_{hex}, respectively. He concluded that the "classical concept of equivalent carbon-carbon bonds must be discarded in favor of one involving unequal bond lengths and angles". The orthorhombic superstructure was also observed in subsequent experiments by Lukesh (1951a, b, c). Laves and Baskin (1956), on the other hand, found no indication of such a superstructure. Furthermore, Freise (1962) pointed out that deformation and twinning in single crystals of graphite provided an explanation for the diffraction effects observed by Lukesh without the necessity of the introduction of an orthorhombic structure.

An entirely different orthorhombic structure has been proposed by Pauling

Table 6-2.   The Lattice Constants of Hexagonal Graphite at 20°C[a]

a, Å	c, Å	Reference
2.47[b]	6.81[b]	Hull, 1917c
2.47[b]	6.80[b]	Hassel & Mark, 1924b
2.45[b]	6.83[b]	Bernal, 1924
2.46[b]	6.74[b]	Mauguin, 1925, 1926
2.46[b]	6.80[b]	Hoffmann, Hoffmann, & Hermann, 1926
2.49[b]	6.79[b]	Ott, 1928
2.460[b]	6.70[b]	Hoffmann & Wilm, 1936
2.4611[c]	6.7085[c]	Trzebiatowski, 1937a
2.4612[c]	6.7087[d]	Nelson & Riley, 1945
2.4614	6.7080[e]	Bacon, 1950a, b
2.461[b]	6.721[b]	Kochanovska, 1953
–	6.7208[f]	Walker, McKinstry, & Wright, 1953
2.468[f]	6.719[f]	Gruhl & Nickel, 1954
2.4589[f]	6.7069[g]	Baskin & Meyer, 1955
–	6.7096[e]	Walker & Imperial, 1957
–	6.7102	Matuyama, 1958
–	6.709[f]	Noda, 1959
–	6.7176[f]	Stewart, Cook, & Kellett, 1960
–	6.7415[h]	Bundy, 1963
	6.7100[i]	Kellett & Richards, 1971
av. 2.4612	6.7090	
±0.0002	±0.0012	

$c/a = 2.7259 \pm 0.0005$

[a]Values for c corrected to 20°C using $\alpha_c = 25 \times 10^{-6}$ deg^{-1} when necessary.
[b]From kX, omitted from the average.
[c]From kX.
[d]From 14.6°C and kX.
[e]From 15°C.
[f]Omitted from the average.
[g]From 24°C.
[h]For a sample melted at 78 kbar, then frozen; omitted from the average.
[i]From 0°C.

(1966), the unit cell having $a = 2.409$, $b = 4.339$, $c = 6.708$ Å, with eight atoms in two sets of positions $\pm(0y\frac{1}{4}, \frac{1}{2}\ \frac{1}{2}+y\ \frac{1}{4})$, with $y_1 = 0.05$ and $y_2 = 0.315 + y_1$. In this structure the layers have quinoid character, with two-thirds of the carbon-carbon bonds longer than the others, 1.453 versus 1.357 Å, as contrasted with all bond lengths equal, at 1.4210 Å, in the hexagonal structure. This new orthorhombic structure has not been tested against the x-ray data.

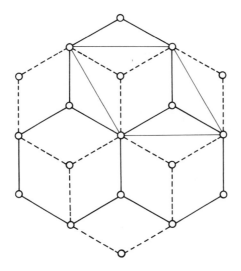

Fig. 6-4. The structure of hexagonal graphite viewed in projection down the c-axis. Solid lines, later at $z = 0$; dashed lines, layer at $z = \frac{1}{2}$; thin lines, outline of unit cell.

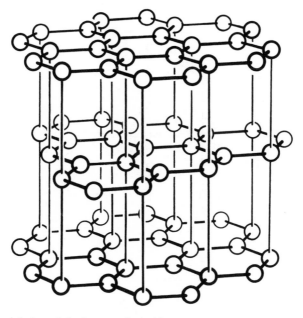

Fig. 6-5. General view of the hexagonal graphite structure.

257

Rhombohedral Graphite

Debye and Scherrer (1916, 1917) proposed a rhombohedral structure for graphite, with two atoms in a unit cell having $a = 3.71$ Å (from kX), $\alpha = 39° 45'$. This structure was rejected by Hull (1922), who stated that new data confirmed the hexagonal structure reported earlier by him (1917c) and could not be reconciled with the structure of Debye and Scherrer. This structure lay dormant for 25 years until Lipson and Stokes (1942a, b) revived it to explain two anomalous reflections which had been observed from graphite by Taylor and Laidler (1940), and indexed by them as $(1\ 0\ \frac{2}{3})$ and $(1\ 0\ \frac{4}{3})$. According to Lipson and Stokes, this form of graphite is rhombohedral, space group $R\bar{3}m$–D_{3d}^5, two atoms per unit cell at $\pm(xxx)$, $a = 3.642$ Å (from kX), $\alpha = 39°29'$. If the x-parameter is exactly $\frac{1}{6}$, the structure consists of flat layers, stacked as illustrated below. The corresponding hexagonal unit cell has $a = 2.460$, $c = 10.061$ Å, $c/a = 4.090$, with atoms at $(000, \frac{2}{3}\frac{1}{3}\frac{1}{3}, \frac{1}{3}\frac{2}{3}\frac{2}{3})$ +$(000, 00\frac{1}{3})$. The c-axis is 3/2 the length of the c-axis in hexagonal graphite, the repeat being after three layers instead of two, as shown in Fig. 6-6. A general view of the structure is shown in Fig. 6-7.

Rhombohedral graphite occurs with hexagonal graphite in both the natural product and artificial powders. Lipson and Stokes found their samples to be 80% hexagonal, 14% rhombohedral, and 6% disordered. Bacon (1950b) found that the proportion of the rhombohedral modification was increased by grind-

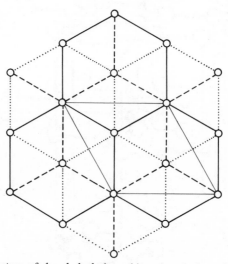

Fig. 6-6.   The structure of rhombohedral graphite, viewed along the hexagonal c-axis. Solid lines, layer at $z = 0$; dashed lines, layer at $z = \frac{1}{3}$; dotten lines, layer at $z = \frac{2}{3}$; thin lines, outline of unit cell.

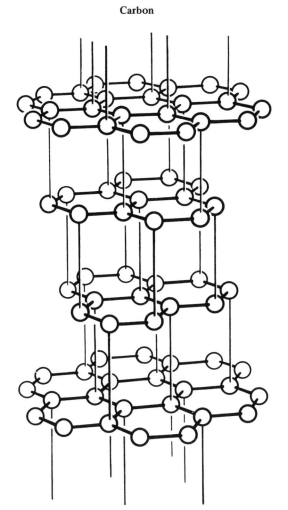

Fig. 6-7.   General view of the rhombohedral graphite structure.

ing, from 8 to 24% for natural graphite, and from 4 to 17% for the artificial. Taylor and Laidler (1940) observed that their anomalous lines disappeared when the sample was heated in a mixture of concentrated nitric-sulfuric acids. Kochanovska (1953) reported hexagonal lattice constants of $a = 2.461$, $c = 10.082$ Å (from kX) for natural rhombohedral graphite, but for artificial rhombohedral graphite prepared by heating coke to $2700°C$ she found the much larger values of $a = 2.508$ Å, with c ranging from 10.221 to 10.296 Å (all from kX). The results of Laves and Baskin (1956) are somewhat different: they found that natural graphite is usually entirely hexagonal, but that some samples are up to 5% rhombohedral. The proportion of the rhombohedral form could be increased by shear, and reduced to zero by heating to 1300°C. Pressures up to 10,000 atm had no effect on either form.

Gibson (1946) also observed extra lines in the diffraction pattern of graphite and suggested that these might be due to alterations of the sequence of the layers, but Nelson and Riley (1947) and Rooksby and Stewart (1947) showed that these could be explained by the presence of small amounts of iron, gold, mercury, and tungsten in the cobalt target used by Gibson, and that the diffraction pattern of the same sample taken with monochromatic radiation showed no anomalous reflections.

The averages of the lattice constants of natural rhombohedral graphite reported by Lipson and Stokes and by Kochanovska correspond to $a = 3.646$ Å, $\alpha = 39°27'$ for the rhombohedral unit cell and $a = 2.461$, $c = 10.072$ Å, $c/a = 4.093$ for the hexagonal unit cell. The carbon-carbon bond distance and interlayer spacings of 1.421 and 3.357 Å are thus essentially identical to those observed in hexagonal graphite, 1.4210 and 3.3545 Å, respectively.

Cubic Carbon, C-24(?)

Aust and Drickamer (1963) reported that they had prepared a new cubic form of carbon by subjecting a single crystal of graphite to pressures above 150 kbar. This form, metastable at atmospheric pressure, gave a powder pattern consisting of seven lines which could be indexed on a cubic unit cell having $a = 5.545 \pm 0.009$ Å. It was estimated that there are 24 atoms in the unit cell, but no other structural details were given. The calculated density is 2.80 g cm^{-3}.

Hexagonal Carbon, H-168(?), a.k.a. chaoite

A new allotropic form of carbon was detected by El Goresey and Donnay (1968) in samples of shock-fused graphite in gneiss from the Ries crater, Bavaria. This new phase was shown to be pure carbon by electron probe analyses. A powder pattern of 22 lines was found to correspond to a hexagonal unit cell having $a = 8.948 \pm 0.009$, $c = 14.078 \pm 0.017$ Å, $c/a = 1.573 \pm 0.003$. The absence of 00ℓ reflections was said to rule out a layer structure. Insufficient amounts of the substance were available for measurement of the density, but it was stated that seven sets of atoms in the twenty-fourfold general position of space group P6/mmm-D_{6h}^1 led to a reasonable calculated density of 3.43 g cm^{-3} (cf. diamond, 3.5123 g cm^{-3}).

This same allotrope was apparently observed by Whittaker and Kintner (1969), who heated pyrolitic graphite to 2700 to 3000°K in an atmosphere of argon at very low pressure, 10^{-4} mm Hg. An electron diffraction pattern showed lines of graphite plus 20 lines which could be indexed as $hk0$ reflections of a hexagonal phase having $a = 8.939 \pm 0.035$ Å. It was stated that for this sample the crystals were oriented with their c-axes parallel to the incident electron beam. In other experiments several crystals were found to be randomly oriented, and these gave 23 additional lines where they were used to calculate hexagonal unit cell dimensions of $a = 8.945 \pm 0.007$, $c = 14.071 \pm 0.011$ Å, $c/a = 1.573 \pm 0.002$. Whittaker and Kintner therefore concluded that their material was the same as that from the Ries crater. Unfortunately, they did not give the relative intensities of the first 20 lines, nor the intensities as well as the indices of the second 23 lines, so it is not possible to assess whether the diffraction patterns are, in fact, the same.

Additional data were then given by Whittaker, Donnay, and Lonsdale (1971). A lower limit of 3.33 g cm^{-3} was found for the density, corresponding to a minimum of 163 carbon atoms per unit cell. In agreement with Whittaker and Kintner, electron probe analyses of the synthetic material showed it to be only carbon, plus, possibly 0 to 2.5 wt% silicon. It was further stated that five 00ℓ reflections having $\ell = 3n$ were observed, but no structural conjectures were made on these meager data, which are insufficient to establish the structure.

Hexagonal Carbon, H(?)–44(?)

Yet another hexagonal form of carbon was reported by Whittaker and Wolten (1972). It was produced, along with H-168(?) carbon, on the surfaces of graphite carbons under free-evaporization conditions at low pressures, with temperatures above 2550°K. Electron diffraction patterns showed 13 lines which could be indexed on a hexagonal unit cell having $a = 5.33 \pm 0.05$, $c = 12.24 \pm 0.18$ Å, $c/a = 2.30 \pm 0.04$. The density was stated to be greater than 2.9 g cm^{-3}. There is thus a minimum of 44 carbon atoms per unit cell. The symmetry was given as trigonal, with P3, P3$_1$, and P3$_2$ as possible space groups. Aside from some numerology concerning the resemblance of the lattice constants to those of hexagonal diamond: $5.33 \approx 2 \times 2.52$ (!) and $12.24 \approx 3 \times 4.12$, and of H-168(?) carbon: $3 \times 5.33 \approx 8.95$ and $12.24 \approx 14.07$, no structural features were proposed.

"Carbon III"

The phase behavior of carbon has been extensively investigated; the results are summarized in Fig. 6-8. Included in this diagram is the region Bundy (1962) labeled "solid III", the existence of which was postulated by analogy with the high pressure behavior of silicon and germanium. This phase has not yet been observed but is estimated to be 15 to 20% denser than diamond and metallic in nature. As yet, of course, there are no diffraction data.

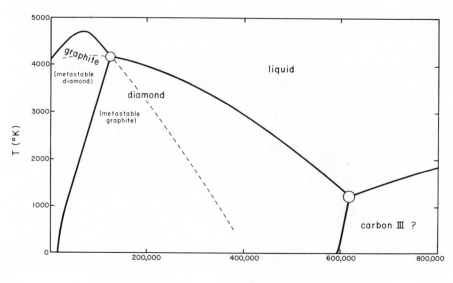

Fig. 6-8.   The phase diagram of carbon (after Bundy, 1962).

SILICON

α-Silicon

The structure of silicon was first determined by Debye and Scherrer (1916). It has the diamond structure, that is, cubic, space group Fd3m–O_h^7, eight atoms per unit cell at (000, $0\frac{1}{2}\frac{1}{2}$, $\frac{1}{2}0\frac{1}{2}$, $\frac{1}{2}\frac{1}{2}0$) ± $\frac{1}{8}\frac{1}{8}\frac{1}{8}$. The various determinations of the lattice constant are presented in Table 14-1. Each silicon atom has four tetrahedral ligands at 2.3515 ± 0.0001 Å, and the atomic volume is 20.020 ± 0.002 Å³.

There is no change in structure up to 1300°C at atmospheric pressure, according to Hoch and Johnston (1953) and Wood (1956); these results contradict an earlier report by Heyd, Kohl, and Kochanovska (1947) that a new form of silicon, possibly hexagonal, could be obtained by vacuum distillation or by quenching from 900°C.

α-Silicon was observed at high pressure by Jamieson (1963a), coexisting with the high pressure forms described below. From the spacings tabulated by Jamieson, excluding that of the first line, a lattice constant of a = 5.278 ± 0.007 Å may be calculated. This value corresponds to a bond distance of 2.285 ± 0.003 Å and an atomic volume of 18.38 ± 0.07 Å³. The pressure at which these values apply was not given.

Table 14-1.   The Lattice Constant of α-Silicon at 25°C[a]

a, Å	Reference	a, Å	Reference
5.47[b]	Debye & Scherrer, 1916	5.43072	Smakula & Kalnajs, 1955
5.44[b]	Hull, 1917b	5.4295[b]	Neff, 1956
5.426[b]	Gerlach, 1921, 1922	5.43050[g]	Vogel & Kempter, 1959
5.4316[b]	Küstner & Remy, 1923	5.43071	Bond & Kaiser, 1960
5.411[b]	Lehmann, 1924	5.43072[c]	Bond, 1960
5.42[b]	Becker, 1926	5.43054	Parrish, 1960
5.421[b]	van Arkel, 1928	5.43057	Ozolins & Ievinš, 1960
5.4279[b]	Jette & Gebert, 1933	5.43035	Hall, 1961
5.43077[c]	Jette & Foote, 1933	5.43070	Straumanis, Borgeaud, & James, 1961
5.4282[b]	Neuberger, 1935a		
5.4299[b]	Straumanis & Ievinš, 1936	5.4304	Dutta, 1962
		5.43074	Beu, Musil, & Whitney, 1962
5.43068[d]	Lipson & Rogers, 1944		
5.41[b]	Hass, 1948	5.43052[h]	Shaw & Liu, 1964
5.43094[e]	Straumanis & Aka, 1952	5.4308[c]	Taylor, Mack, & Parrish, 1964
5.4301[f]	Swanson & Fuyat, 1953	5.43042[f]	Middleton & Gazzara, 1969

[a]Values corrected to 25°C where necessary, $\alpha = 3.1 \times 10^{-6}$ deg^{-1}.
[b]From kX, not included in the average.
[c]From kX.
[d]From kX and 18.5°C.
[e]From kX and 20°C.
[f]Omitted from the average.
[g]Average of five different samples.
[h]From kX and 30°C.

There do not appear to have been any structural studies at low temperature.

High Pressure Forms of Silicon

Jamieson (1962) reported that at 195 kbar and room temperature silicon "inverted" from the diamond structure to the tetragonal white tin structure, space group $I4_1/amd-D_{4h}^{19}$, four atoms at $(000, \frac{1}{2}\frac{1}{2}\frac{1}{2})$ $\pm(0\frac{1}{4}\frac{3}{8})$. This form was characterized as "metallic", in contradistinction to the "semiconducting" ordinary form. The lattice constants were later reported (Jamieson, 1963a) as $a = 4.686$, $c = 2.585$ Å, $c/a = 0.5516$, but the pressure to which these correspond was not stated. In this structure (depicted in Fig. 50-3, p. 276) each silicon atom has four nearest neighbors at 2.430 Å and two next nearest neighbors at 2.585 Å, and the atomic volume is 14.19 Å³. Following the usage with respect to tin, this tetragonal form of silicon is termed herein β-silicon.

Wentorf and Kasper (1963) reported a new form of silicon, which they prepared by subjecting silicon to pressures of about 110 to 160 kbar at 25°C and at 200 to 1000°C. Although the resistivity of this form is several orders of magnitude less than that of normal silicon, it was not certain whether it, which was recovered at room temperature and atmospheric pressure after the above treatment, is the same form which exists at high pressure. This new form, termed γ-silicon here, was found to be body centered cubic, with 16 atoms per unit cell; the structure was later described in more detail by Kasper and Richards (1964). The unit cell has $a = 6.636 \pm 0.005$ Å, with 16 atoms in position 16c of space group $Ia3-T_h^7$, $(000, \frac{1}{2}\frac{1}{2}\frac{1}{2}) \pm (xxx)(\frac{1}{2}+x \ \frac{1}{2}-x \ \bar{x})$ $(\bar{x}\frac{1}{2}+x \ \frac{1}{2}-x)(\frac{1}{2}-x \ \bar{x} \ \frac{1}{2}+x)$, with $x = 0.1003 \pm 0.0008$. A view of this structure is shown in Fig. 14-1. Each silicon atom has four nearest neighbors, one at

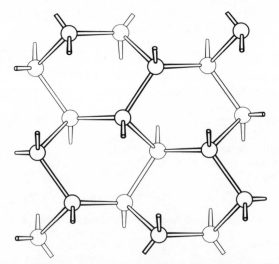

Fig. 14-1.   The cubic form of high pressure silicon discovered by Wentorf & Kasper (1963), viewed in projection down one of the cubic axes. Only part of the structure is shown. The three-dimensional network is obtained by translation of this structure by $\frac{1}{2}\frac{1}{2}\frac{1}{2}$, with bonds between them forming at the appropriate places.

2.306 ± 0.024 Å and three at 2.392 ± 0.015 Å (these values differ slightly
from those given by Kasper and Richards), so the average bond distance of
2.371 Å is not very different from that in normal silicon. There are, however,
substantial angular distortions from tetrahedral, three bond angles being 99.2°,
and three being 117.9°. The decrease in atomic volume of 8.8% to 18.26 $Å^3$
is reflected by the occurrence of four next-nearest neighbors, one at 3.441 Å
and three at 3.575 Å, which have no counterpart in normal silicon, where the
next-nearest neighbors are four at 3.840 Å. There does not appear to be any
simple mechanism, according to Kasper and Richards, of transforming the
normal α-silicon structure into the high pressure γ structure, for example,
by shearing.

This body centered cubic form which Bundy and Kasper observed at room
temperature and atmospheric pressure was also probably present at high pressure in Jamieson's (1963a) experiments. Two powder lines which could not
be attributed to either α (diamond)-silicon or β (white tin)-silicon were
assigned by Bates, Dachille, and Roy (1965) to the γ form. These lines are
the two strongest in the pattern reported by Kasper and Richards; from the
spacings given by Jamieson a lattice constant of $a = 6.405 ± 0.006$ Å may be
calculated. The atomic volume is 16.42 $Å^3$, or 10.1% smaller than that observed at atmospheric pressure.

A fourth form of silicon, termed here δ-silicon, was reported by Wentorf
and Kasper (1963), who found that when the above body centered cubic dense
form was heated at 200 to 600°C for between 30 minutes and 3 days the
powder lines of the dense form disappeared, being replaced by a new pattern
which could be indexed on a hexagonal unit cell with $a = 3.80$, $c = 6.28$ Å,
$c/a = 1.653$. This was assumed to correspond to a wurtzite type structure
(as in the case of hexagonal "diamond", see above). The atomic volume is
19.63 Å, or 1.9% (error unknown) less than that of normal silicon. The ideal
values for the lattice constants are $a = 3.840$ and $c = 6.271$ Å.

The phase diagram of silicon, according to Bundy (1964) is presented in
Fig. 14-2. Some structural data for the various forms are summarized in
Table 14-2.

Table 14-2. Structural Data for Allotropes of Silicon

Allotrope	Structure	Temperature	Pressure	Bond Distance, Å	Atomic Volume, Å³
α	Diamond	25°C	1 atm	2.352	20.020 ± 0.002
		Room	"High"	2.285	18.38 ± 0.07
β	White tin	Room	"High"	4 at 2.430 2 at 2.585	14.19 ± ?
γ	bcc(16)	Room	1 atm (quenched)	1 at 2.306 3 at 2.392	18.26 ± 0.04
		Room	"High"	1 at 2.225[a] 3 at 2.308[a]	16.42 ± 0.05
δ	Wurtzite	Room	1 atm	3 at 2.330 1 at 2.355	19.63 ± ?

[a]Distance calculated assuming no change in the value of the positional parameter at high pressure.

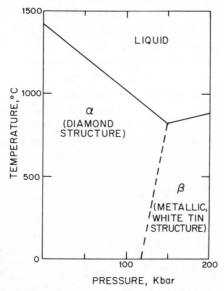

Fig. 14-2. The phase diagram of silicon (after Bundy, 1964). Not shown: The positions of the metastable γ and δ forms (see text), which are uncertain.

GERMANIUM

α-Germanium

The structure of germanium was first determined by Hull (1922) and by Kolkmeijer (1922). It is isostructural with diamond, that is, cubic, space group $Fd3m-O_h^7$, eight atoms per unit cell at $(000, \frac{1}{2}\frac{1}{2}0, \frac{1}{2}0\frac{1}{2}, 0\frac{1}{2}\frac{1}{2}) \pm \frac{1}{8}\frac{1}{8}\frac{1}{8}$. Various determinations of the lattice constant are presented in Table 32-1. Each germanium atom has four tetrahedral ligands at 2.4497 ± 0.0001 Å, and the atomic volume is 22.634 ± 0.001 Å^3.

There is no structure change at atmospheric pressure down to -253°C (Nitka, 1937) nor up to 890°C (Shaw and Liu, 1964).

Table 32-1.   The Lattice Constant of Germanium at 20°C[a]

a, Å	Reference	a, Å	Reference
5.64[b]	Hull, 1922	5.65737[d]	Smakula & Kalnajs, 1955
5.62[b]	Kolkmeijer, 1922	5.65702[e]	Blum & Durif, 1956
5.658[b]	Goldschmidt, 1926	5.65735	Mack, 1957, 1958
5.6594[b]	Nitka, 1937	5.65736[d]	Smakula & Kalnajs, 1957
5.65742[c]	Straumanis & Aka, 1950	5.65739[f]	Cooper, 1962
5.65747[c]	Straumanis & Aka, 1952	5.6577[e]	Shaw & Liu, 1964
5.65733[d]	Greiner, 1952	5.6576[e]	Singh, 1968
5.65743[d]	Swanson & Tatge, 1953		
av. 5.65739 ± 0.00005			

[a]Correction to 20°C made using $\alpha = 5.9 \times 10^{-6}$ deg^{-1} where necessary.
[b]From kX, omitted from the average.
[c]From kX.
[d]From 25°C.
[e]Omitted from the average.
[f]From kX and 24.6°C.

Germanium at High Pressure

A transition in germanium at room temperature above 120 kbar to a conduct-
ing form was found by Minomura and Drickamer (1962). Jamieson (1962)
reported a transition at 120 kbar and stated that the high pressure metallic
form had the white tin structure, space group $I4_1/\text{amd}-D_{4h}^{19}$, four atoms at
$(000, \frac{1}{2}\frac{1}{2}\frac{1}{2}) \pm (0\frac{1}{4}\frac{3}{8})$ (see Figs. 50-3 and 50-4). By analogy with white tin, this
form is termed here β-germanium. It has sometimes been called germanium
II. The lattice constants were later reported as $a = 4.884$, $c = 2.692$ Å, $c/a =$
0.5512 (Jamieson, 1963a), at an unstated high pressure. In this structure
each germanium atom has four nearest neighbors at 2.533 Å and two next-
nearest neighbors at 2.692 Å. The atomic volume is 16.05 Å^3'

Bundy and Kasper (1963) prepared a new form by compressing germanium
to excess of 120 kbar and decompressing to atmospheric pressure. This form,
which they labeled germanium III, but which is termed here γ-germanium,
had the resistivity of a semiconductor. The change in the resistance of the
sample during a typical run is shown in Fig. 32-1. γ-Germanium was found to
be tetragonal, with $a = 5.93$, $c = 6.98$ Å, $c/a = 1.177$. The space group is
$P4_12_12-D_4^4$ and the enantiomorphic $P4_32_12-D_4^8$, and there are 12 atoms per
unit cell. The atomic volume is 20.45 Å^3, or 9.6% less than that of normal
germanium. The refinement of this structure was later carried out by Kasper
and Richards (1964). Four Ge(I) atoms lie in position $4a$, at $(xx0)$, $(\bar{x}\bar{x}\frac{1}{2})$,

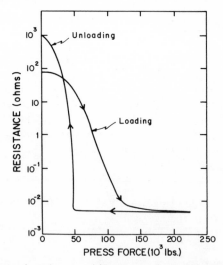

Fig. 32-1.   The change in resistance with pressure of germanium (after Bundy & Kasper,
1963).

$(\frac{1}{2}-x\ \frac{1}{2}+x\ \frac{3}{4})$, $(\frac{1}{2}+x\ \frac{1}{2}-x\ \frac{1}{4})$ with $x = 0.0912 \pm 0.0060$, and eight Ge(II) atoms
lie in position $8b$, at $\quad (xyz)$, $(\frac{1}{2}+y\ \frac{1}{2}-x\ \frac{1}{4}+z)$, $(\bar{x}\ \bar{y}\ \frac{1}{2}+z)$, $(\frac{1}{2}-y\ \frac{1}{2}+x\ \frac{3}{4}+z)$,
$\quad\quad\quad (yx\bar{z})$, $(\frac{1}{2}+x\ \frac{1}{2}-y\ \frac{1}{4}-z)$, $(\bar{y}\ \bar{x}\ \frac{1}{2}-z)$, $(\frac{1}{2}-x\ \frac{1}{2}+y\ \frac{3}{4}-z)$,

with $x = 0.1730 \pm 0.0037$, $y = 0.3784 \pm 0.0051$, $z = 0.2486 \pm 0.0048$. Each
atom of type II has four nearest neighbors, two at 2.489 ± 0.025 Å and one
each at 2.486 ± 0.040 Å and 2.479 ± 0.042 Å; each atom of type I also has
four nearest neighbors, two at 2.486 ± 0.040 Å and two at 2.479 ± 0.042 Å.
The average bond distance is 2.485 Å, which is slightly greater than the value
of 2.450 Å in α-germanium. These nearest neighbors are shown in Fig. 32-2.
The next shortest distances which are shorter than the value of 4.000 Å found
in α-germanium lie in the range 3.455 to 3.957 Å. These may be identified in
Fig. 32-3 as those associated with the bond angles of less than 110°. The
angular distortions from tetrahedral are rather large, as may be seen in Fig.
32-3. The bond angles average to 109.1°, but the root mean square deviation
from the mean is 15.7°. A general view of the structure is presented in Fig.
32-4.

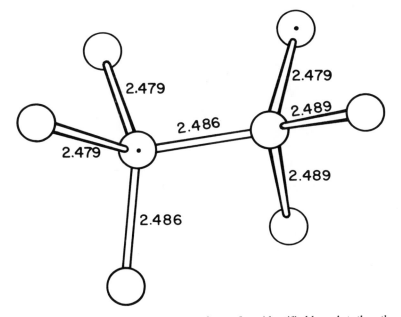

Fig. 32-2.   Bond distances in γ-Ge. Atoms of type I are identified by a dot; the others
are type II.

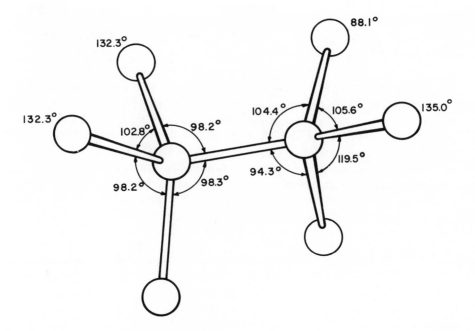

Fig. 32-3.   Bond angles in γ-germanium.

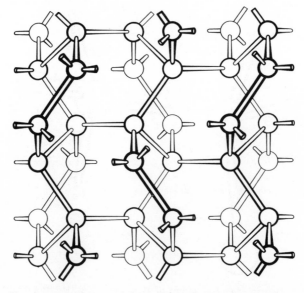

Fig. 32-4.   The structure of γ-germanium, projected along the c-axis.

270

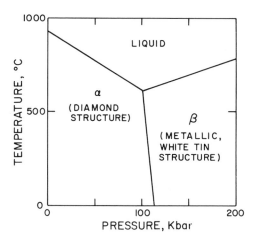

Fig. 32-5.   The phase diagram of germanium (after Bundy, 1964).

A more extensive study of the phase behavior was carried out by Bundy (1964), whose phase diagram is presented in Fig. 32-5. This diagram is in disagreement with that given shortly afterward by Bates, Dachille, and Roy (1965), which is presented in Fig. 32-6. Bates et al. state that the transition α-Ge⇌γ-Ge takes place at about 25 kbar at 25°C. Their powder pattern of γ-Ge (which they erroneously term body centered tetragonal) agrees well with that of Kasper and Richards. This pattern, however, was apparently recorded at atmospheric pressure. It is thus difficult to reconcile the phase diagram of Bates et al. (Fig. 32-5) with the phase changes observed by Bundy and Kasper (Fig. 32-1).

The positive slope of the α-Ge:γ-Ge boundary of Fig. 32-6 led Bates et al. to predict that γ-Ge should be stable below –135°C. They sought it at dry ice temperature under shear conditions of 10 to 20 kbar, but failed to find it. As mentioned previously, Nitka (1937) observed no structure change down to –253°C.

Bates et al. also reported that samples held at 110 to 130 kbar at room temperature and then quenched disclosed four lines which could not be accounted for by γ-Ge. (This result is also at variance with that of Bundy and Kasper.) These lines were indexed on a cubic unit cell with lattice constant $a = 6.92 \pm 0.02$ Å. They termed this new form Ge-IV and, by analogy with silicon, assumed that it was isostructural with γ-Si. It is termed here δ-germanium. The atomic volume is thus 20.71 ± 0.18 Å^3, or 8.5% less than that of normal germanium, the same decrease observed for the corresponding forms of silicon.

Some structural data for the various forms of germanium are summarized in Table 32-2.

Group IV

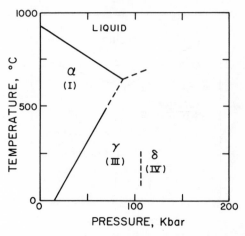

Fig. 32-6. The phase diagram of germanium according to Bates et al. (1965). The struc-
tures assigned are: α, diamond structure; γ, tetragonal, 12 atoms per unit cell; δ, body centered cubic, 16 atoms per unit cell. The position of the β (white tin) form was not determined.

Table 32-2.   Structural Data for Allotropes of Germanium

Allotrope	Structure	Tempera-ture	Pressure	Bond Distance, Å	Atomic Volume, Å³
α	Diamond	25°C	1 atm	2.4497	22.634
β	White tin	Room	"High"	4 at 2.533 2 at 2.692	16.05
γ	Tetragonal $Z=12$	Room	1 atm (quenched)	2.479 2.486 2.489	20.45
δ	bcc(16)	Room	1 atm (quenched)	Not known	20.71

TIN

There are two forms of tin at atmospheric pressure. Below about 18°C, and
down to at least –130°C (Thewlis and Davey, 1954), the stable form is α, or
gray, tin, which has the diamond structure. Above 18°C, and up to at least
212°C (Deshpande and Sirdeshmukh, 1962), the stable form is β, or white,
tin, which is tetragonal. The $\alpha \rightarrow \beta$ transition is slow at room temperature, so
that α-tin may be studied at those conditions, and the $\beta \rightarrow \alpha$ transition is also

sluggish, proceeding rapidly only at temperatures well below 18°C. α-Tin is nonmetallic, β-tin is metallic.

(A third form of tin which was proposed to exist below the melting point of 232°C on the basis of discontinuities of various properties was sought, but not found, in diffraction experiments by Mason and Pelissier, 1939).

α-Tin (gray tin)

The structure of α-tin was first determined by Bijl and Kolkmeijer (1918b). It has the diamond structure, space group Fd3m–O_h^7, eight atoms per unit cell at $(000\ \frac{1}{2}\frac{1}{2}0\ \frac{1}{2}0\frac{1}{2}\ 0\frac{1}{2}\frac{1}{2}) \pm \frac{1}{8}\frac{1}{8}\frac{1}{8}$. Various determinations of the lattice constant are presented in Table 50-1. There is no apparent explanation for the discrepancy between the equally precise values of Brownlee and of Thewlis and Davey; the result of Swanson and Fuyat favors the latter. Each tin atom is tetrahedrally bonded to four others at 2.810 Å, and the atomic volume is 34.157 Å^3. The structure is shown in Fig. 50-1.

β-Tin (white tin)

The structure of β-tin was first reported to be tetragonal, with *three* atoms per unit cell, by Bijl and Kolkmeijer (1918a), who later gave lattice constants a = 5.85, c = 2.38 Å (from kX), c/a = 0.406 (Bijl and Kolkmeijer, 1918b, 1919a). This unit cell corresponds to a calculated density of 7.26 g cm^{-3} as compared with the observed value of 7.29 g cm^{-3}. It is, nevertheless, wrong, as was shown by Mark, Polanyi, and Schmid (1923).

The correct structure is tetragonal, space group I4$_1$/amd–D_{4h}^{19}, *four* atoms

Table 50-1. The Lattice Constant of α (Gray)-Tin at Room Temperature

a, Å	Reference
6.47[a]	Bijl & Kolkmeijer, 1918b, 1919b
6.4912	Brownlee, 1950
6.489	Swanson & Fuyat, 1953
6.4892	Thewlis & Davey, 1954
Preferred value: 6.4892 Å at 20°C	

[a]From kX.

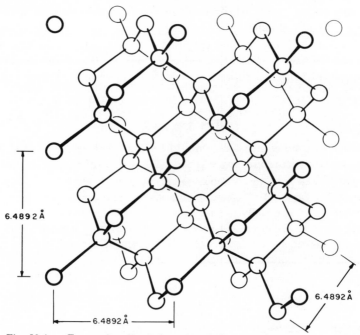

Fig. 50-1.   Four unit cells of the α (gray)-tin structure.

per unit cell at $(000 \; \frac{1}{2}\frac{1}{2}\frac{1}{2}) \pm 0\frac{1}{4}\frac{3}{8}$. (This origin is on a center of symmetry; the structure is often described with a different origin, with the atoms lying at $(000 \; \frac{1}{2}\frac{1}{2}\frac{1}{2}) + 0\frac{1}{2}\frac{1}{4} \; \frac{1}{2}0\frac{3}{4}$.) Various determinations of the lattice constants are presented in Table 50-2.   The value of a is close to that reported by Bijl and Kolkmeijer, while the value of c is 4/3 the old value.

Each tin atom has four nearest neighbors at 3.022 Å; these lie at the vertices of a flattened tetrahedron, as shown in Fig. 50-2.   There are two more near neighbors, above and below, at 3.182 Å, so the true coordination number is six.   The atomic volume is 27.049 Å^3.   A general view of the structure is shown in Fig. 50-3.

Another view of the β-tin structure is presented in Fig. 50-4.   In this view the unit cell is face centered, with $a' = \sqrt{2}a$.   The relationship of the α- and β-tin structures may be seen by comparing Fig. 50-4 with Fig. 50-1.   Expansion of the a- and b-axes of the α structure by 27.1%, plus contraction of the c-axis by 51.0% gives the β structure.   The overall change is a volume contraction of 20.8% in going from α to β.   It is, however, unlikely that the transition proceeds in this simple way, for although Kuo and Burgers (1956) observed that in 12 out of 20 experiments a single crystal of β-tin transformed

Table 50-2. The Lattice Constants[a] of β (white)-Tin at 25°C[b]

a, Å	c, Å	Reference
5.84[c]	3.17[c]	Mark, Polanyi, & Schmid, 1923
5.85[c]	3.16[c]	Mark & Polanyi, 1923, 1924
5.81[c]	3.18[c]	van Arkel, 1924
5.835[c]	3.171[c]	Phebus & Blake, 1925
5.830[c]	3.180[c]	Solomon & Morris-Jones, 1931
5.830[c]	3.183[c]	Bowen & Morris-Jones, 1931
5.8317[d]	3.1822[c]	Stenzel & Weerts, 1932
5.8284[c]	3.1792[c]	Jette & Gebert, 1933
5.819[c]	3.180[c]	Hägg & Hybinette, 1935
5.8313[e]	3.1814[e]	Jette & Foote, 1935
5.8307[c]	3.1811[c]	Kossolapov & Trapeznikov, 1936
5.8315[e]	3.1813[e]	Ievinš, Straumanis, & Karlsons, 1938
5.8315[e]	3.1814[e]	Carapella & Hultgren, 1941
5.8313[e]	3.1814[e]	Fink, Jette, Katz, & Schnettler, 1945
5.828[c]	3.181[c]	Nial, 1947
5.8315[e]	3.1813[e]	Straumanis, 1949
5.831[f]	3.182[f]	Swanson & Tatge, 1953
5.8317[e]	3.1813[e]	Lee & Raynor, 1954
5.8318	3.1818	Vogel & Kempter, 1959
5.8318	3.1819	Deshpande & Sirdeshmukh, 1961
5.8318[g]	3.1813[g]	Deshpande & Sirdeshmukh, 1962
av. 5.8316	3.1815	
±0.0002	±0.0002	

$c/a = 0.54556 \pm 0.00004$

[a]Values reported for the face centered cell transformed to the body centered cell when necessary.

[b]Values converted to 25°C with $\alpha_\perp = 17 \times 10^{-6}$ deg^{-1} and $\alpha_\parallel = 32 \times 10^{-6}$ deg^{-1} when necessary.

[c]From kX, omitted from the average.

[d]From kX and 20°C.

[e]From kX.

[f]Omitted from the average.

[g]From 33°C.

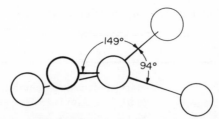

Fig. 50-2. The flattened tetrahedral configuration of the four nearest neighbors of a tin atom in the β (white)-tin structure.

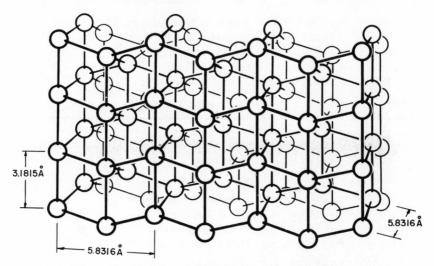

3.1815Å

5.8316Å

5.8316Å

Fig. 50-3. Nine unit cells of the β (white)-tin structure. The c-axis is vertical.

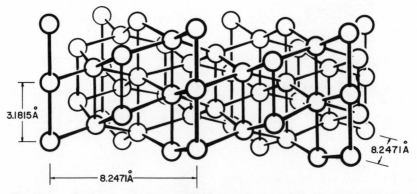

3.1815Å

8.2471Å

8.2471Å

Fig. 50-4. Four unit cells (face centered setting) of the β-tin structure. (Cf. Fig. 50-1; see text.)

to a single crystal of α-tin, at –20°C, the relative orientations of the two lattices were completely random.

γ-Tin

The results of studies of tin at high pressure are conflicting. It is agreed that a new form appears, termed γ-tin in the following discussion; the designation Sn-II for this form has been used in some places.

Results of work on the phase behavior of tin are summarized in Fig. 50-5. The triple point is in the vicinity of 33 kbar and 310°C, but there is disagreement concerning the placement of the β-γ boundary, as well as on the structure of γ-tin. Jamieson (1962; point 3 of Fig. 50-5) reported that it is body centered cubic (no lattice constant given) at room temperature and 130 kbar, while Barnett, Bennion, and Hall (1963b; point 5 of Fig. 50-5) stated that it is body centered tetragonal, $a = 3.811$, $c = 3.483$ Å, $c/a = 0.914$, two atoms per unit cell, atomic volume 25.29 Å^3, at 39 kbar and 308 to 320°C. This same structure was reported by Barnett, Bean, and Hall (1966; point 6 of Fig. 50-5), who found, at 98 kbar and 25°C, $a = 3.70$, $c = 3.37$ Å, $c/a = 0.911$, atomic volume 23.07 Å^3. This result is in disagreement with that of Stager, Balchan, and Drickamer (1962; point 2 of Fig. 50-5), who stated that the transition at 25°C does not occur until a pressure of 113 to 115 kbar.

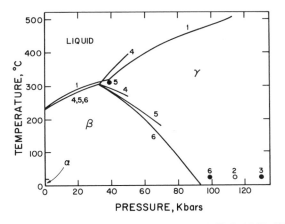

Fig. 50-5.   The phase behavior of tin. (1) Dudley & Hall, 1960; (2) Stager, Balchen, & Drickamer, 1962; (3) Jamieson, 1962; (4) Kennedy & Newton, 1963; (5) Barnett, Bennion, & Hall, 1963b, and Barnett and Hall, 1964; (6) Barnett, Bean, & Hall, 1966. At point 3 the structure was stated to be body centered cubic, and points 5 and 6, body centered tetragonal, while at point 2 it was stated that the β structure was still retained. See text.

According to Barnett, Bean, and Hall (1966), the decrease in volume at the transition at 92 kbar and room temperature is $1.1 \pm 0.2\%$.

LEAD

The structure of lead was first determined by Vegard (1916b); it is cubic close packed. Various determinations of the lattice constant are presented in Table 82-1. There is no change in structure down to $20°K$ (Keesom & Kammerlingh-Onnes, 1926) nor up to $320°C$ (Stokes and Wilson, 1941).

At room temperature and 130 kbar a transition to the hexagonal close packed structure takes place (Takahashi, Mao, and Bassett, 1969). At 139 kbar the lattice constants are $a = 3.265 \pm 0.004$, $c = 5.387 \pm 0.007$ Å, $c/a = 1.650 \pm 0.003$. The change in the atomic volume on transition, as determined by Takahashi et al., on samples where the two phases coexisted in the high pressure cell, is $-0.30 \pm 0.10 Å^3$, or -1.2%. The lattice constant of the cubic close packed phase is thus calculated to be 4.652 ± 0.006 Å.

Table 82-1.   The Lattice Constant of Lead at 25°C[a]

a, Å	Reference
4.92[b]	Vegard, 1916b
4.993[b]	Owen & Preston, 1923a
4.930[b]	Davey, 1924, 1925, 1926
4.943[b]	Levi, 1924, 1925
4.952[b]	Phebus & Blake, 1925
4.934[b]	Halla & Staufer, 1928
4.94[b]	Solomon & Morris-Jones, 1930, 1931
4.953[b]	Zintl & Harder, 1931
4.93[b]	Zeidenfeld, 1931
4.948[b]	Darbyshire, 1932
4.9505[c]	Owen & Iball, 1932
4.9485[b]	Jette & Gebert, 1933
4.9496[d]	Obinata, 1933
4.9506[e]	Owen & Yates, 1933a
4.9492[e]	Ölander, 1934
4.9505[e]	Straumanis & Ievinš, 1936
4,9500[f]	Lu & Chang, 1941
4.0505[g]	Foote & Jette, 1941a
4.950[c]	Fricke, 1941
4.9503[g]	Stokes & Wilson, 1941
4.9503[g]	Klug, 1946
4.9500[g]	Straumanis, 1949
4.9505	Swanson & Tatge, 1953
4.9500	Tyzack & Raynor, 1954
av. 4.9502 ± 0.0005	

[a]Values corrected to 25°C with thermal expansion coefficient of 29×10^{-6} deg^{-1} when necessary.
[b]From kX, omitted from the average.
[c]From kX and 16.5°C.
[d]From kX and 20°C.
[e]From kX and 18°C.
[f]From kX and 21°C.
[g]From kX.

Chapter 8

Group V

At atmospheric pressure there is a transition in solid nitrogen,

$$\alpha \xrightarrow{\ 35.6°K\ } \beta$$

A third form (γ) exists only at high pressure. The phase diagram, as determined by Swenson (1955a), is shown in Fig. 7-1.

α-Nitrogen

The structure of α-nitrogen is the subject of an as yet unsettled controversy. The first work on this substance is that of de Smedt and Keesom (1925a, b), who were unable to index powder photographs taken at 20°K and concluded that it was of low symmetry. Vegard (1929a), however, found that it is cubic, with lattice constant 5.66 Å (from kX) at 20°K [later revised to 5.67 Å (from kX); Vegard, 1929c]. A structure was proposed by Vegard (1929b, c), based on 25 observed powder lines. It consists of eight nitrogen atoms in two sets of position $4a$ of space group T^4–$P2_13$, $(xxx, \frac{1}{2}+x\,\frac{1}{2}\,x\bar{x}, \bar{x}\frac{1}{2}+x\frac{1}{2}-x, \frac{1}{2}-x\,x\,\bar{x}\,\frac{1}{2}+x)$, with x_I = 0.0695 and x_{II} = –0.039. These give a bond distance of 1.064 Å. In this structure the N_2 molecules lie on the threefold axes, and the molecular centers are displaced by 0.150 Å from the origin and the centers of the faces of the unit cell. In addition to the bond distance, each nitrogen-I has three neighbors at 3.470 Å (II), six at 3.584 Å (I), and three at 3.742 Å (II), and each nitrogen-II has three at 3.470 Å (I), six at 3.736 Å (II), and three at 3.742 Å (I).

De Smedt, Keesom, and Mooy (1929a, b), at about the same time, reported

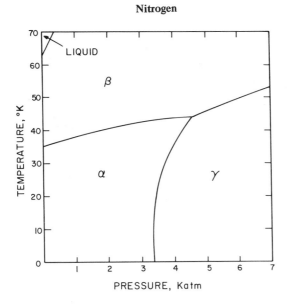

Fig. 7-1.   The phase diagram of nitrogen (after Swenson, 1955a).

that α-nitrogen at 20°K is either cubic, with a = 5.667 Å (from kX) or tetragonal, with a = 4.007, c = 5.667 Å (from kX), c/a = $\sqrt{2}$, preferring the latter
because of the observation that the solid was birefringent.

Ruhemann (1932), also working with samples at 20°K, observed 16 powder
lines but did not observe two lines, 110 and 310, observed by Vegard, and
concluded that the true space group is probably Pa3-T_h^6. The lines 110 and
310 are forbidden in space group Pa3. In this structure the eight nitrogen
atoms in the unit cell are equivalent, occupying position $8c$, $\pm(xxx, \frac{1}{2}+x\ \frac{1}{2}-x\ \bar{x},$
$\bar{x}\ \frac{1}{2}+x\ \frac{1}{2}-x, \frac{1}{2}-x\ \bar{x}\ \frac{1}{2}+x)$. Ruhemann reported the value 5.68 ± 0.02 Å (from
kX) for the lattice constant. In this space group the molecules lie on centers
of symmetry. This structure is seen in Fig. 7-2.

Bolz, Boyd, Mauer, and Peiser (1959) collected data in the region 4.2 to
20°K. They reported a = 5.644 ± 0.005 Å at 4.2°K, and a thermal expansion
coefficient of 2×10^{-4} deg^{-1} for this temperature range. These yield a =
5.662 Å at 20°K.

Hörl and Marton (1961), in an electron diffraction investigation, observed
20 lines and also did not observe the 110 and 310 reflections. They found,
at 20°K, the value a = 5.661 ± 0.008 Å. They also calculated intensities for
the Pa3 structure, assuming a bond distance of 1.094 Å, as observed in gaseous
N_2 (Herzberg, 1955); this corresponds to a value for the positional parameter
of 0.0558. The intensity agreement is satisfactory.

Donohue (1961b), who also assumed the Pa3 structure, refined the intensity

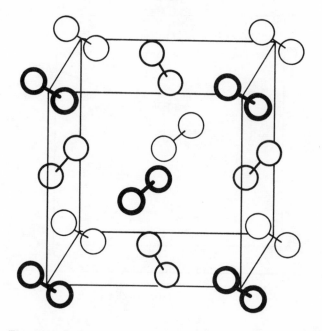

Fig. 7-2.   The unit cell of. α-nitrogen for the case of space group Pa3, with the molecules lying on symmetry centers.

data of Hörl and Marton by least squares, obtaining the value 0.0530 ± 0.0020 for the positional parameter. This yields 1.039 ± 0.040 Å for the bond distance, a result not significantly different from the spectroscopic value.

Jordan, Smith, Streib, and Lipscomb (1964), in more than 20 attempts, succeeded only once in obtaining a single crystal of α-nitrogen by cooling a single crystal of β-nitrogen through the transition point. With this one crystal they confirmed the m3–T_h symmetry of the diffraction pattern and observed 49 unique reflections, many of them several times by virtue of symmetrical equivalents. Two of these reflections, 501 and 520, are forbidden by space group Pa3; the former was observed at five different points, the latter at three different points, and a careful search revealed no other violations of Pa3 symmetry. Least squares refinements were carried out in both Pa3 and $P2_1 3$ on two sets of data: the 49 observed reflections, and the 116 observed and unobserved reflections. Their results are summarized in Table 7-1. On the basis of the better agreement obtained with the $P2_1 3$ structure, Jordan et al. preferred it over the centrosymmetric Pa3 structure and, moreover, favored the results obtained in the refinement based on the observed reflections only. This structure is quite close to that obtained 35 years earlier by Vegard, described

Table 7-1.   Results of Two Different Refinements of α-N_2 by Jordan et al. (1964)[a]

	$P2_1 3$	$Pa3$
49 observed $F_{hk\ell}$		
R_F	13.4%	20.6%
R_{F^2}	24.5%	34.0%
R_{wF^2}	25.4%	31.3%
x_I	0.0735	0.0546
x_{II}	-0.0388	(-0.0546)
116 observed and unobserved $F_{hk\ell}$		
R_F	24.6%	36.0%
R_{F^2}	33.0%	38.9%
R_{wF^2}	33.6%	38.6%
x_I	0.0648	0.0548
x_{II}	-0.0464	(-0.0548)

$$^a R_F = \frac{\Sigma ||F_o| - |F_c||}{\Sigma |F_o|}$$

$$R_{F^2} = \frac{\Sigma |F_o^2 - F_c^2|}{\Sigma |F_o^2|}$$

$$R_{wF^2} = \sqrt{\frac{\Sigma w(F_o^2 - F_c^2)^2}{\Sigma w F_o^4}}; \quad w_{hk\ell} = \sigma^{-2}_{hk\ell}$$

above. The bond distance is 1.099 Å, and the neighbors of nitrogen-I are three II at 3.428 Å, six I at 3.556 Å, and three II at 3.737 Å, while those of nitrogen-II are three I at 3.428 Å, six II at 3.724 Å, and three I at 3.737 Å. The molecular centers are displaced by 0.17 Å from those in the Pa3 structure. Jordan et al. speculate that the lowering of the symmetry from Pa3 to $P2_1 3$ may be associated with the tendency for each nitrogen atom to avoid having as many as 12 nearest nonbonded neighbors in the range 3.57 to 3.63 Å, but they do not elaborate on the source of this "tendency".

It should further be pointed out that six reflections forbidden in Pa3 and not observed by Jordan et al. have calculated F values in the $P2_1 3$ structure *greater than* the corresponding minimum threshold intensities. There is thus a contradiction in that two reflections forbidden in Pa3 were observed, but six reflections allowed in $P2_1 3$, and calculated as observable, were not detected; the latter include 110 and 310 reported by Vegard, but by no subsequent investigators, including Ruhemann (1932), Bolz et al. (1959), and Hörl and Marton (1961). The evidence that the structure of α-nitrogen is based on space group $P2_13$ instead of Pa3 is accordingly far from unexceptionable.

Additional evidence which argues against the noncentric $P2_13$ structure is provided by the work of Schuch and Mills (1970), who investigated crystalline nitrogen at various temperatures and pressures. In particular, at 3785 atm and $19.6°K$ they found no evidence of the $P2_13$ structure. They observed 14 powder lines, and did not observe 110, the calculated intensity of which in $P2_13$ was almost three times above the limit of their observation. They concluded "we are unable to prove the existence of the $P2_13$ structure".

The results of an electron diffraction investigation by Venables (1970) of both powders and single crystals also supports the Pa3 structure.

If the structure is the noncentric $P2_13$, then coincidences should be found in the Raman and infrared spectra, but none have been detected, for example, by Brith, Ron, and Schnepp (1969) or Anderson, Sun, and Donkersloot (1970), although the former remark that "presumably the distortion [from Pa3 to $P2_13$] is quantitatively too small to be manifested in the spectra".

Theoretical reasons for the distortions of the $P2_13$ structure have not been forthcoming. In fact, crystal potential models and calculation of the lattice dynamics agree on the undistorted Pa3 model (Schnepp and Ron, 1969; Anderson et al., 1970).

On the other hand, evidence in favor of the $P2_13$ structure has been found by Brookeman and Scott (1972), who observed piezoelectric signals from solid nitrogen at $4.2°K$. If these signals were not artifacts—and Brookeman and Scott stated that careful checks were made to ensure that they were indeed produced by the sample—then the correct space group cannot be the centric Pa3 but must be $P2_13$.

More detailed least squares refinement of the original data of Jordan et al. was then carried out by LaPlaca and Hamilton (1972), who assumed the space group was $P2_13$ and used only the observed values of F^2. They obtained the values $x_I = 0.0699 \pm 0.0021$ and $x_{II} = -0.0378 \pm 0.0024$, not too different from those obtained previously, but because they introduced four thermal vibration parameters as additional variables, were able to reduce the R_F index to 10.5%. They did not present any calculated F values, including, of course, those for the unobserved reflections, but did remark that a set of more precise diffraction data is very much needed.

The weight of evidence would thus appear to favor the centrosymmetric Pa3 structure for α-nitrogen. The forbidden reflections observed by Jordan et al., however, remain unexplained. Various determinations of the lattice constant are presented in Table 7-2. The average, at $20°K$ and 1 atm, when combined with the bond distance of 1.0976 Å in gaseous N_2 (Stoicheff, 1954),

Table 7-2.   The Lattice Constant of α-Nitrogen at
20°K and Atmospheric Pressure

a, Å	Reference
5.66[a]	Vegard, 1929a
5.67[a]	Vegard, 1929c
5.667[b]	de Smedt, Keesom, & Mooy, 1929a, b
5.68[a]	Ruhemann, 1932
5.662[c]	Bolz, Boyd, Maurer, & Peiser, 1959
5.661	Hörl & Marton, 1961
5.649[d]	Schuch & Mills, 1970
5.656[e]	Schuch & Mills, 1970
av. 5.659 ± 0.007	

[a]From kX, omitted from the average.
[b]From kX.
[c]From 4.2°K and a thermal expansion coefficient of 2×10^{-4} deg^{-1}.
[d]X-ray result.
[e]From p-V-T data.

yields a positional parameter of 0.05599, and the following environment of each nitrogen atom, in addition to the intramolecular distance: six neighbors at 3.582 Å and six at 3.637 Å. These correspond to the 12 nearest neighbors in the range 3.57 to 3.63 Å referred to by Jordan et al. The molecular volume at 20°K and 1 atm is 45.31 ± 0.17 Å^3.

The lattice constant at high pressure has been measured by Schuch and Mills (1970). At the extreme of 3785 atm (at 20°K) they found $a = 5.433$ Å and a molecular volume of 40.09 Å^3.

A packing drawing of the α-nitrogen structure, viewed along one of the cubic axes, is presented in Fig. 7-3, and a view along one of the threefold axes is presented in Fig. 7-4.

β-Nitrogen

The hexagonal nature of β-nitrogen was first determined by Ruhemann (1932), who found two molecules in a hexagonal unit cell having $a = 4.042$, $c = 6.601$ Å (from kX), $c/a = 1.633$ at 39°K. He observed 16 powder lines. The space group P6$_3$/mmc–D$_{6h}^4$ was preferred, with the nitrogen atoms in position 4f, $\pm (\frac{1}{3} \frac{2}{3} z, \frac{2}{3} \frac{1}{3} \frac{1}{2}+z)$. In this structure the molecules lie on crystallographic

Fig. 7-3.   The structure of α-N_2, viewed along one of the cubic axes.

centers of symmetry and are directed parallel to the c-axis.

A different structure was proposed by Vegard (1932, 1934). At 45°K he observed 10 powder lines and found $a = 4.047$, $c = 6.683$ Å (from kX), $c/a = 1.651$. The structure is based on space group P$\bar{3}$m1–D$_{3d}^3$, with the four nitrogen atoms in two sets of position $2d$, $\pm$ ($\frac{1}{3}$ $\frac{2}{3}$ z); the molecules do not lie on symmetry centers.

Bolz, Boyd, Mauer, and Pieser (1959) reported $c/a = 1.627$ and a molecular volume of 48.1 Å³ for β-nitrogen at an unspecified temperature. From these data lattice constants of $a = 4.087$, $c = 6.650$ Å may be deduced.

Streib, Jordan, and Lipscomb (1962) observed 14 unique reflections from single crystals of β-nitrogen at 50°K. They gave crude values for the lattice constants, $a = 3.93 \pm 0.16$, $c = 6.50 \pm 0.51$ Å, $c/a = 1.65 \pm 0.15$. They considered the proposed structure of Vegard to be physically unrealistic and chose P6$_3$/mmc as the correct space group. The data were found to agree equally well with two nearly physically indistinguishable models, in both of which the centers of the molecules lie on the points of an hexagonal close packed lattice. In the first model the molecules are precessing about axes parallel to the c-axis passing through their centers at a fixed angle between c and the N–N bond; in the second model the molecules are statistically distributed among six orientations with the molecular axes at the same fixed angle to c. This angle was

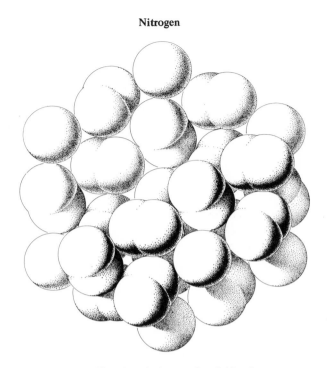

Fig. 7-4.   The structure of α-N_2, viewed along a threefold axis.

found to be $54.5 \pm 2.5°$, but in a subsequent emendation by Jordan, Smith, Streib, and Lipscomb (1964) it was corrected to $56.0 \pm 2.5°$. These two models are far superior to both the fixed atom model of Ruhemann and one in which there is complete free rotation of the molecules, that is to say, complete, or spherical disorder.

Various determinations of the lattice constant are presented in Table 7-3. The averages give a molecular volume of 46.99 ± 0.34 $Å^3$ at $43°K$ and 1 atm. The lattice constants at high pressure have been measured by Schuch and Mills (1970). At the extreme of 4125 atm (at $49°K$) they reported $a = 3.861$, $c = 6.265$ Å, $c/a = 1.623$, and a molecular volume of 40.44 $Å^3$.

γ-Nitrogen

The structure of γ-nitrogen was first determined by Mills and Schuch (1969), who observed 10 powder lines. The structure was also described by Schuch and Mills (1970). It is tetragonal, space group $P4_2/mnm$-D_{4h}^{14}, $a = 3.957$, $c = 5.109$ Å at 4015 atm and $20.5°K$. There are two molecules per unit cell, the four atoms lying in position $4f$, $\pm(xx0, \frac{1}{2}+x\ \frac{1}{2}-x\ \frac{1}{2})$, with $x = 0.098$. Each nitrogen atom has, in addition to the intramolecular bond of 1.097 Å, two neighbors at 3.275 Å and eight at 3.448 Å. A packing drawing of several unit

Table 7-3.   The Lattice Constants of β-Nitrogen at Atmospheric Pressure

	a, Å	c, Å	c/a	T, °K	Reference
	4.042[a]	6.601[a]	1.633	39	Ruhemann, 1932
	4.047[a]	6.683[a]	1.651	45	Vegard, 1932
	4.087[b]	6.650[b]	1.627	?	Bolz, Boyd, Mauer, & Pieser, 1959
	3.93[b]	6.50[b]	1.65	50	Streib, Jordan, & Lipscomb, 1962
	4.050	6.604	1.631	46	Schuch & Mills, 1970
av.	4.046	6.629	1.638	43	
	±0.004	±0.047	±0.012	±4	

[a]From kX.
[b]Omitted from the average.

cells is presented in Fig. 7-5, and a view of the structure down the c-axis in Fig. 7-6.

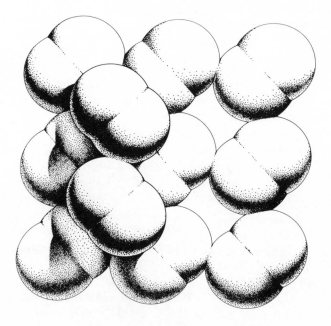

Fig. 7-6.   The structure of γ-nitrogen viewed along the c-axis.

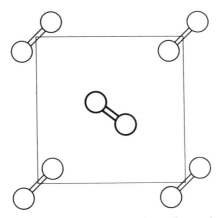

Fig. 7-5.   A projection of the γ-nitrogen structure down the c-axis.

PHOSPHORUS

The structural chemistry of elemental phosphorus is a Comedy of Errors on which the final curtain has not yet descended. Some of the allotropes which have been proposed are, as will be seen, of dubious validity. In the discussion which follows the various forms are considered in order of increasing confusion with regard to their structures.

There are four allotropes of phosphorus the structures of which have been well established by diffraction methods: (1) black (orthorhombic); (2) monoclinic; (3) rhombohedral; (4) metallic (cubic). In addition, two forms of white phosphorus have been well characterized thermodynamically and by x-ray diffraction, but detailed structures for them have not yet been determined. Remaining to be resolved is the interpretation of no less than nine different x-ray powder patterns obtained from various preparations, all termed "red phosphorus".

Black Phosphorus

The discovery of black phosphorus is generally attributed to Bridgman (1914), who obtained it irreversibly from white phosphorus by treatment at 200°C and 12,000 kg cm^{-2}. It is characterized by a metallic black graphitelike appearance and a density of 2.691 g cm^{-3} (Bridgman). Frost (1930) prepared what he called violet phosphorus following the directions of Hittorf (1865, 1866). Samples obtained by crystallization from bismuth at 570 to 580°C or from lead at 525 to 590°C gave closely similar powder x-ray diffraction patterns which Frost was unable to index. These patterns are, however, closely similar to one reported by Hultgren, Gingrich, and Warren (1935)*, who exam-

*It is curious that this concordance was not mentioned by Hultgren et al., who presumably were familiar with the work of Frost, which they cited.

ined some of the preparations of Bridgman. (The details of this work are described below.) It would thus appear that black phosphorus was discovered by Hittorf. There are two discrepancies, though, for Hittorf characterized his preparations as black in reflected light and red in transmitted light, while Frost gave the density as 2.34 to 2.36 g cm^{-3}. It was recently verified by Brown and Rundqvist (1965) that black phosphorus could be obtained by crystallization from liquid bismuth. Their powder pattern matches those of both Frost and Hultgren et al. (The results of their single crystal x-ray study are given below).

The first x-ray study of black phosphorus was that of Linck and Jung (1925), who reported that the structure consists of phosphorus atoms at the lattice points of two interpenetrating face centered rhombohedral lattices, separated by 2.47 Å, having a = 5.97 Å (from kX), α = 60°47', as deduced from 21 observed powder lines. The calculated density is 2.68 g cm^{-3}.

This arrangement, in which each phosphorus atom has one nearest neighbor at 2.47 Å plus six at 3.01 Å, was considered improbable by Hultgren and Warren (1935a, b), who redetermined the structure from new powder data. The details of this work were reported by Hultgren, Gingrich, and Warren (1935). Their powder pattern also consisted of 23 lines but was not the same as that reported by Linck and Jung. The structure is orthorhombic, with a = 3.32, b = 10.52, c = 4.39 Å (from kX), space group Cmca–D_{2h}^{18}, eight atoms in position 8f, (000, $\frac{1}{2}\frac{1}{2}$0) ± (0yz, $\frac{1}{2}$ y $\frac{1}{2}$-z), with y = 0.098 and z = 0.090. The calculated density is 2.68 g cm^{-3}. (The interatomic distances as obtained in a later refinement of the structure are given below.)

This unit cell was confirmed by Thiel (1956), who observed 42 powder lines but did not derive new values for the lattice constants. He noted that his powder pattern was also in disagreement with the earlier one of Linck and Jung. The sample used by Thiel had been prepared by an entirely different method (Krebs, Weitz, and Worms, 1955). Some impurity must have been present in the sample used by Thiel, because five of the observed lines he listed violate an extinction rule for the space group.

The structure was then refined by Brown and Rundqvist (1965) in a single crystal study. The revised values of the lattice constants are a = 3.3136 ± 0.0005, b = 10.478 ± 0.001, c = 4.3763 ± 0.0005 Å, and of the positional parameters the values are y = 0.10168 ± 0.00009, z = 0.08056 ± 0.00028. The calculated density is 2.708 g cm^{-3}.

The structure consists of puckered layers parallel to the ac plane. The projection of two unit cells along the b-axis is presented in Fig. 15-1. Within each layer each phosphorus atom is bonded to three other phosphorus atoms, two at 2.224 ± 0.002 Å and one at 2.244 ± 0.002 Å, average, 2.234 ± 0.014 Å; the bond angles are 96.3° (one) and 102.1° (two). The shortest distance between layers is 3.592 Å, two per atom. A perspective view of portions of three of the layers is presented in Fig. 15-2.

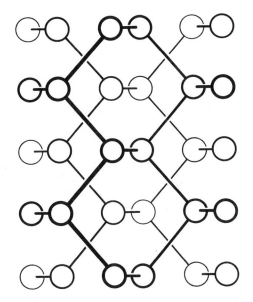

Fig. 15-1. A portion of the structure of black phosphorus showing two of the puckered layers; these layers are parallel to the *ac* plane.

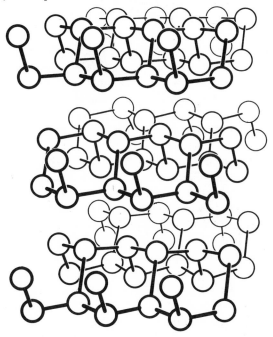

Fig. 15-2. Perspective view of a portion of the black phosphorus structure; *b*-axis vertical, *a*-axis horizontal.

Monoclinic (Hittorf's) Phosphorus

This form of phosphorus is usually named after its first preparator (Hittorf, 1865, 1866), but now that it has been fully characterized structurally this somewhat fusty nomenclature best be abandoned. The improved preparation scheme is rather complicated: 1 g of white phosphorus and 30 g of lead are heated slowly in a sealed tube to 630°C and held for a short time at that temperature. The solution is then cooled at the rate of 10° per day to 520°C, and cooled rapidly to room temperature thereafter. It is next electrolyzed in a solution of 2 kg of lead acetate in 8 liters of 6% acetic acid, and the phosphorus is collected in a watch glass placed under the anode. Nearly square tabular crystals, about 0.2 × 0.2 × 0.05 mm, are obtained in this way (Thurn, 1967).

It is obvious that there is another discrepancy here, for, as noted previously, Frost (1930), following the method of Hittorf, obtained from both lead and bismuth solutions samples, which are now identified as black (orthorhombic) phosphorus, a result confirmed by Brown and Rundqvist (1965), who also carried out a complete single crystal x-ray study. A possible, but unsatisfactory explanation of this situation is that both varieties are formed under the experimental conditions, and that Frost as well as Brown and Rundqvist overlooked the monoclinic polymorph, while Thurn overlooked the orthorhombic polymorph.

Pakulla (1953) found that the crystals were monoclinic, with $a = 9.27$, $b = 9.17$, $c = 22.61$ Å (from kX), $\beta = 106.18°$. She further stated that the space group was either Pc–C_s^2 or P2/c–C_{2h}^4, and the number of atoms in the unit cell was in the range 82.1 to 83.5. (Actually, the above lattice constants, combined with the observed density of 2.30 to 2.34 g cm^{-3}, correspond to a range in Z of 82.6 to 84.0.) The same data were also given by Krebs, Müller, Pakulla, and Zürn (1955).

The structure was later determined by Thurn and Krebs (1969) (see also Thurn and Krebs, 1966, and Thurn, 1967). They found slightly different lattice constants of $a = 9.21$, $b = 9.15$, $c = 22.60$ Å, $\beta = 106.1°$, and established the space group as P2/c–C_{2h}^4 from intensity statistics. A total of 2845 observed reflections was refined to an R of 5.8%. There were no spurious peaks in the final difference map. The unit cell contains 84 atoms, giving a calculated density of 2.361 g cm^{-3}. The atoms lie in 21 sets of position 4g of space group P2/c, $\pm (xyz)(x \; \bar{y} \; \frac{1}{2}+z)$. The positional parameters, as obtained from Fourier and least squares refinements, are presented in Table 15-1.

The structure consists of cagelike P_8 and P_9 groups, which are linked alternately by pairs of phosphorus atoms to form tubes of pentagonal cross section. Parallel tubes form double layers in which tubes in different layers are approximately perpendicular to each other; there are occasional cross-linkages between these two layers. These bonds are arranged so that the double layer consists

Table 15-1.   Positional Parameters in Monoclinic Phosphorus

Atom	x	y	z
1	0.30089	0.20127	0.18147
2	0.17387	0.03262	0.11695
3	0.05014	−0.05231	0.18035
4	−0.07589	−0.21901	0.11634
5	−0.20537	−0.32128	0.17380
6	−0.31537	−0.48468	0.10402
7	−0.43399	−0.55068	0.17224
8	−0.57576	−0.72259	0.11672
9	0.04120	0.39067	0.07245
10	−0.00092	0.15881	0.04497
11	−0.21153	0.13878	0.07346
12	−0.25140	−0.09081	0.04464
13	−0.46426	−0.12736	0.06842
14	−0.49167	−0.35276	0.03304
15	−0.69485	−0.36285	0.06617
16	−0.74959	−0.59445	0.04420
17	0.14600	0.38905	0.17219
18	−0.13962	0.10055	0.17357
19	−0.40394	−0.17616	0.16940
20	−0.58144	−0.35419	0.16732
21	−0.05418	0.32296	0.20060

of two interpenetrating systems of tubes with no chemical bonds between them. There are no chemical bonds between two adjacent double layers. A schematic, general view of the structure is shown in Fig. 15-3. Some of the van der Waals distances between layers, for example, from P(14) at (xyz) to P(14′) at ($1-x$ $1-y$ $\bar{z}$) and P(16′) at ($1-x$ $1-y$ $\bar{z}$) 3.06 Å and 3.22 Å, respectively, and from P(6) at (xyz) to P(21′) at ($1+x$ y z) of 3.28 Å, are surprisingly short in comparison to the interlayer distances of 3.59 Å observed in black phosphorus.

The bond distances and bond angles in the tubes are seen in Figs. 15-4 and 15-5. The bond distances fall into three groups. As pointed out by Thurn and Krebs, there is a correlation between the bond distance and the torsion angle about the bond. In the eclipsed systems—those where the central bond is P(2)-P(10), P(4)-P(12), P(6)-P(14), P(8)-P(16), or P(19)-P(20) (see Fig. 15-4) — the P-P bond distances are significantly longer than the others, and average being 2.278 ± 0.014 Å. In the staggered systems, those where the central bond is P(5)-P(19), P(7)-P(20), P(13)-P(19), or P(15)-P(20), the distances are intermediate, at 2.238 ± 0.002 Å. The remaining systems are all *trans*, in

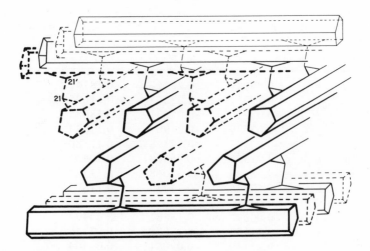

Fig. 15-3.   A schematic representation of the structure of monoclinic phosphorus (after Thurn and Krebs, 1969).

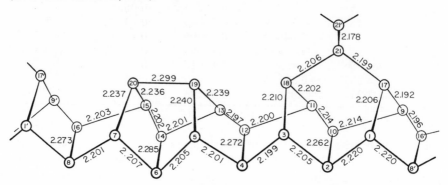

Fig. 15-4.   Bond distances in monoclinic phosphorus (after Thurn & Krebs, 1969).

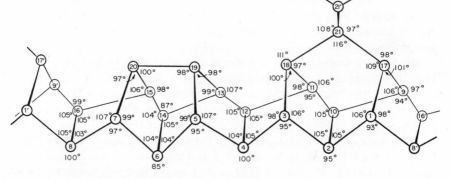

Fig. 15-5.   Bond angles in monoclinic phosphorus.

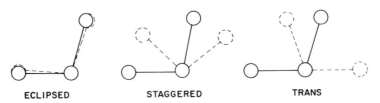

ECLIPSED   STAGGERED   TRANS

Fig. 15-6.   The three kinds of conformation in monoclinic phosphorus, viewed along a
P–P bond.

part, with the shortest P-P bond distances, average 2.203 ± 0.008 Å. Typical
geometries of these three kinds of systems are shown in Fig. 15-6.

Rhombohedral Phosphorus

When black phosphorus at room temperature is subjected to high pressure,
two transitions occur (Jamieson, 1963b). The pressure at the first transition
is about 50 kbar (Bridgman, 1948). According to Jamieson, who observed
22 powder lines of this phase, it has the arsenic structure (see Fig. 33-1, p. 304),
space group $R\overline{3}m$–D_{3d}^{5}, two atoms per unit cell at ± (xxx). The lattice constants at about 100 kbar are a = 3.524 Å, α = 57.25°, and for the corresponding hexagonal cell a = 3.377, c = 8.806 Å. The calculated density is 3.56
g cm^{-3}. The value of x was not determined but the intensities were said to
be consistent with a value near 0.21 to 0.22. If the value 0.215 is assumed
each phosphorus atom has three neighbors within a layer at 2.13 Å, plus three
interlayer contacts of 3.27 Å. The ratio nonbonded:bond distance of 1.54:1
is thus only somewhat smaller than that in black phosphorus of 1.607:1, but
in the latter there are only two nonbonded interlayer contacts instead of three.
The P–P–P bond angle within the layers is 105°, as compared with the average
value of 100.2° in black phosphorus.

It is interesting that transformation of black phosphorus under pressure to
a form with the arsenic structure had been predicted earlier by Parthé (1961).

Metallic Phosphorus

The second transition in black phosphorus at high pressure occurs at 111 ±
9 kbar, as reported by Jamieson (1963b). He observed 12 powder lines which
could be indexed with a primitive cubic lattice having a = 2.377 Å (at about
120 kbar). With one atom per unit cell, the calculated density is 3.83 g cm^{-3}.
The space group is thus Pm3m–O_{h}^{1}, with one atom per unit cell, at (000).
Each phosphorus atom has six nearest neighbors, at 2.377 Å.

As pointed out by Jamieson, the simple cubic structure may be regarded as
a special case of an arsenic-type structure having α = 60° and $x = \frac{1}{4}$. The
transition is thus exceedingly simple, a situation which doubtless explains
its easy reversibility.

White Phosphorus

White (yellow) phosphorus is formed by condensing phosphorus vapor. It is generally agreed that both the gas and the solid consist of P_4 molecules. A transition in the solid at $-77°C$ and atmospheric pressure was discovered by Bridgman (1914), who, from the appearance of the crystals, speculated that the low temperature form is hexagonal, in contrast to the well-known cubic symmetry of the high temperature form. The phase diagram for white phosphorus is shown in Fig. 15-7. The observed density at room temperature is 1.83 g cm^{-3}.

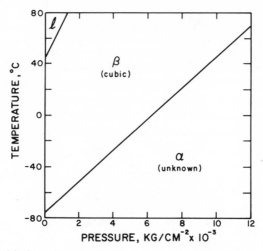

Fig. 15-7.   Partial phase diagram of white phosphorus (after Bridgman, 1914).

α-White Phosphorus (stable below –77°C at 1 atm)

The α form of white phosphorus was first studied by Natta and Passerini (1930), who obtained from a sample at $-170°C$ a powder pattern which they were unable to index. They accordingly concluded that the substance is not cubic, but of low symmetry.

In a later report, Natta and Passerini (1936) tentatively indexed this powder pattern, which consisted of 36 lines, on a hexagonal unit cell with $a = 8.57$, $c = 13.70$ Å (from kX), $c/a = 1.60$. If this unit cell contains nine molecules of P_4, the calculated density is 2.12 g cm^{-3}. Natta and Passerini made no attempt at a structure determination, and even remarked that the symmetry was perhaps only pseudo-hexagonal. (To avoid confusion, it should be pointed out that Natta and Passerini termed the low temperature form of white phosphorus β.)

Sugawara, Sakamoto, and Kanda (1949) reported photographing the α form at $-100°C$ (the $100°C$ in their paper is an obvious misprint), but they were unable to obtain a good diffraction pattern.

Sugawara and Kanda (1949) took photographs with the sample at $-83°$ and below but were unable to index them as cubic, hexagonal, or tetragonal. They concluded "α-yellow phosphorus will be orthorhombic or monoclinic, with a large unit cell". They did not publish their observed spacings.

The structure of α-white phosphorus thus remains unknown.

β-White Phosphorus (stable $-77°C$ to melting point, at atmospheric pressure)

The first x-ray study of β-white phosphorus is that of Jung (1926b), who stated that it changes to the red form upon x-irradiation (CuK_α). He stated that the observed density corresponds to a lattice constant of 3.83 Å if the cubic unit cell contains two atoms, and to a lattice constant of 4.82 Å if it contains four atoms. Later developments (described below) strikingly bring out the danger of proposing a partial structure solely on the basis of an observed density.

Natta and Passerini (1930), on the other hand, obtained a powder photograph of 22 lines (iron radiation) from a sample at $-35°C$, which they indexed as cubic, with $a = 7.18$ Å (from kX). They proposed that the unit cell contains four molecules of P_4, giving a calculated density of 2.22 g cm^{-3}.

In their later report, Natta and Passerini (1936) presented a table of the spacings they observed with the sample at $-55°C$, with chromium (12 lines), iron (11 lines), and copper (8 lines) radiation. They stated that no special precautions had been made in their previous (1930) work to prevent the formation of ice, were unable to confirm the earlier unit cell reported by them, and added that the unit cell must be very large and the lattice very complex.

Sugawara, Sakamoto, and Kanda (1949) obtained powder photographs showing "only a few rings" when the samples were held at 10 and $-40°C$. From Laue and oscillation photographs of single crystals at an unstated temperature they determined that β-white phosphorus is cubic, with a lattice constant of 18.8 Å. Taking the value 1.82 g cm^{-3} for the density, they calculated that there are about 236 atoms per unit cell. From the absences observed on the oscillation photographs they stated that the space group is either $I432\text{-}O^5$, $I\bar{4}3m\text{-}T_d^3$, or $Im3m\text{-}O_h^9$, but no structure was proposed.

Sugawara and Kanda (1949), in continuing the above work, noted that the existence of P_4 molecules allowed only 232, 236, or 240 atoms per unit cell, and that 59 molecules (236 atoms) is not possible in these space groups. The number of molecules per unit cell must thus be either 58 or 60. They stated:

. . .58 and 60 P_4 molecules were placed on every possible sets of equivalent positions belonging to T_d^3, O^5 or O_h^9 and corresponding scattering intensities F^2 were calculated and compared with observations. The only case which was in fair agreement with the observations and satisfied the van der Walls radious of about 3.8 Å for P atom was as follows:

Space group T_d^3

Types of molecules	Positions of molecular centers		Positions of atoms	
P_4 I	2(a)		8(c)	making
P_4 II	8(c)	x_{II} = 0.32	8(c) + 24(g)	tetrahedrons
P_4 III	24(g)	x_{III} = 0.36, z_{III} = 0.04	24(g) + 24(g) + 48(h)	around the positions of
P_4 IV	24(g)	x_{IV} = 0.09, z_{IV} = 0.28	24(g) + 24(g) + 48(h)	the centers

The agreement is given in a figure, an examination of which reveals that the basis of this complicated, many-parameter structure rests on only 12 observed and 4 unobserved reflections! Furthermore, since they were unable to fix the positions of many of the atoms, it is not clear how they calculated the intensities.

It is remarkable that the positions of the molecular centers given by Sugawara and Kanda are exactly those (rounded off to two decimal places) given by Bradley and Thewlis (1927) for the positions of the atoms in α-manganese, but this rather astonishing coincidence was not mentioned by them.

Contrary to the experience of Natta and Passerini (1930, 1936), but in agreement with that of Jung (1926b), Sugawara and Kanda found that white phosphorus was transformed into the red modification when irradiated by x-rays (CuK_α).

Corbridge and Lowe (1952) obtained single crystal photographs of β-white phosphorus at –30 to –35°C. These indicated a body centered cubic lattice with a = 18.51 ± 0.03 Å, and space group either $I\bar{4}3m$-T_d^3, $I432$-O^5, or $Im3m$-O_h^9, in agreement with the results of Sugawara et al. However, the measured density at –25°C of 1.84 g cm^{-3} yields 56 P_4 molecules per unit cell instead of 58 (calculated value 56.7). Corbridge and Lowe made preliminary structural studies which indicated that 56 molecules of P_4 could not be accommodated in the unit cell without appreciable overlap if the van der Waals radius of the phosphorus atom is assigned its usual value of 1.9 Å. This result differs from that described by Sugawara and Kanda. Corbridge and Lowe eliminated $Im3m$, on unstated general crystallographic grounds, and thought $I\bar{4}3m$ a more likely choice than $I432$.

The structure of β-white phosphorus is thus also unknown, but it is probable that the space group is I$\bar{4}$3m with lattice constant 18.51 Å at $\sim$ -30°C, but the number of molecules per unit cell remains to be established, as do the positions of the individual atoms.

Red Phosphorus (and its progeny)

Red phosphorus is obtained by heating white phosphorus in an inert atmosphere. It may also be obtained from the white variety by the action of light. The color, density, melting point, and vapor pressure depend on the method of preparation. In particular, densities of 2.0 to 2.4 g cm^{-3} have been reported. Commercial red phosphorus is amorphous. X-ray patterns consisting of discrete lines are obtained only if the samples have been subjected to heat, pressure, or various other treatments. The following discussion is chronological.

1. Olshausen (1925) reported a powder pattern of six lines from what he called red phosphorus, but the method of preparation was not given. This pattern was indexed on cubic unit cell having a = 7.346 Å (from kX) and containing 16 atoms, giving a calculated density of 2.075 g cm^{-3}.

2. Jung (1926b) reported a powder pattern of 10 lines from what *he* called red phosphorus, the sample having been obtained from a solution in molten lead. This pattern is not the same as that of Olshausen and could not be indexed as characteristic of a cubic substance. Jung referred to his preparation as Hittorf's phosphorus, but his diffraction pattern does not correspond to what would be expected from monoclinic phosphorus.

3. Frost (1930) also prepared, in addition to what we now know was black phosphorus, a different form which was obtained by slow cooling of the melt from 600 to 580°C, followed by annealing at this temperature for 6 hours. The powder pattern of this preparation is quite different from those reported by both Olshausen and Jung.

4. Hultgren, Gingrich, and Warren (1935), in addition to carrying out the first determination of the structure of black phosphorus (see above), also examined red phosphorus from several different sources. Commercial red phosphorus showed only diffuse rings in powder patterns and was therefore termed "amorphous". On the other hand, a sample of red phosphorus which had been subjected to 12 katm and samples of black phosphorus which had been heated to 550°C in evacuated quartz [sic] tubes gave identical x-ray lines which they said checked those of Frost (1930) but Hultgren et al. neglected to state which of the patterns of Frost their own patterns matched. Radial distribution functions calculated from both the amorphous and crystalline samples were closely similar, and it was concluded that the amorphous specimen was simply extremely fine crystals. These functions indicated three closest neighbors at 2.28 Å and about twelve at 3.6 Å.

5. Hultgren et al. also examined a sample prepared by Birch (1935). It had been obtained by treatment of red phosphorus at 300°C and 8 katm. This sample was deep red, and, unlike black phosphorus, showed no flakiness. Its x-ray diffraction pattern was said to be different from any of the other forms of phosphorus. Unfortunately, they did not have the time to investigate it further.

6. Klein (1947) examined red phosphorus which had been heated to 600°C. The powder pattern consisted of 26 lines, which were indexed on a cubic unit cell having a = 11.31 Å (from kX) and containing 66 atoms, giving a calculated density of 2.35 g cm^{-3}, as compared with the observed value of 2.38 g cm^{-3}. There were no systematic absences, so Klein concluded that the space group was one of the five (uncentered) in the cubic system which contain neither glide planes nor screw axes, but no further speculations were made.

7. Roth, DeWitt, and Smith (1947) made extensive investigations of red phosphorus which had been subjected to various types of heat treatment, but the results of their work only compounded the confusion rather than contributing to its resolution. Samples prepared by heating white phosphorus to 280°C, or by heating it with a tungsten wire at 900°C, behaved similarly. Heating curves obtained from these samples gave the following results:

$$\text{I} \xrightarrow{460°C} \text{II} \xrightarrow{520°C} \text{III} \xrightarrow{540°C} \text{IV} \xrightarrow{> 550°C} \text{V}$$

The first three of these transitions were found to be exothermic, the fourth was not detected by thermal analysis. The results of x-ray and micrographic analyses of these five forms are presented in Table 15-2. The five x-ray diffractions are all different. Roth et al. stated that their pattern of form V corresponded to those Frost obtained from phosphorus crystallized from lead or

Table 15-2.   Summary of the Results of Roth, DeWitt and Smith (1947) on Red Phosphorus

Form	X-Ray Diffraction Pattern[a]	Photomicrographic Results
I	1 diffuse ring	Amorphous
II	7 diffuse rings	Hexagonal?
III	7 diffuse rings	Hexagonal?
IV	13 sharp lines	Tetragonal?
V	34 sharp lines	Triclinic?

[a]None of these patterns was indexed.

bismuth. This conclusion is difficult to reconcile with the fact that there are very few points of correspondence between the two sets of data. In his summary van Wazer (1966) equates form IV of Roth et al. with the foregoing preparations of Frost. This observation is equally difficult to reconcile with the great differences between the two respective diffraction patterns.

8. Rice, Potocki, and Gosselin (1953) prepared yet another form of phosphorus by vaporizing white phosphorus at $1000°C$ and condensing at liquid nitrogen temperature. A dark brown deposit was obtained under these conditions. This deposit was observed to change irreversibly to a mixture of red and white phosphorus on heating. No diffraction data were taken.

The summary of the incomplete results on the structural chemistry of elemental phosphorus is presented in Table 15-3.

Table 15-3.   Unsolved Problems of Phosphorus Allotropy[a]

Reference	Remarks
Olshausen, 1925	Red P, method of preparation not stated, gave x-ray pattern indicating a cubic structure with $a = 7.346$ Å and 16 atoms per unit cell
Linck & Jung, 1925	Black P, powder pattern and structure different from those usually attributed to black phosphorus
Jung, 1926b	Red P, from solutions in lead
Frost, 1930	Red P, from annealing a slowly cooled melt
Hultgren et al., 1935	Red P, heated to $300°C$ at 8 katm
Klein, 1947	Red P, heated to $600°C$; x-ray pattern indexed as cubic, $a = 11.31$ Å, 66 atoms per unit cell
Roth et al., 1947	Red P, untreated; amorphous
Roth et al., 1947	Red P, heated to $460°C$; hexagonal?
Roth et al., 1947	Red P, heated to $520°C$; hexagonal?
Roth et al., 1947	Red P, heated to $540°C$; tetragonal?
Roth et al., 1947	Red P, heated above $550°C$; triclinic?
Rice et al., 1953	Condensing vapor at $1000°C$ at $78°K$; no diffraction data; product brown, unstable when heated

[a]All of these have different x-ray diffraction patterns, which are different from those of other well-characterized varieties of phosphorus.

ARSENIC

There are six named forms of arsenic: α (gray, metallic, rhombohedral, ordinary), β, γ, δ, ϵ (orthorhombic, arsenolamprite), and yellow (cubic). The β, γ, and δ varieties are amorphous and will not be treated in detail here. Their methods of preparation and properties have been discussed by Stöhr (1939), Krebs and Schultze-Gebhardt (1956), and Krebs, Holz, Lippert, and Worms (1956, 1957); they all convert to α-arsenic on heating.

α-Arsenic

The structure of α-arsenic was first determined by Bradley (1924a). His and subsequent determinations of the lattice constants are presented in Table 33-1. The space group is $R\bar{3}m-D_{3d}^5$. The structure may be described in various ways: (1) primitive rhombohedral unit cell with α somewhat less than 60°, two atoms at $\pm (xxx)$, with x a little less than $\frac{1}{4}$; (2) hexagonal unit cell, six atoms at $(000\ \frac{2}{3}\frac{1}{3}\frac{1}{3}\ \frac{1}{3}\frac{2}{3}\frac{2}{3}) \pm (00z)$ with z having the same value as x, above; and (3) face centered rhombohedral unit cell with α somewhat less than 90°, eight atoms at $(000\ 0\frac{1}{2}\frac{1}{2}\ \frac{1}{2}0\frac{1}{2}\ \frac{1}{2}\frac{1}{2}0) \pm (\frac{x}{2}\frac{x}{2}\frac{x}{2})$, x as above. There have been three different determinations of the value of the x-parameter at room temperature: Bradley (1924a) obtained 0.226, Olshausen (1925), 0.2325, and Schiferl and Barrett (1969), 0.22707; the last value is the most accurate and is used here.

The structure consists of puckered sheets of covalently bonded arsenic atoms stacked in layers perpendicular to the hexagonal c axis (Fig. 33-1). Each arsenic atom has three ligands in the same layer at 2.517 Å and three next nearest neighbors in an adjacent layer at 3.120 Å. The As–As–As bond angles in the layers are 96°39'.

The relationship between the two different rhombohedral unit cells is illustrated in Fig. 33-2.

Lattice constants and values of the positional parameter at low temperatures have been reported by Schiferl and Barrett (1969), and lattice constants at high temperatures have been reported by Taylor et al. (1965). These results are summarized in Fig. 33-3 and in Table 33-2. It is interesting that the length of the hexagonal a-axis remains constant over the entire temperature range of 4.2 to 677°K. Furthermore, the variations of x and the length of the hexagonal c-axis with temperature combine to give the result that only the interlayer nonbonded distances increase with temperature (from 3.095 to 3.166 Å in the above range), both the bond distances and bond angles within the layers remaining constant.

Table 33-1. The Lattice Constants of α-Arsenic at 25°C[a]

	Rb		Hc		Fd		Reference
	a, Å	α, °	a, Å	c, Å	a, Å	α, °	
	4.150[e]	54.12[f]	3.776	10.594	5.611	84.59	Bradley, 1924a
	4.159[e]	53.72[f]	3.758	10.645	5.605	84.20	Olshausen, 1925
	4.159[e]	53.82[f]	3.764	10.638	5.610	84.30	Jung, 1926b
	4.143[e]	54.13[f]	3.770	10.575	5.601	84.61	Willott & Evans, 1934
	4.131[e]	54.10[f]	3.757	10.547	5.584	84.58	Hägg & Hybinette, 1935
	4.1318[e]	54.08[f]	3.7564	10.551	5.5842	84.55	Trzebiatowski & Bryjak, 1938
	4.131[e]	54.17[f]	3.762	10.542	5.587	84.67	Stöhr, 1939
	4.1322	54.12	3.760	10.549	5.5865	84.59	Swanson, Fuyat, & Ugrinic, 1954
	4.1318	54.13	3.7598	10.547	5.5867	84.61	Taylor, Bennett, & Heyding, 1965
	4.1319	54.12	3.7598	10.5475	5.5860	84.59	Schieferl & Barrett, 1969
av.	4.1320	54.12	3.7599	10.5478	5.5864	84.60	
	±0.0002	±0.01	±0.0001	±0.0010	±0.0004	±0.01	

[a] Average temperature of the last three determinations.
[b] Primitive rhombohedral unit cell.
[c] Hexagonal cell.
[d] Face centered rhombohedral cell.
[e] From kX, omitted from average (for all three cells).
[f] Omitted from average (for all three cells).

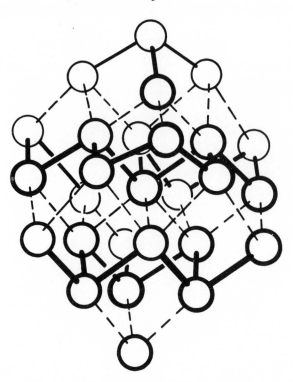

Fig. 33-1.   One unit cell, face centered rhombohedral description, of α-arsenic. In the corresponding hexagonal description, the c-axis is vertical. Covalent intralayer bonds shown by solid lines and interlayer contacts as dashed lines.

ε-Arsenic

Jung (1926b) examined powder samples of the mineral arsenolamprite from Copiapó, Chile, and observed 32 lines, only 14 of which corresponded to those of α-arsenic. He suggested that another modification of arsenic, isostructural with metallic (orthorhombic) phosphorus, was present in the sample. The pattern of another sample, from Marienburg, Saxony, was similar to the first. This mineral has sometimes been classed as an allotropic form of arsenic (e.g., by Palache, Berman, and Frondel, 1944).

Krebs, Holz, Lippert, and Worms (1956, 1957) prepared ε-arsenic by heating one of the amorphous varieties at 125 to 175°C in the presence of mercury. They stated that it is isostructural with black phosphorus, with lattice constants $a = 3.62$, $b = 4.48$, $c = 10.85$ Å.

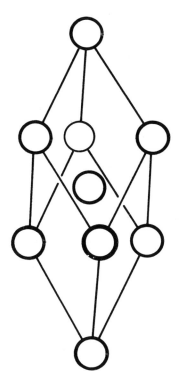

Fig. 33-2.   Some of the atoms of Fig. 33-1, showing the primitive rhombohedral cell of α-arsenic.

Johan (1959) examined samples of arsenolamprite from Černy Důl, Czecho-slovakia. He also stated that it is isostructural with black phosphorus, with $a = 3.63$, $b = 4.45$, $c = 10.96$ Å, values close to those of Krebs et al.

The details of the structure of this allotrope have yet to be worked out. Indeed, additional chemical analyses are needed to establish whether the materials studied are in fact pure arsenic.

Yellow Arsenic

Yellow arsenic is formed as a sublimation product. It is cubic, and presumably consists of As_4 molecules, but structural data have not been obtained because it is unstable and decomposes in the x-ray beam (Jung, 1926b).

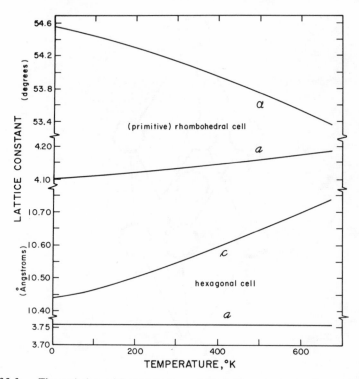

Fig. 33-3.   The variation with temperature of the lattice constants of α-arsenic (calculated from the data of Taylor, Bennett, & Heyding, 1965, and Schiferl & Barrett, 1969).

Table 33-2.   Data for α-Arsenic at Various Temperatures

	4.2°K	78°K	298°K	677°K
a, Å	4.1018	4.1063	4.1321	4.1860
α, °	54.55	54.48	54.13	53.37
x	0.2276	0.2275	0.2271	0.2260[a]
Bond distance, Å	2.517	2.516	2.516	2.517
Bond angle, °	96.73	96.73	96.73	96.73
Interlayer distance, Å	3.095	3.099	3.120	3.166
Atomic volume, Å^3	21.30	21.33	21.52	21.91

[a]Extrapolated value.

ANTIMONY

Six forms of antimony have been named: α (metallic, ordinary), yellow, black, explosive, and two high pressure forms, II and III. The yellow form was prepared by Stock and Guttmann (1906); it is unstable above -90°C and has not been investigated structurally. The black form, obtained from the yellow form or by cooling gaseous antimony, reacts rapidly with air and has also not been the subject of structural investigation; on heating it converts to the α form. Explosive antimony, prepared electrolytically, was said by Kersten (1932) to be amorphous, while Krebs, Schultz-Gebhardt, and Thees (1955) characterized it as a mixed polymer; Kersten found that the powder pattern of the explosion product was identical with that of α-antimony.

α-Antimony

The structure of α-antimony was first determined by James and Tunstall (1920). Their and subsequent determinations of the lattice constants are presented in Table 51-1. It is isostructural with α-arsenic, space group $R\bar{3}m-D_{3d}^5$, two atoms per unit cell at $\pm (xxx)$. (For description of the alternate face centered rhombohedral and hexagonal unit cells see α-arsenic, above.) The value of the positional parameter obtained in 1920 by James and Tunstall of 0.232 is very close to the more accurate value of 0.2335 obtained recently by Barrett, Cucka, and Haefner (1963).

A portion of this layer structure is shown in Fig. 51-1. Each antimony atom has three intralayer ligands at 2.908 Å, and the Sb–Sb–Sb bond angle is 95°35'. The shortest interatomic distance between layers, three per antimony atom, is 3.355 Å. The atomic volume is 30.201 Å^3.

α-Antimony has been studied down to 4.2°K by Barrett et al. (1963). Some data obtained from this work are presented in Table 51-2.

Antimony at High Pressure

Kabalkina and Mylov (1964) found that at 50 kbar antimony had a simple cubic structure, one atom in a unit cell having a = 2.966 ± 0.010 Å (from kX). The corresponding atomic volume is 26.09 ± 0.27 Å^3. This structure is obtained from the normal rhombohedral structure by an increase of the angle α from 57.1 to 60° together with small shifts in the atomic positions so that each atom has six equidistant neighbors instead of three short and three longer. (The equivalent description of this change in structure in terms of the face centered rhombohedral unit cell is an increase of the angle α from 87.4 to 90°, plus the atomic shifts, as may easily be visualized by referring to Fig. 51-1; in the hexagonal cell the axial ratio becomes equal to $\sqrt{6}$.) Kabalkina and

Table 51-1. The Lattice Constants of Antimony at Room Temperature

R[a]		H[b]		F[c]		Reference
a, Å	α, °	a, Å	c, Å	a, Å	α, °	
4.508[d]	56.62[d]	4.275	11.316	6.213	86.97	James & Tunstall, 1920
4.511[d]	57.08[d]	4.311	11.287	6.239	87.40	Persson & Westgren, 1928
4.504[d]	57.08[d]	4.304	11.270	6.230	87.40	Jette & Gebert, 1933
4.505[d]	57.20[d]	4.313	11.263	6.237	87.51	Hägg & Hybinette, 1935
4.5067[e]	57.11	4.3083	11.2743	6.2346	87.42	Jette & Foote, 1935
4.50	57.1	4.29	11.26	6.23	87.3	Dorn & Glockler, 1937
4.5069[e]	57.12	4.3091	11.2741	6.2354	87.43	Trzebiatowski & Bryjak, 1938
4.5063[e]	57.11	4.3079	11.2732	6.2344	87.42	Lu & Chang, 1941
4.5060	57.10	4.307	11.273	6.2333	87.25	Swanson, Fuyat, & Ugrinic, 1954
4.507[d]	57.06[d]	4.305	11.28	6.233	87.37	Tatarinova, 1955
4.5067	57.11	4.3084	11.274	6.2348	87.42	Barrett, Cucka, & Haefner, 1963
4.504[d]	57.12[d]	4.306	11.267	6.231	87.43	Vershchagin & Kabalkina, 1964
av. 4.5065	57.11	4.3081	11.2737	6.2345	87.42	
±0.0004	±0.01	±0.0008	±0.0006	±0.0008	±0.01	

[a] Primitive rhombohedral cell.
[b] Hexagonal cell.
[c] Face centered rhombohedral cell.
[d] From kX, omitted from the average (for all three cells).
[e] From kX.

309

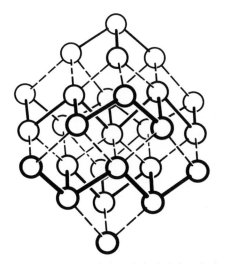

Fig. 51-1. One unit cell, face centered rhombohedral description, of α-antimony. Covalent intralayer bonds shown as solid lines, interlayer contacts as dashed lines. (Cf. Figs. 33-1 and 33-2, α-arsenic.)

Table 51-2. Data for α-Antimony at Various Temperatures

	4.2°K	78°K	298°K
a, Å	4.4898	4.4927	4.5067
α, °	57.23	57.20	57.11
x	0.2336	0.2336	0.2335
Bond distance, Å	2.902	2.903	2.908
Bond angle, °	95.63	95.60	95.60
Interlayer distance, Å	3.343	3.345	3.355
Atomic volume, Å³	29.96	29.99	30.21

Mylov were not certain whether this change was abrupt at some undetermined minimum pressure or continuous with pressure increase.

At 90 kbar they found that antimony has the hexagonal close packed structure, with a = 3.34, c = 5.28 Å (from kX), c/a = 1.58. The corresponding atomic volume is 25.5 Å³.

This work was extended by Vershchagin and Kabalkina (1964), who took powder photographs of antimony at various pressures up to 46 kbar, at 50 kbar, and at 90 kbar. Their data may be used to calculate the atomic volume at various pressures, as in Fig. 51-2, where it is seen that the structure change occurs with zero volume change, suggesting that the change is continuous. On

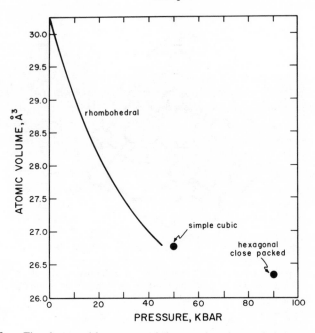

Fig. 51-2.　The change with pressure of the atomic volume of antimony (calculated from the data of Vershchagin & Kabalkina, 1964).

the other hand, when similar calculations are made giving the rhombohedral angle α as a function of pressure, as in Fig. 51-3, the change is apparently abrupt. The lattice constants reported by Vershchagin and Kabalkina are: at 50 kbar (simple cubic) $a = 2.992 \pm 0.009$ Å (from kX), and at 90 kbar (hexagonal close packed) $a = 3.376$, $c = 5.341$ Å (from kX), $c/a = 1.58$.

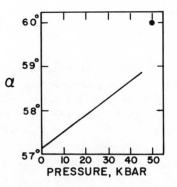

Fig. 51-3.　The change with pressure of the rhombohedral angle α for antimony. At $\alpha = 60°$, the structure is simple cubic. (Calculated with the data of Vershchagin & Kabalkina, 1964).

BISMUTH

α-Bismuth

The structure of bismuth was first correctly determined by Ogg (1921). The determinations of the lattice constants are presented in Table 83-1. Bismuth is isostructural with α-arsenic and α-antimony, space group $R\bar{3}m–D_{3d}^5$, two atoms per unit cell at ± (xxx). (For alternate descriptions of this structure as face centered rhombohedral or hexagonal, see α-arsenic, above.) The Bi–Bi bond distance of 2.92 kX given by Ogg corresponds to the value 0.223 for x. Values ranging from 0.224 to 0.237 have been reported by some of the investigators listed in Table 83-1, but the most recent and most accurate is the value 0.2339 reported by Cucka and Barrett (1962).

A portion of the structure is seen in Fig. 83-1. Each bismuth atom has three intralayer ligands at 3.072 Å, and the bond angle is $95°27'$. The shortest interatomic distance between layers, three per bismuth atom, is 3.529 Å.

There is no change in structure, at atmospheric pressure, up to 265°C (Goetz and Hergenrother, 1932b) nor down to $4.2°$K (Barrett, 1959, 1960; Cucka and Barrett, 1962). Some data obtained from the low temperature work are presented in Table 83-2.

Bismuth at High Pressure

Work on the behavior of bismuth at high pressure at room temperature and above includes that of Bridgman (1935, 1941, 1942), Butuzov and Ponyatovski (1956), Bundy (1958), Ponyatovski (1960), Klement, Jayaraman and Kennedy (1963c), and Kennedy and Newton (1963). The phase diagram, after Bundy, is presented in Fig. 83-2. There is disagreement concerning whether there is a triple point liquid-β-γ or liquid-β-η. The nomenclature of the various phases has not been consistent, as shown in Table 83-3.

Low temperature studies at high pressure were carried out by Ilina and Itskevich (1967), who reported the appearance of a new phase below 0°C, as shown in Fig. 83-3. However, Compy (1970), who investigated the same region as did Ilina and Itskevich, did not observe the new phase reported by them.

β-Bismuth

Jamieson, Lawson, and Nachtrieb (1959) published a powder pattern consisting of about eight lines which they obtained from a sample of bismuth at about 27 kbar; they did not index this pattern.

Table 83-1. The Lattice Constants of α-Bismuth at 25°C

R[a] a, Å	R[a] α, °	H[b] a, Å	H[b] c, Å	F[c] a, Å	F[c] α, °	Reference
4.72[d]	57.27[d]	4.52	11.79	6.53	87.57[e]	Ogg, 1921
4.75[d]	57.25[d,e]	4.55	$\frac{1}{2}\times$11.84	6.58	87.56	Kahler, 1921
4.74[d]	57.27[d,e]	4.54	11.84	6.57	87.57[e]	James, 1921
4.736[d]	57.27[d,e]	4.539	11.83	6.559	87.57[e]	McKeehan, 1923c
4.743[d]	57.30[d]	4.548	11.85	6.571	87.60	Davey, 1924, 1925
4.75[d]	57.27[d,e]	4.55	11.87	6.58	87.57[e]	Hassel & Mark, 1924a
4.759[d]	57.27[d,e]	4.561	11.892	6.592	87.57[e]	Ehret & Fine, 1930
4.736[d]	57.28	4.540	11.832	6.561	87.58	Parravano & Caglioti, 1930
4.731[d]	57.27[d,e]	4.534	11.82	6.553	87.57[e]	Solomon & Morris-Jones, 1931
4.745[d]	57.22[d]	4.5445	11.860	6.569	87.53	Jette & Foote, 1932
–	–	–	11.860	–	–	Goetz & Hergenrother, 1932a
4.745[d]	57.23[d]	4.544	11.860	6.570	87.52	Jette & Gebert, 1933
4.7460[f]	57.24	4.5464	11.8620	6.5723	87.53	Jette & Foote, 1935
4.7459[f]	57.23	4.5459	11.8623	6.5718	87.53	Ievinš, Straumanis, & Karlsons, 1938; Straumanis, 1949
4.7458[f]	57.23	4.5459	11.8620	6.5718	87.53	Chiswik & Hultgren, 1940
4.745[g]	57.15	4.546	11.860	6.571	87.55	Swanson, Fuyat & Ugrinic, 1954
4.7461	57.23	4.5461	11.8629	6.5721	87.53	Cucka & Barrett, 1962
av. 4.7460	57.23	4.5461	11.8623	6.5720	87.53	
±0.0001	±0.01	±0.0002	±0.0004	±0.0002	±0.01	

a Primitive rhombohedral cell.
b Hexagonal cell.
c Face centered rhombohedral cell.
d From kX, omitted from averages.

e Goniometric value.
f From kX.
g Omitted from average.

Fig. 83-1.   One unit cell, face centered rhombohedral description, of α-bismuth. Co-valent intralayer bonds shown as solid lines, interlayer contacts as dashed lines. (Cf. Figs. 33-1 and 33-2, α-arsenic, and Fig. 51-1, α-antimony.)

Table 83-2.   Data for α-Bismuth at Various Temperatures

	4.2°K[a]	78°K	298°K
a, Å	4.723	4.729	4.746
α, °	57.33	57.30	57.23
x	0.2341	0.2340	0.2339
Bond distance, Å	3.061	3.064	3.072
Bond angle, °	95.48	95.47	95.45
Interlayer distance, Å	3.511	3.516	3.529
Atomic volume, Å³	34.96	35.07	35.39

[a]Value of the length of the hexagonal a-axis used in these calcu-lations obtained by extrapolation.

Jaggi (1964a) said that the foregoing pattern corresponded to a simple cubic structure with a = 3.177 ± 0.009 Å. No details were given. This struc-ture is the same as that of the first high pressure form of antimony, and was later said to be verified on theoretical grounds by Jaggi (1964b).

Meanwhile, Piermarini and Weir (1962) tabulated eight spacings of a pat-tern observed from a sample of bismuth at about 28 kbar. It is quite different from that of Jamieson et al. Piermarini and Weir did not index their pattern, and not only strongly suspected that there may have been some of the normal low pressure phase still present but also did not disallow the possibility that two of the high pressure phases were present.

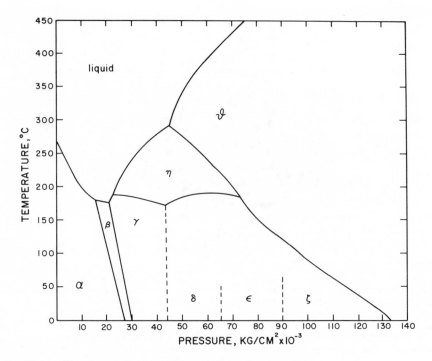

Fig. 83-2.   The phase behavior of bismuth (after Bundy, 1958).

Table 83-3.   **Designation of the Phases of High Pressure Forms of Bismuth by Various Authors**

Bridgman (1935, 1942)	Bundy (1958)	Ponyatovski (1960)	Kennedy & Newton (1963)	Klement et al. (1963c)	Fig. 83-2
I	I	α	I	I	α
II	II	β	II	II	β
III	III	γ	III	III	γ
IV	IV				δ
V	V				ϵ
VI	VI				ζ
	VII	δ	VII	IV	η
	VIII		VI	V	θ

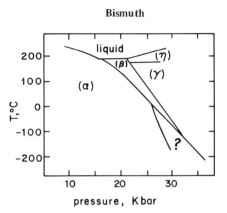

Fig. 83-3. A portion of the phase diagram of bismuth (according to Ilina and Itskevich, 1967). See text.

Brugger, Bennion, and Worlton (1967), on the other hand, published a powder pattern of 40 lines obtained from bismuth at 26 kbar. It is also quite different from that of Jamieson et al. as well as from that of Piermarini and Weir. Brugger et al. indexed their pattern on the basis of a monoclinic unit cell containing four atoms with $a = 6.674$, $b = 6.117$, $c = 3.304$ Å, $\beta = 110.33°$; they further stated that the space group was $C2/m\text{-}C_{2h}^3$, the four atoms lying in position $4i$, (000), $(\frac{1}{2}\frac{1}{2}0) \pm (x0z)$, with $x = \frac{1}{4}$ and $z = \frac{1}{8}$. This gives a structure in which the bismuth atoms form puckered layers parallel to the ab plane. One of these layers is shown in Fig. 83-4. Each bismuth atom is bonded to three others within a layer, at 3.147 Å (one) and 3.168 Å (two). These are somewhat longer than the three bonds of 3.072 Å formed by α-bismuth. However, the bonds between layers are shorter and more numerous: 3.304 Å (two), 3.396 Å (one), and 3.559 Å (one), as compared with 3.529 Å (three) in the α phase. The atomic volume is of course smaller, 31.62 Å³, or 10.7% less than the value for α-bismuth.

Further work is needed, however, to resolve the differences between the two published powder patterns and to establish the structure of β-bismuth.

ζ-Bismuth

In a preliminary note Schaufelberger, Merx, and Contré (1972) reported that they had found the "highest pressure modification of bismuth. . .discovered by Bridgman (1942)" to have a body centered cubic structure, with two atoms per unit cell and a lattice constant of 3.800 Å, at 90 kbar and room temperature. This gives 3.291 Å for the (eight) nearest neighbor distance, and an atomic volume of 27.44 Å³.

(The nomenclatural confusion continues: Schaufelberger et al. refer to this

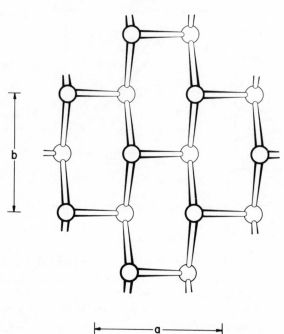

Fig. 83-4.   Part of one "layer" of the β-bismuth structure proposed by Brugger et al., (1967).

phase as "bismuth V", although Bridgman (1942) unambiguously terms the form stable above 90,000 kg cm^{-2} "bismuth VI". The pressure of 90 kbar, or 91,800 kg cm^{-2}, is above the transition $\epsilon \rightarrow \zeta$ (V→ VI), as may be seen in Fig. 83-2.)

Chapter 9
Group VI

There are four different forms of crystalline oxygen (O_2). Three are related by the following transitions:

$$\alpha \xrightarrow{\ 23.9^\circ K\ } \beta \xrightarrow{\ 43.6^\circ K\ } \gamma \xrightarrow{\ 54.4^\circ K\ } \text{liquid}$$

The fourth form, termed α', occurs in thin films deposited from the gas phase at a low flow rate (Hörl, 1969); its diffraction pattern is quite different, according to Hörl, from that of the α form, which is obtained by cooling the β form.

Much of the early work on solid oxygen was carried out at the temperature of boiling hydrogen (20.4°K), and conflicting results were reported. The opinion has been expressed that in all of this work mixtures of α- and β-oxygen were examined, and not the pure α phase (Hörl, 1962; Alikhanov, 1964a).

α-Oxygen

De Smedt and Keesom (1925a,b) concluded that oxygen at 20.4°K was neither cubic, tetragonal, hexagonal, nor trigonal, on the basis of four powder lines. They did not, however, publish the values of the observed spacings.

McLennan and Wilhelm (1927) reported that at 21°K oxygen is body centered orthorhombic with $a = 5.51$, $b = 3.83$, $c = 3.45$ Å (from kX), two molecules per unit cell. The corresponding molecular volume is 36.4 Å^3. However, neither Ruhemann (1932) nor Vegard (1935b) could explain the observed data with this unit cell, and Hörl (1962) said that the d-values of McLennan and Wilhelm agree partly with his α-pattern and partly with his β-pattern.

317

Mooy (1932) reported that at 20.4°K oxygen is hexagonal, with a = 5.76, c = 7.61 Å (from kX), c/a = 1.32, six molecules per unit cell, molecular volume, 36.4 Å³. Ten powder lines were observed. Hörl (1962) remarked that this pattern is very similar to his β-pattern except for one line which corresponds to his α-pattern.

Alikhanov (1963) indexed the powder pattern on the basis of a monoclinic unit cell having a = 4.28, b = 3.45, c = 5.08 Å, $β$ = 110°, but further details were not given. The volume of this unit cell is 70.5 Å³, and thus contains two molecules.

The monoclinic unit cell of Alikhanov was confirmed by Barrett, Meyer, and Wasserman (1967b), who found a = 5.403, b = 3.429, c = 5.086 Å, $β$ = 132.53°, at 23°K, two molecules per cell, molecular volume 34.72 Å³. (These result from a different choice in the a-axis than was made by Alikhanov, whose values convert to a = 5.41 Å, $β$ = 132°, b and c unchanged, in good agreement with those of Barrett et al.) A total of 26 lines was observed, six of which are composite. Barrett et al. deduced that the structure is based on space group $C2/m$–C_{2h}^3, with four oxygen atoms per unit cell at $(000 \frac{1}{2}\frac{1}{2}0)$ ± $x0z$, x = 0.089 ± 0.010, z = 0.153 ± 0.015.

The arrangement of the molecules, as viewed down the b-axis, is presented in Fig. 8-1. Each oxygen atom has six nearest neighbors in adjacent molecules, four at 3.20 ± 0.12 Å and two at 3.25 ± 0.18 Å. The next-nearest neighbors are at 3.35 Å (one), 3.40 Å (four), and 3.43 Å (two). A packing drawing prepared taking 1.60 Å as the van der Waals radius is presented in Fig. 8-2. The molecules lie in distorted close packed layers parallel to the ab plane, as shown in Fig. 8-3. The molecules are within experimental error of being perpendicular to this plane. A packing drawing of this view is presented in Fig. 8-4.

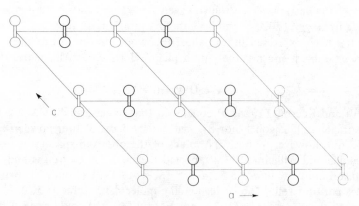

Fig. 8-1.   Four unit cells of the α-oxygen structure in projection along the b-axis; lighter molecules at y = 0, darker molecules at $y = \frac{1}{2}$.

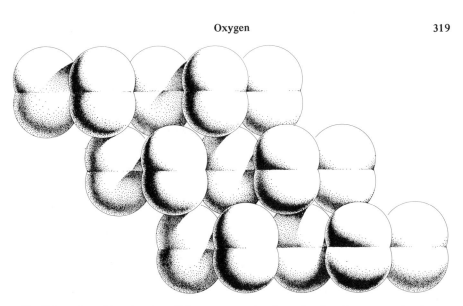

Fig. 8-2. A packing drawing of the α-oxygen structure, showing the same molecules appearing in Fig. 8-1.

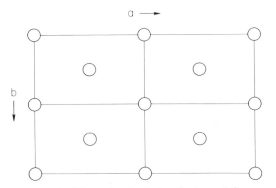

Fig. 8-3. The arrangement of the molecules in the *ab* plane of the α-oxygen structure.

The parameters and their uncertainties give a bond distance of 1.15 ± 0.12 Å, not significantly different from the value of 1.2074 Å in gaseous oxygen.

Collins (1966) observed two extra lines in a neutron diffraction invistigation of α-oxygen powder. These were interpreted by Barrett et al. as implying that this antiferromagnetic substance consists of antiparallel moments in the molecules at 000 and $\frac{1}{2}\frac{1}{2}$0.

α′-Oxygen

As mentioned previously, the diffraction pattern of oxygen deposited below

Fig. 8-4.   A packing drawing of the α-oxygen structure showing the same molecules appearing in Fig. 8-3.

23.9°K as thin films is different from that obtained from cooled samples of bulk crystals of the β phase (Hörl, 1969). This pattern, of eight lines, could not be indexed as cubic, tetragonal, or hexagonal.

This form was also observed by Bostanjoglo and Lischke (1967), who termed it α, the label also used for it in Hörl (1962).

β-Oxygen

Ruhemann (1932) obtained a powder pattern (12 lines) of β-oxygen at 29°K, but could not interpret it.

Vegard (1935a) indexed his powder pattern of nine lines on the basis of a rhombohedral cell with a = 6.20 Å (from kX), α = 99.1° at 33°K, six molecules per unit cell. (The corresponding hexagonal unit cell has a = 9.44, c = 8.88 Å, c/a = 0.941, and 18 molecules.) The molecular volume is 38.1 Å^3.

The pattern published by Black et al. (1958) matches the first three lines in Vegard's table.

An entirely different rhombohedral unit cell was found by Hörl (1962), who reported that at 28°K, a = 4.210 Å, α = 46°16′, one molecule per unit cell, on the basis of 11 observed lines, eight of which match those of Vegard. The space group is $R\bar{3}m$–D_{3d}^5, with atoms at $\pm(xxx)$, where x was taken as 0.0536 corresponding to the observed bond distance of 1.2074 Å in the gas. The hexagonal description is a = 3.307, c = 11.256 Å, c/a = 3.404, six oxygen atoms per cell at $(000\ \frac{2}{3}\frac{1}{3}\frac{1}{3}\ \frac{1}{3}\frac{2}{3}\frac{2}{3}) \pm (00z)$, with z = 0.0536. The molecular volume is 35.54 Å^3.

There is no apparent relationship between the unit cells of Vegard and Hörl, but the latter pointed out the the former had used an incorrect formula for calculating the spacings, a circumstance which may account for the satisfactory, but spurious agreement found by Vegard.

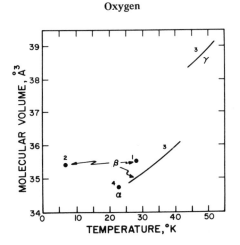

Fig. 8-5. The molecular volume of the various allotropes of oxygen. (1) Hörl, 1962; (2) Curzon & Pawlowicz, 1965; (3) Barrett, Meyer, & Wasserman, 1967a; (4) *idem.*, 1967b.

The structure of Hörl was confirmed by Alikhanov (1964a, b), who reported the identical lattice constants.

The lattice constants of the hexagonal cell *at 7°K* were determined by Curzon and Pawlowicz (1965), who found a = 3.296, c = 11.291 Å, c/a = 3.426; the corresponding rhombohedral values are a = 4.217 Å, α = 46°0'. The molecular volume is 35.41 Å^3. It was not explained how the β structure was retained at so low a temperature, 17° below the transition temperature. Barrett, Meyer, and Wasserman (1967a) measured the lattice constants over a range of temperatures, as shown in Fig. 8-5.

In a neutron diffraction study Collins (1966) observed 10 powder lines which agreed with the structure of Hörl. No extra lines were observed, and β-oxygen is thus not antiferromagnetic.

The structure may be thought of as distorted cubic close packed, the distortion being brought about by the fact that the molecules are not spherical, the elongation reducing the rhombohedral angle from 60 to 46.27°. Each oxygen atom has three nearest neighbors at 3.18 Å plus six more at 3.31 Å, as shown in Fig. 8-6. A packing drawing of a portion of the structure is presented in Fig. 8-7.

γ-Oxygen

Ruhemann (1932) obtained powder data from γ-oxygen, was unable to interpret them, but suggested that it might be hexagonal close packed.

Vegard (1935a, b) and Keesom and Taconis (1936a, c, d), on the other hand, agreed that γ-oxygen is cubic, with a = 6.84 Å (from kX), eight mole-

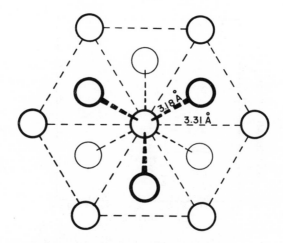

Fig. 8-6.   The structure of β-oxygen projected down the c-axis.

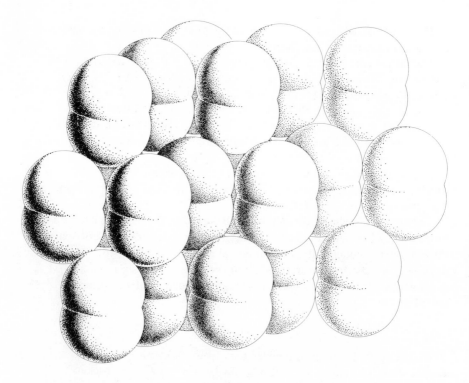

Fig. 8-7.   Part of the β-oxygen structure (c-axis vertical).

cules per unit cell, space group $Pa3-T_h^6$, with rotating, possibly dimeric, molecules.

In a single crystal study, however, Jordan, Streib, Smith, and Lipscomb (1964), found that the space group is $Pm3n-O_h^3$ and that the dimeric hypothesis was probably untenable. At 50°K the lattice constant is 6.83 ± 0.05 Å and molecular volume 39.82 ± 0.89 Å^3. The structure, which is disordered, is the same as that of β-fluorine. The unit cell contains two kinds of molecule, two of type I, centered at $000\ \frac{1}{2}\frac{1}{2}\frac{1}{2}$, and six of type II, centered at $\pm (\frac{1}{4}\frac{1}{2}0\ \frac{1}{2}0\frac{1}{4}\ 0\frac{1}{4}\frac{1}{2})$. The molecules of type I are approximately spherically disordered and those of type II show an oblate spheroidal distribution (Fig. 8-8). Each molecule of type I has 12 neighboring molecules at 3.82 Å between molecular centers; each of the other molecules has two at 3.42 Å, four at 3.82 Å, and eight at 4.18 Å. The short distance of 3.42 Å is responsible for the oblate spheroidal distribution: were the molecular axes along this line of contact, oxygen atoms in neighboring molecules would be only 2.22 Å apart.

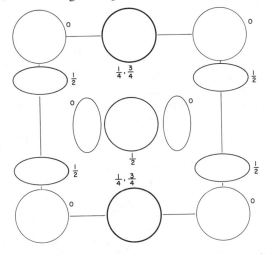

Fig. 8-8.   The structure of γ-oxygen. Out-of-plane coordinates are indicated for each molecular center. The molecules at $(000,\ \frac{1}{2}\frac{1}{2}\frac{1}{2})$ are approximately spherically disordered. The other six, at $\pm(0\frac{1}{4}\frac{1}{2},\ \frac{1}{2}0\frac{1}{4},\ \frac{1}{4}\frac{1}{2}0)$, show an oblate spheroidal distribution of electron density such that the ratio of major to minor axis is about two. Minor axes are aligned along the shortest intermolecular contacts. (After Jordan, Streib, Smith, & Lipscomb, 1964).

Ozone

Solid ozone is capriciously explosive, and no diffraction data have been reported. Even the density was not measured until 1959 (Streng and Gross). The observed value at 77.4°K is 1.728 ± 0.002 g cm^{-3}. The corresponding value of 15.37 Å^3 per oxygen *atom* is 21.0% smaller than that found for γ-oxygen at 50°K.

SULFUR

CONTENTS

A recent attempt (Donohue and Meyer, 1965) to untangle the structural chemistry of sulfur was not an unqualified success. Of all of the elements sulfur presents the most confusion and complexity in this respect. Only part of this situation is caused by the uniquely large number of molecular species which have been observed; these are S_n, with n = 1, 2, 3, 4, 5, 6, 7, 8, 9, 10, 12, 18, and ∞. Of these, however, only S_6, S_7, S_8, S_9, S_{10}, S_{12}, S_{18}, and S_∞ have been observed in the crystalline state. They are treated in that order below, followed by the discussion of provisional and doubtful forms.

A. Allotrope of Cyclohexasulfur, S_6

Names used for this include rhombohedral sulfur, trigonal sulfur, Engel's sulfur, Aten's sulfur, ϵ-sulfur, and ρ-sulfur. The term Aten's sulfur has also been applied to a hypothetical component of liquid sulfur which has also been termed π-sulfur, and the term ρ-sulfur has also been applied to the molecular species cyclo-S_6.

This form of sulfur was discovered by Engel (1891) and shown to be a hexamer by Aten (1914), who also reported improved directions for the preparation of it. The crystallization from cold toluene gives both rhombohedral and ordinary orthorhombic sulfur simultaneously. The former are orange hexagonal prisms, as contrasted with the yellow elongated octahedra of the latter. Unless special precautions are taken to remove small amounts of impurities, the hexagonal prisms, on exposure to air, convert in an hour or so to a mixture of orthorhombic and amorphous sulfur. This difficulty led to an incomplete x-ray study by Frondel and Whitfield (1950), who were able to determine hexagonal lattice constants of a = 10.9, c = 4.27 Å (from kX), c/a = 0.392. They further found that vertical planes of symmetry were absent but were unable to determine whether the lattice type was hexagonal or rhombohedral. This observation, combined with the crystal habit, limits the possible space groups to either $P\bar{3}$-C_{3i}^1 or $R\bar{3}$-C_{3i}^2. With three molecules of S_6 per unit cell the calculated density is 2.18 g cm^{-3}, as compared with the value 2.14 g cm^{-3} observed by Engel.

Donnay (1955) then pointed out that on purely morphological grounds the lattice was rhombohedral, the space group $R\bar{3}$-C_{3i}^2, and the hexagonal unit cell triply primitive. These conclusions were verified in a second single crystal x-ray study by Donohue, Caron, and Goldish (1958), who gave preliminary results which included lattice constants of a = 10.83, c = 4.26 Å, c/a = 0.393, and a structure consisting of staggered S_6 molecules having S–S bond lengths of 2.0 Å and S–S–S bond angles of about 100°. More precise lattice constants, based on the spacings of 31 observed powder lines, were later given by Caron and Donohue (1960), of a = 10.818 ± 0.002, c = 4.280 ± 0.001 Å, c/a = 0.3956 ± 0.0002. These correspond to a calculated density of 2.209 g cm^{-3}, much higher than observed for any of the other forms of sulfur.

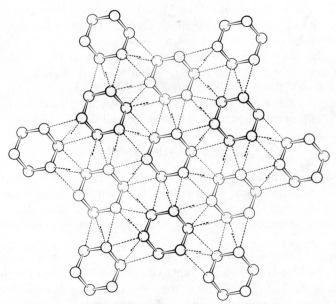

Fig. 16-1.   Projection of the rhombohedral sulfur structure down the *c*-axis. The inter-molecular contacts of 3.50 and 3.53 Å shown as dashed lines.

The crystallographic description of this structure is: space group, $R\bar{3}$-C_{3i}^2; 18 atoms per hexagonal unit cell at $(000)(\frac{1}{3}\frac{2}{3}\frac{2}{3})(\frac{2}{3}\frac{1}{3}\frac{1}{3}) \pm (xyz)(y$–$x$ $\bar{x}$ $z)(\bar{y}$ x–y $z)$. Fourier and least-squares refinement of 25 observed ($hk0$) and 16 observed ($0k\ell$) reflections were carried out by Donohue, Caron, and Goldish (1961), who reported the values for the three positional parameters of $x = 0.1454 \pm 0.0012$, $y = 0.1882 \pm 0.0016$, $z = 0.1055 \pm 0.0020$. These yield the following values for the molecule: S–S bond length $= 2.057 \pm 0.018$ Å S–S–S bond angle $= 102.2 \pm 1.6°$ S–S–S–S torsion angle $= 74.5 \pm 2.5°$.

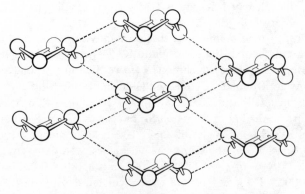

Fig. 16-3.   Projection of the rhombohedral sulfur structure perpendicular to *c*. Inter-molecular contacts as in Fig. 16-1.

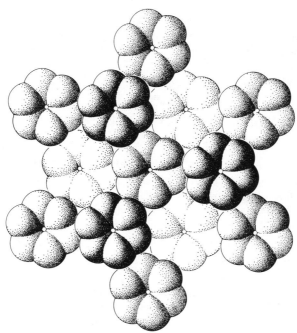

Fig. 16-2.   Packing drawing of the view of Fig. 16-1.

The molecular packing is very efficient. Each molecule has short intermolecular contacts with 12 neighboring molecules, as shown in Fig. 16-1. Each atom has two neighbors at 3.50 Å and one at 3.53 Å; there are thus 18 short contacts per molecule. Figure 16-2 is a packing drawing of the view in Fig. 16-1. The small "hole" in the center of the molecule, plus the efficient packing, accounts for the high density. Figures 16-3 and 16-4 show the structure as viewed perpendicular to the c-axis.

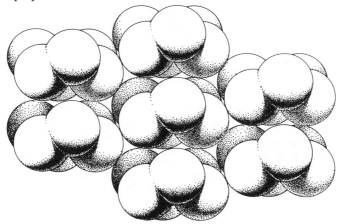

Fig. 16-4.   Packing drawing of the view of Fig. 16-3.

The hypothesis of Soklakov (1962) that rhombohedral sulfur consists of spirally wound zigzag chains of sulfur atoms was thoroughly demolished by Donohue and Goodman (1964).

B. Allotrope of Cycloheptasulfur, S_7

Cycloheptasulfur was first prepared by Schmidt, B. Block, H. D. Block, Köpf, and Wilhelm (1968), who carried out the following reaction:

$$(C_5H_5)_2 TiS_5 + S_2Cl_2 \rightarrow (C_5H_5)_2 TiCl_2 + S_7$$

The substance forms pale yellow needles that decompose rapidly at room temperature. An x-ray study of the crystals at $-80°C$ by Kawada and Hellner (1970) gave the following results: $a = 21.77 \pm 0.04$, $b = 20.97 \pm 0.04$, $c = 6.09 \pm 0.01$ Å, $\alpha = \beta = \gamma = 90°$; the experimental density ($-80°C$) of 2.090 g cm^{-3} compares with the value 2.144 g cm^{-3} calculated for 16 molecules of S_7 per unit cell. The structure was solved only in projection down the c-axis, so structural details are not yet available.

C. Allotropes of Cyclooctasulfur, S_8

1. Orthorhombic Sulfur

Names used for this allotrope, the ordinary form stable at room temperature and atmospheric pressure, include rhombic sulfur, Muthmann's sulfur I, α-sulfur, and orthorhombic sulfur; the last is preferable, but the term α-sulfur may also be used for brevity.

Orthorhombic sulfur was among the first substances to be examined with x-ray diffraction by Bragg (1914), who determined the lattice constants, which, however, because of the unrecognized nature of the extinctions, were all reported as one-half of their true values. Several subsequent investigators made the same error. Various determinations of the lattice constants are presented in Table 16-1. The most recent values of Cooper (1962) are the most precise and are used here.

The correct space group was first reported by Mark and Wigner (1924), but the structure itself was not determined until 1935, by Warren and Burwell, who assumed a staggered ring conformation for the S_8 molecule, thus reducing the problem to one of only two variables. They obtained satisfactory qualitative agreement between observed and calculated intensities for 41 observed and 11 unobserved reflections. The space group is Fddd–D_{2h}^{24}, with the 128 sulfur atoms lying in four sets of position $32(h)$, at $(000 \ 0\frac{1}{2}\frac{1}{2} \ \frac{1}{2}0\frac{1}{2} \ \frac{1}{2}\frac{1}{2}0) \pm (xyz)$, $(x \ \frac{1}{4}\text{-}y \ \frac{1}{4}\text{-}z)$, $(\frac{1}{4}\text{-}x \ y \ \frac{1}{4}\text{-}z)$, $(\frac{1}{4}\text{-}x \ \frac{1}{4}\text{-}y \ z)$. [In this description the origin lies on a center of symmetry; in an often used alternate description the origin lies $(\frac{1}{8}\frac{1}{8}\frac{1}{8})$ from the above, at the intersection of three twofold axes.] The 128

Table 16-1.   The Lattice Constants of Orthorhombic (α) Sulfur[a]

a, Å	b, Å	c, Å	Reference
5.22	6.42	12.22	Bragg, 1914
5.35	6.47	12.29[b]	Mark, Weissenberg, & Gonell, 1923
10.63	12.90	24.61	Mark & Wigner, 1924
10.50	12.95	24.60	Warren & Burwell, 1935
5.2	6.4	12.2	Trillat & Oketani, 1937
10.48	12.92	24.55	Ventriglia, 1951
10.437	12.845	24.369	Abrahams, 1955
10.45	12.84	24.46	de Wolff, 1958
10.468	12.870	24.49	Swanson, Cook, Isaacs, & Evans, 1960
10.467[c]	12.870	24.493	Caron & Donohue, 1961
10.4646[d]	12.8660	24.4860	Cooper, 1962; Cooper, Bond, & Abrahams, 1961

[a]All values before 1950 from kX.
[b]The value of c in this paper is misprinted as 15.27 kX, instead of 12.27 kX.
[c]At 25°C.
[d]At 24.8°C; values used here.

atoms form 16 S_8 molecules, the centers of which lie in position 16(g), at $(000 \; 0\frac{1}{2}\frac{1}{2} \; \frac{1}{2}0\frac{1}{2} \; \frac{1}{2}\frac{1}{2}0) \pm (\frac{1}{8}\frac{1}{8}z), (\frac{1}{8} \; \frac{1}{8} \; \frac{1}{4}+z)$, on twofold axes parallel to c, This structure was confirmed with more extensive data by Ventriglia (1951), but the first accurate structure determination is that of Abrahams (1955), who refined 669 observed structure factors by Fourier and least squares methods. The average S–S bond length was found to be 2.037 ± 0.005 Å. In a subsequent least squares refinement Abrahams (1961) re-refined these data using an improved scattering factor for sulfur, allowing for anisotropic thermal motion, and including the 377 unobserved reflections. The average S–S bond length was essentially unchanged, at 2.041 ± 0.003 Å; the three root mean square thermal displacements, but not their directions, were given for each of the atoms.

The observed structure factors of Abrahams (1955) were then subjected to a second least squares re-refinement by Caron and Donohue (1965), who omitted 18 highly discrepant reflections and made an empirical correction for extinction on the remaining 651 data. The resulting positional parameters were scarcely changed from those given by Abrahams (1955), the average difference being 0.005 Å. The thermal parameters for the individual atoms were found to be consistent with a rigid-body motion of the entire molecule. The librational amplitudes obtained in this way were then used to obtain corrected values for the positional parameters.

The complete, uncorrected data set of Abrahams (1955) was also refined by least squares somewhat later by Pawley and Rinaldi (1972), who were obviously unaware of the work of Caron and Donohue, and, not surprisingly, obtained positional and thermal parameters virtually identical to those of the earlier work. Pawley and Rinaldi also investigated the effect of refining the structure when (a) the molecular symmetry was constrained to be D_{4d}-$\bar{8}2m$ and (b) the thermal parameters were constrained to be those of a rigid molecule. Somewhat contradictory results were obtained in that the results of refinement a indicated that a model having unrestrained, that is, distorted, molecules gave significantly better agreement, while refinement b showed that allowance for individual atomic thermal motions gave no better agreement than the model with rigid-body molecules. Pawley and Rinaldi imply that this matter may be resolved when new, more accurate (and presumably more extensive) diffraction data become available, a conclusion which may also be inferred from the note of Caron and Donohue. Pawley and Rinaldi were also apparently unaware of Abrahams' observation (1965) that the data set in question may contain systematic errors, an opinion which reinforces the need for collection of new data.

The various values for the positional parameters are presented in Table 16-2.

Table 16-2. Positional Parameters in Orthorhombic (α) Sulfur (origin on $\bar{1}$)

Atom		Warren & Burwell, 1935	Abrahams, 1955	Caron & Donohue, 1965		Pawley & Rinaldi, 1972	
				a	b	c	d
1	x	−0.142	−0.1446	−0.1437	−0.1437	−0.1439	−0.1418
	y	−0.042	−0.0474	−0.0471	−0.0465	−0.0472	−0.0467
	z	−0.053	−0.0484	−0.0484	−0.0489	−0.0485	−0.0483
2	x	−0.219	−0.2156	−0.2155	−0.2158	−0.2152	−0.2175
	y	0.036	0.0301	0.0305	0.0315	0.0305	0.0301
	z	0.075	0.0763	0.0762	0.0763	0.0763	0.0771
3	x	−0.292	−0.2931	−0.2924	−0.2934	−0.2927	−0.2911
	y	−0.020	−0.0205	−0.0201	−0.0195	−0.0200	−0.0178
	z	0.000	0.0040	0.0039	0.0037	0.0041	0.0037
4	x	−0.219	−0.2138	−0.2140	−0.2145	−0.2142	−0.2154
	y	−0.097	−0.0927	−0.0922	−0.0920	−0.0920	−0.0946
	z	0.125	0.1290	0.1296	0.1300	0.1296	0.1289

[a]Uncorrected.   [c]Unrestrained.
[b]Corrected for libration.   [d]Molecules restrained to symmetry D_{4d}-$\bar{8}2m$.

The molecular quantities were calculated with the lattice constants of Cooper (1962) and the corrected positional parameters of Caron and Donohue (1965). Until the new data sets mentioned by Pawley and Rinaldi are refined these are the best available.

The shortest intermolecular S· · ·S distances are presented in Fig. 16-5. In keeping with the complexity of the structure, there does not appear to be any regularity of these, and it may also be noted that the environments of the four nonequivalent sulfur atoms are rather different.

Distances and angles within a single molecule are presented in Fig. 16-6 and 16-7. The average values are:

$$\begin{array}{lll}
\text{S–S bond length} & = 2.060 \pm 0.003 \text{ Å} \\
\text{S–S–S bond angle} & = 108.0 \pm 0.7° \\
\text{S–S–S–S torsion angle} & = 98.3 \pm 2.1°
\end{array}$$

Although the crystal symmetry imposes an exact twofold axis on the molecule,

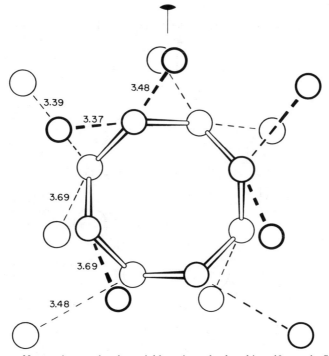

Fig. 16-5.   Nearest intermolecular neighbors in orthorhombic sulfur; only S· · ·S distances less than 3.70 Å are shown.

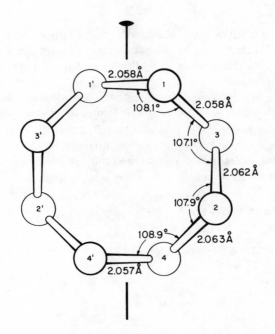

Fig. 16-6.   Bond distances and angles in the S₈ molecule in orthorhombic sulfur.

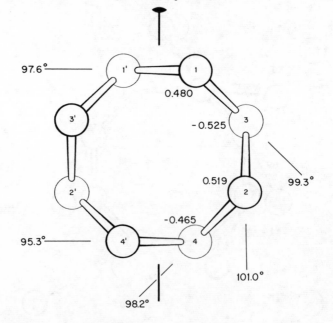

Fig. 16-7.   Torsion angles and distances from the mean molecular plane in the S₈ molecule in orthorhombic sulfur.

the higher symmetry of $\bar{8}2m$–D_{4d} is closely approximated. The libration correction raises the average bond length from 2.045 Å but has no significant changes on any of the angles.

The molecular packing is very complex. Views of the structure in projection down the *a*-, *b*-, and *c*-axes are shown in Figs. 16-8 through 16-16. A

Fig. 16-8. Projection of the orthorhombic sulfur structure along the *c*-axis, top half of the unit cell. The complete structure is the superposition of this figure with Fig. 16-10.

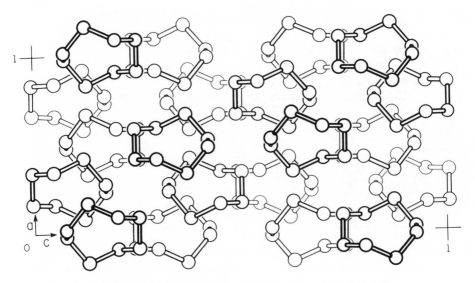

Fig. 16-9.　Projection of the orthorhombic sulfur structure along the *a*-axis.

Fig. 16-10.　Projection of the orthorhombic sulfur structure along the *c*-axis, bottom half of the unit cell.

334

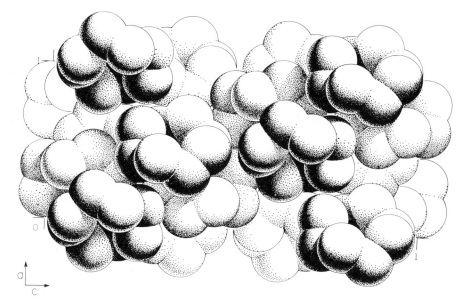

Fig. 16-11.   Packing drawing of the view of Fig. 16-9.

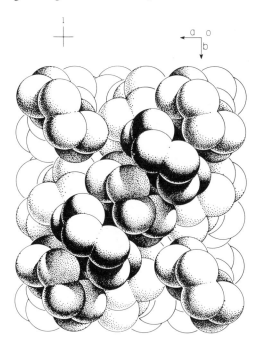

Fig. 16-12.   Packing drawing of the view of Fig. 16-10. Note the "crankshaft" running from the lower left corner to the upper right corner.

portion of the structure, viewed perpendicular to the mean plane of half of the molecules, a direction which is not rational with respect to the crystal axes, is shown in Figs.16-13 and 16-14. From these two figures it is apparent why the molecular packing in this crystal is sometimes referred to as the crankshaft arrangement. As may be seen, these crankshafts extend in two different directions, a result leading to the great complexity of the complete crystal structure. The depiction of the orthorhombic sulfur structure has often been made as one in which the rings stack *directly* over one another

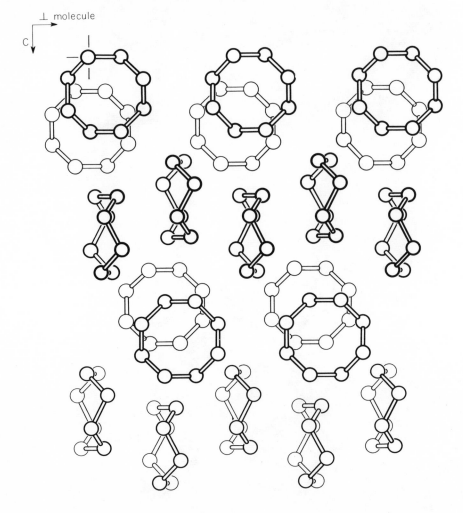

Fig. 16-13.   Projection of the orthorhombic sulfur structure perpendicular to the mean plane of half of the molecules, showing two aspects of the "crankshafts".

(e.g., Pinkus, Kim, McAtee, and Concilio, 1959a, b; Pinkus and McAtee, 1960; Cotton and Wilkinson, 1966). Although this picture has the advantage of simplicity, and might quickly be arrived at by the manipulation of space-filling molecular models, it is clearly incorrect, as is easily seen from Figs. 16-13 and 16-14. This misconception has been pointed out earlier by Donohue and Caron (1961).

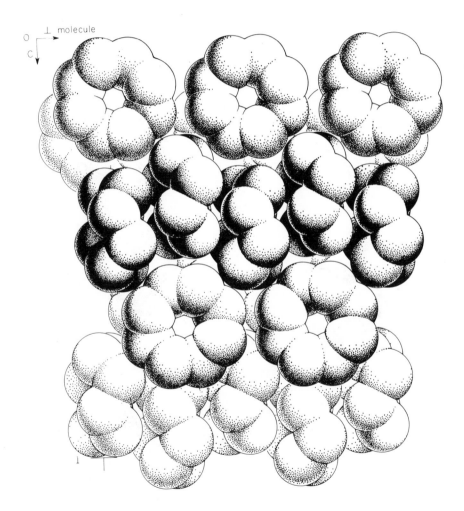

Fig. 16-14. Packing drawing of the view of Fig. 16-13.

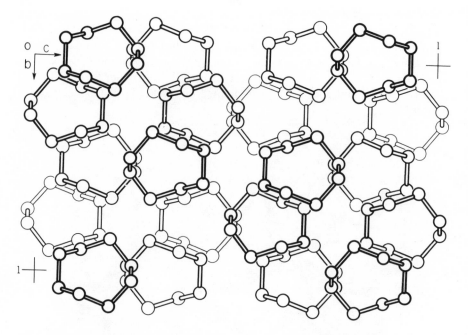

Fig. 16-15. Projection of the orthorhombic sulfur structure along the *a*-axis.

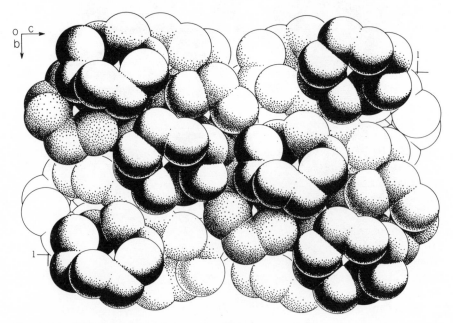

Fig. 16-16. Packing drawing of the view of Fig. 16-7.

338

2. β-Monoclinic Sulfur

At least five different preparations said to be forms of sulfur have been described as having monoclinic symmetry, so that that term used by itself is
highly ambiguous. Following previous usage the appelation used here for the
allotrope of sulfur which is formed by crystallization from the melt at atmospheric pressure is β-monoclinic sulfur, or, more simply, β-sulfur.

The first diffraction experiment on β-sulfur appears to be that of Trillat
and Forestier in 1932. They reproduced a rotation photograph of a single
crystal of the substance but did not attempt any interpretation of it.

The first unit cell and space group data were obtained by Burwell (1937)
from single crystals, grown from the melt, which were maintained at about
103°C during oscillation and Weissenberg photography. The lattice constants
(from kX) are $a = 10.92$, $b = 10.98$, $c = 11.04$ Å, $\beta = 96°44'$. With 48 atoms
of sulfur per unit cell the calculated density is 1.94 g cm^{-3}. The observed absences of $(0k0)$ with k odd and $(h0\ell)$ with h odd lead uniquely to the space
group $P2_1/a-C_{2h}^5$. If, as is generally assumed, β-sulfur also consists of S_8
rings, then the unit cell contains six such molecules, and at least two of them
must lie on crystallographic centers of symmetry, a conclusion incompatible
with the crown conformation found in α-sulfur. Burwell suggested that a
statistical center of symmetry might be achieved by oscillation or rotation
of the molecules about their fourfold axes.

Additional work on this structure was reported 26 years later by Sands
(1965), in a preliminary note. Sands described a unit cell containing four
ordered molecules and two disordered molecules; the statistical centers of symmetry are thus achieved by having the latter in one of two possible orientations related by centers, at random. The atoms of the ordered molecules lie
in eight sets of position $4(e)$, at $\pm(xyz)$, $(\frac{1}{2}+x \ \frac{1}{2}-y \ z)$, giving four S_8 molecules
centered in a set of positions $4(e)$ with $x = 0.072$, $y = 0.356$, and $z = 0.367$,
and those of the disordered molecules also lie in eight sets of position $4(e)$,
but with occupancy factors of one-half, giving superimposed double molecules
at (000), $(\frac{1}{2}\frac{1}{2}0)$. The values of the positional parameters are presented in
Table 16-3. These must be regarded as provisional because additional refinement was implied, and "the details. . .will be published later". At the present
stage the agreement index is 13.1% for the 1270 observed reflections.

The parameters of Table 16-3, which were determined with the crystal at
room temperature, combined with the lattice constants of $a = 10.778$, $b =
10.844$, $c = 10.924$ Å, $\beta = 95.80°$ at room temperature (Sands, 1971), lead to
the bond distances and angles of Figs. 16-17 and 16-18. These are also, obviously, subject to revision. A view of the structure along the b-axis is shown
in Fig. 16-19.

Doubt on the above structure was cast by Boon (1962), who was stated to
have shown that the absence of $(0k0)$ reflections with k odd was not system-

Table 16-3. Positional Parameters in β-Monoclinic Sulfur, According to Sands (1965); Space Group Orientation $P2_1/a$

	x	y	z
Atoms of the ordered molecules			
1	0.031	0.525	0.234
2	0.006	0.356	0.152
3	−0.101	0.252	0.256
4	0.014	0.144	0.374
5	0.037	0.228	0.537
6	0.208	0.315	0.553
7	0.176	0.498	0.512
8	0.202	0.525	0.332
Atoms of the disordered molecules			
9	0.081	−0.108	0.182
10	0.206	−0.069	0.052
11	0.188	0.113	0.004
12	0.079	0.126	−0.152
13	−0.093	0.171	−0.119
14	−0.193	0.009	−0.129
15	−0.201	−0.048	0.053
16	−0.073	−0.185	0.094

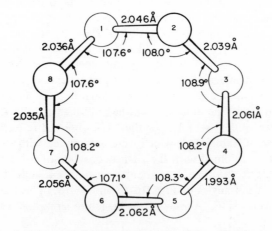

Fig. 16-17.   Bond distances and angles in the ordered S_8 molecules in β-monoclinic sulfur.

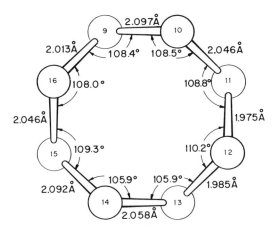

Fig. 16-18.   Bond distances and angles in the disordered S$_8$ molecules in β-monoclinic sulfur.

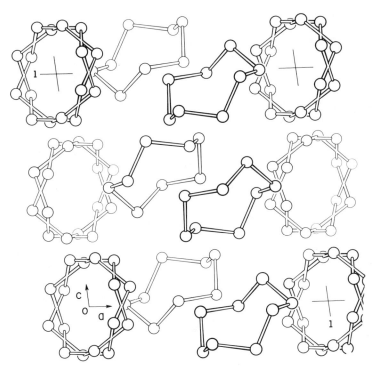

Fig. 16-19.   Projection of the β-monoclinic sulfur structure along the b-axis, showing the disorder of one-third of the molecules.

atic. The space group thus would be Pc, rather than $P2_1/c$. The structure of
Boon was said to have features in common with that of Sands, but no details
were given. However, this conclusion was found to be incorrect by Van de Loo
(Tuinstra, 1971), who, working with 1458 reflections, proved that the structure proposed by Sands is essentially correct, but the details are not yet
available. Meanwhile, the values in Table 16-3 and Figs. 16-18 and 16-19 will
have to suffice.

3. γ-Monoclinic Sulfur

A second form of sulfur having monoclinic symmetry has been variously termed
Muthmann's sulfur III, nacreous sulfur, and γ-sulfur. It may be produced from
an undercooled melt, or by slow cooling of a nearly saturated solution of
sulfur in ethanol which had been refluxed for several hours. The crystals are
stable when kept in contact with the solution, but most of them revert to α-
sulfur when dry. It also occurs as the mineral rosickyite.

The crystallographic literature of γ-monoclinic sulfur is confusing because
three different conventions have been used in labelling the axes, as shown in
Fig. 16-20. The various determinations of the lattice constants are presented
in Table 16-4. There is also confusion regarding the space group. White (1944)
gives it as P2/a for the cell outlined with solid lines in Fig. 16-20, but all
succeeding authors found that this cell was B-centered. DeHaan (1958) states
that the cell outlined with the dashed lines has space group P2/n, but the
systematic absences he gives correspond to P2/c. Strunz and Herda (1961)
give P2/m as the space group of the dashed-line cell (this would then also be
the space group of the cell outlined with dotted lines), but Strunz (1962)
then gives P2/n for the dashed-line cell and P2/c for the dotted-line cell (these
correspond to B2/c for the solid-line cell).

DeHaan claims to have determined the structure and presented two drawings
of it, showing a "sheared penny roll" arrangement of S_8 molecules, but gave
neither the positional parameters nor the observed and calculated structure
factors. The complete structure determination, said to be in progress, has not
yet been published.

The structure of γ-monoclinic sulfur must also thus be regarded as not yet
established.

D. Cycloenneasulfur, S_9

Schmidt and Wilhelm (1970) prepared cycloenneasulfur in accord with the
overall reaction: $(C_5H_5)_2TiS_5 + S_4Cl_2 \xrightarrow{HCl} (C_5H_5)_2TiCl_2 + S_9$.
It forms intense yellow needles which were said to be more stable towards
light and heat than cyclohexasulfur. No diffraction work has been reported.

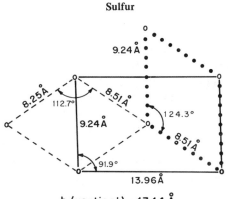

b (vertical) = 13.14 Å

Fig. 16-20.   The relation among the three different unit cells reported for γ-monoclinic sulfur. See Table 16-4.

Table 16-4.   The Lattice Constants of γ-Monoclinic Sulfur[a]

	White, 1944	deHaan, 1958	Strunz, 1962	Average
b	13.22 Å	13.05 Å	13.16 Å	13.14 Å
a_1	13.95	13.96	(13.96)	13.96
a_2	9.31	9.26	(9.29)	9.29
a_3	(8.26)	(8.23)	8.27	8.25
a_4	(8.51)	(8.52)	8.50	8.51
$\angle a_1 a_2$	91.9°	92.2°	(91.7°)	91.9°
$\angle a_3 a_4$	(112.6°)	(112.9°)	112.7°	112.7°
$\angle a_2 a_4$	(125.0°)	(125.2°)	(124.8°)	125.0°

[a]Values in parentheses calculated from the published values.

E. Cyclodecasulfur, S_{10}

Cyclodecasulfur was also prepared by Schmidt et al. (1968), who used the reaction

$$2(C_5H_5)_2TiS_5 + 2SO_2Cl_2 \rightarrow 2(C_5H_5)_2 + 2SO_2 + S_{10}$$

No structural work has been carried out.

F. Allotrope of Cyclododecasulfur, S_{12}

A new molecular species of sulfur was reported by Schmidt and Wilhelm (1966), who prepared cyclododecasulfur by the reaction of S_xCl_2 and H_2S_y, $x + y = 12$. The crystal structure was reported at the same time by Kutoglu and Hellner (1966). The pale yellowish crystals are orthorhombic, with $a = 4.730$, $b = 9.104$, $c = 14.574$ Å. With 24 atoms of sulfur, two molecules of

S_{12}, per unit cell, the calculated density is 2.036 g cm^{-3}; The space group is Pnnm-D_{2h}^{12}, and the atoms are placed as follows: four atoms, S(4), in position 4(e) at $\pm(00z)(\frac{1}{2}\ \frac{1}{2}\ \frac{1}{2}+z)$; four atoms, S(3), in position 4(g) at $\pm(xy0)$ $(\frac{1}{2}-x\ \frac{1}{2}+y\frac{1}{2})$; sixteen atoms, S(2) and S(1), in two sets of position 8(h) at $\pm(xyz)(x\bar{y}z)(\frac{1}{2}+x\ \frac{1}{2}-y\ \frac{1}{2}-z)\ (\frac{1}{2}+x\ \frac{1}{2}-y\ \frac{1}{2}+z)$. Numerical values of the parameters (Hellner, 1967) are presented in Table 16-5; these were obtained by least squares refinement of 634 visually estimated intensities to an R of 11%.

Individual values of the independent bond distances, angles, and torsion angles are shown in Fig. 16-21; The average values are

$$
\begin{aligned}
\text{S-S bond length} \quad &= 2.053 \pm 0.007 \text{ Å} \\
\text{S-S-S bond angle} \quad &= 106.5 \pm 1.4^\circ \\
\text{S-S-S-S torsion angle} \quad &= 86.1 \pm 5.5^\circ
\end{aligned}
$$

The crystal symmetry of a molecule is 2/m-C_{2h}, but the individual molecules closely approximate symmetry $\bar{3}$m-D_{3d}. The observed conformation is probably assumed because it allows more favorable values for the torsion angles. In the crown conformation, such as is observed for the S_6 and S_8 molecules, a 12-membered ring with symmetry $\overline{12}$m2-D_{6d} and bond angles of 106° would have torsion angles of 129°. Moreover, the hole in the center of the molecule would be far larger than is observed, as may be seen in Fig. 16-22, which is a packing drawing of one molecule. The observed torsion angles in the S_{12} ring are intermediate between those found in the S_6 and S_8 molecules.

A view of the structure in projection along the x-axis is presented in Fig. 16-23, which shows the 24 nearest neighbors of the central molecule. The distance of 3.34 Å between two atoms of type S(2) of different molecules is the shortest intermolecular distance yet observed in any of the allotropes of sulfur.

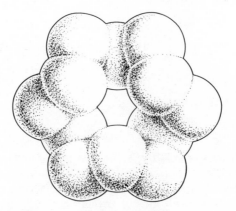

Fig. 16-22.   Packing drawing of one molecule of S_{12}.

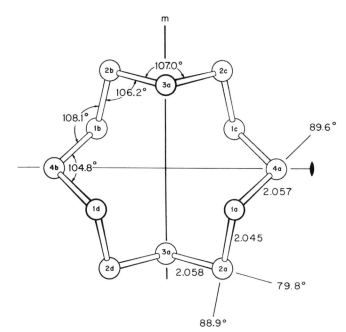

Fig. 16-21. Bond distances, bond angles, and torsion angles in the S_{12} molecule.

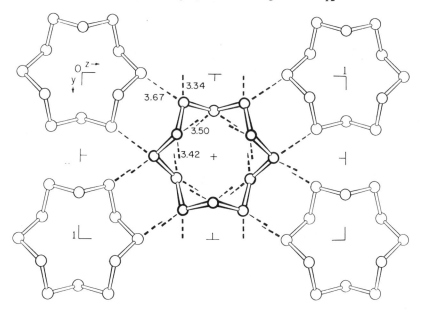

Fig. 16-23. Projection of the S_{12} structure down the a-axis. The 24 shortest $S\cdots S$ inter-molecular distances are shown as dashed lines.

Table 16-5.   Positional Parameters in Cyclododecasulfur

	x	y	z	Position
1	0.2383	0.1293	0.1402	8(h)
2	0.0152	0.3169	0.1135	8(h)
3	−0.2293	0.2725	−	4(g)
4	−	−	0.2263	4(e)

G. Cyclooctadecasulfur, S_{18}

Schmidt and Wilhelm (1970), referring to their own unpublished work, state they have prepared cyclooctadecasulfur, but no further details were given.

H. Allotrope of Catenapolysulfur, S_∞

Crystalline samples of this form of sulfur, often called fibrous or plastic sulfur, may be prepared by stretching quenched liquid sulfur from the boiling point, or by drawing filaments from hot liquid sulfur. After several hours at room temperature the material becomes hard and brittle and it was generally assumed that this was caused by conversion to microcrystalline α-sulfur, perhaps with some amorphous material admixed. Trillat and Forestier (1931) first obtained diffraction data from stretched samples of fibrous sulfur, but made no attempt to interpret the pattern aside from reporting a 9.35 Å repeat along the fiber axis. A series of photographs published by them (1932) showed that a fresh sample gave a pattern typical of a fiber, while a three-week-old sample gave a pattern identical to that of powdered α-sulfur.

In order to avoid the crystallization problem Meyer and Go (1934) devised an ingenious method for continuously generating fresh filaments of fibrous sulfur from a melt for passage through the x-ray beam. They found that the diffraction pattern could be indexed on a monoclinic unit cell having a = 26.5, b = 9.28 (fiber axis), c = 12.34 Å (from kX), and $β$ = 79°15′. The observed density of 2.01 g cm^{-3} corresponds to 112 atoms per cell (calc., 2.00 g cm^{-3}). They proposed a structure consisting of flat chains (Fig. 16-24), with 14 such chains in the unit cell.

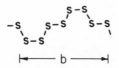

Fig. 16-24.   The flat chain structure for fibrous sulfur, S_∞, proposed by Meyer and Go (1934).

Now follows a curious interlude. In 1942 Meyer apparently considered, as an alternative to the 14 chain-8 atom structure, a 16 chain-8 atom structure, in spite of the fact that the calculated density of this structure is 2/14 larger than that of the first structure, or 2.29 g cm^{-3}. Huggins (1945), using the analogy with metallic selenium and tellurium, which consist of "uniform spirals with three chain atoms per identity distance", preferred an 18 chain-6 atom structure, stating that this seemed to be in as good or better agreement with the published data. In this case the calculated density is 1.93 g cm^{-3}, but why this value is in "as good or better agreement" with 2.01 than 2.00 g cm^{-3} is mysterious. Huggins suggested that a repeat after six atoms, rather than three, could be accounted for by assuming a reversal of direction of the helix after each complete rotation, but he did not investigate how, or even whether, such irregular helices could fit into the unit cell. Pauling (1949), accepting 112 as the number of atoms in the unit cell, rejected the first structure of Meyer and Go because the chains were flat and accordingly had unfavorable values for the S-S-S-S torsion angles, and also rejected the Huggins structure because it led to unsatisfactory bond angles and because 112 is not a multiple of six. Pauling found that a helix with seven atoms in two turns, with a repeat of 9.26 Å, had acceptable values for the bond and torsion angles and would account for the unusual occurrence of a multiple of seven atoms in the unit cell, with 16 chains per cell. Meyer (1950) later cited Pauling's proposed structure without comment.

It is noteworthy that all of the foregoing structures were based on the meagerest of information, with the emphasis on one number, the number of atoms in the unit cell. None of them was tested by comparison of the observed intensities with those calculated from a set of atomic coordinates derived from a model, nor was the molecular packing explored in detail.

A considerable contribution toward the solution to this problem was then made by Prins, Schenk, and Hospel (1956), who discovered that there was no difference among the diffraction patterns of freshly prepared fibers and those which had been aged for more than a year. Of even more significance was the discovery that extraction of the fibers with carbon disulfide caused a number of diffraction maxima to disappear, with the rest keeping their exact positions and gaining in strength. In their opinion the new pattern was the true one for long chain sulfur, and the spots which disappeared were attributed to crystals of some form of S_8. They further suggested that the long chains were spirals with 10 atoms in three turns, as opposed to the seven atom-two turn structure of Pauling.

Confirmation of some of the preceding points was soon provided by Prins, Schenk, and Wachters (1957), who showed that the previously accepted pattern of fibrous sulfur (termed by them φ-sulfur) was, in fact, due to the superposition of two constitutents, catenapolysulfur, which they termed ψ-sulfur, and γ-monoclinic sulfur. Crystals of the latter, oriented with a 9.2 Å axis (a_2 of Table 16-4) parallel to the fiber axis, were found to occur in the fibers; that part of the pattern due to this phase disappears after treatment with carbon disulfide.

The indexing of Meyer and Go thus pertains to *two different* substances, and all structures based upon it are clearly invalidated.

The model proposed by Prins, Schenk, and Wachters consists of close-packed helices the axes of which lie 4.7 Å apart, each helix having 10 atoms in three turns, giving a repeat distance of 13.70 Å. The helical radius was stated to be 0.92 Å, and bond distances, bond angles, and torsion angles of 2.04 Å, 107°, and 87°, respectively. Comparison of the observed and calculated intensities was not given.

Liquori and Ripamonti (1959) stated that their photographs of stretched fibers of plastic sulfur could be interpreted on the basis of an ideal helical structure having 10 atoms in three turns, thus supporting the foregoing result.

However, Prins and Tuinstra (1936a, b) modified the model on the basis of new and improved data, by formulating alternating right- and left-handed helices, packed together as shown in Fig. 16-25. The unit cell of this arrangement is monoclinic, a = 8.88, b = 9.20, c = 13.7 Å, β = 114°, and contains 40 atoms. They also stated that there was evidence for a superstructure, with a repeat distance of at least 78 Å along the fiber axis. This superstructure was thought of as a slow twist, a full turn of which is completed after 78 Å, superimposed on the ten atom-three turn helix. Again, no intensity comparisons were presented.

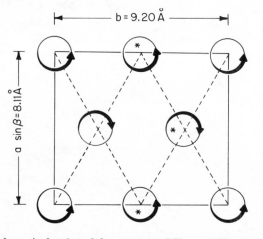

Fig. 16-25.   Schematic drawing of the structure of fibrous sulfur according to Prins and Tuinstra (1963a, b), viewed along the hilical axes. The asterisks indicate that a "chain has been 'screwed up' half an atom (rotation by 18°, translation about $\frac{1}{2}$·1.37 Å in c-direction)". The two helices in the center are displaced from the intersections of the dashed lines by about 0.6 Å.

A Diversion Concerning Helices. All subsequent proposals for the structure of catenapolysulfur use homoatomic regular helices. Such helices are defined by a number of parameters:

r = the radius
P = the pitch, which is the distance along the axis after one turn
p = the translation along the axis between successive atoms
γ = the rotation between successive atoms
k = the number of atoms per turn
I = the repeat distance along the axis
n_a = the number of atoms in the repeat distance (an integer)
n_t = the number of turns in the repeat distance (another integer) (n_a and n_t relatively prime)

These parameters are, obviously, not independent, save for r. Thus the following relations obtain:

$$P/p = k = n_a/n_t$$
$$\gamma = 360°/k$$
$$I = n_a p = n_t P$$

The simplest helices have $n_t = 1$, such as the simple threefold helix of metallic selenium and tellurium used by Huggins in his model for fibrous sulfur. The Pauling structure is only slightly more complicated, having $n_a/n_t = 7/2 = 3.50$, while the Prins, Schenk, and Wachters structure goes up in complexity, with $n_a/n_t = 10/3 = 3.33\cdots$.

Prins and Tuinstra's structure was later modified by Tuinstra (1966), who obtained improved data from strongly stretched fibers which had been heated for 40 hours at 80°C. The unit cell was now given as orthorhombic, with $a = 8.11$, $b = 9.20$ Å, c "indeterminate". Refinement of the 65 observed reflections gave the following dimensions for the helices: $r = 0.96$ Å, $P = 4.60$ Å, $p = 1.37$ Å. P and p were said to be incommensurate, a statement incompatible with the fact that $P/p = 460/137 = 3 \ 49/137 \ (= 3.358)$. Strictly speaking, this would imply a helix containing 460 atoms in 137 turns, but there are far less complicated ways of expressing this ratio. No errors on either P or p were given, but Tuinstra gave γ as $107°27' \pm 45'$, from which it may be calculated that $k = 3.350 \pm 0.021$, a value within experimental error of the simpler helix having $n_a/n_t = 10/3$, as well as numerous other simpler structures, for example, a 47 atom-14 turn helix ($k = 3.357$) or a 76 atom-20 turn helix ($k = 3.350$). The slow twisting superstructure of Prins and Tuinstra was apparently abandoned. The proposed packing of the helices is indicated in Fig. 16-26. No interhelical packing distances were reported.

Tuinstra's method of indexing has been severely criticized by Geller and Lind (1969), and an unacceptable discrepancy between an observed and cal-

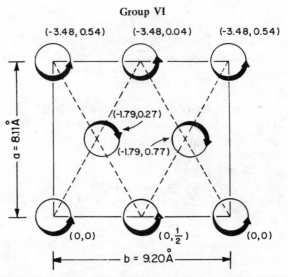

Fig. 16-26.   Modification of the fibrous sulfur structure according to Tuinstra (1966). The numbers in parentheses give, first, the translation (in Å) of a helix along its axis, and, second, the screw motion as a fraction of the axial distance between successive atoms.

culated intensity was pointed out by Donohue, Goodman, and Crisp (1969). It is thus not unlikely that although some features of this model are correct it cannot be accepted *in toto*. The molecular structural details for one helix given by Tuinstra are:

$$
\begin{aligned}
\text{S–S-bond length} \quad &= 2.069 \pm 0.014 \text{ Å} \\
\text{S–S–S bond angle} \quad &= 106.0 \pm 1.7° \\
\text{S–S–S–S torsion angle} \quad &= 84.2 \pm 1.0°
\end{aligned}
$$

At about the same time Geller (1966) reported that single crystals of a form of sulfur prepared at 27 kbar, but which did not transform at 1 atm, gave a diffraction pattern "very close to, but not precisely the same as" that of fibrous sulfur prepared by the method of Prins, Schenk, and Hospel. This resemblance was later emended by Lind and Geller (1969), who stated that a photograph of fibrous sulfur superposed *exactly* on one of their photographs of pressure-induced sulfur. It thus appears that if there are structural differences between the two preparations they are minor, and it is not unlikely that differences in the diffraction patterns, if real, may arise from twinning in crystals of the pressure induced material and orientation effects in the stretched fibers.

Geller described the crystals as orthorhombic, with $a = 13.8$, $b = 32.4$, $c = 9.25$ Å, space group either Cc2m–C_{2v}^{16} or Ccm2$_1$–C_{2v}^{12}, 160 atoms per unit cell in 16 ten atom-three turn helices, in a nearly close packed arrangement. This unit cell is closely similar to that of Prins and Tuinstra (1963a):

Prins & Tuinstra	Geller

$$c = 13.7 \text{ Å (fiber axis)} \quad a = 13.8 \text{ Å (helix axis)}$$
$$a \sin \beta = 8.11 \text{ Å} \quad b = 4 \times 8.10 \text{ Å}$$
$$b = 9.20 \text{ Å} \quad c = 9.25 \text{ Å}$$

Both space groups were said to require a certain amount of disorder of the helices about their axes, with right- and left-handed helices present in both. Details of the molecular structure were not given, but a complete structure analysis was promised.

Tuinstra (1967b) soon after gave what was apparently a slightly modified model of his structure. For one helix the values given were $P = 4.60 \pm 0.03$ Å and $k = 3.350 \pm 0.003$. The helix radius was given as 0.96 ± 0.02 Å. The difference is now the specification of k rather than p, but the subtle relationships among the standard errors on these quantities should be explored, in particular, the limit or error of 0.003 on k, a value only one-seventh as large as that obtained via the limit on γ given in Tuinstra (1966). Somewhat different structural results were given:

S–S bond length	$= 2.067 \pm 0.032$ Å
S–S–S bond angle	$= 105.73 \pm 1.83°$
S–S–S–S torsion angle	$= 84.20 \pm 1.73°$

It is mysterious that while these quantities differ, albeit slightly, from those given earlier by Tuinstra (1966), the calculated intensities in the two publications are identical.

Lind and Geller (1969) then described the results of their complete analysis. The true unit cell was now given as monoclinic, with $a = 17.6$, $b = 9.25$, $c = 13.8$ Å, $\beta = 113°$. The relation between this cell and the previous end-centered orthorhombic one is seen in Fig. 16-27. The pseudo-orthorhombic symmetry was stated to be the result of multiple twinning.

The 429 observed F values were refined by least squares to an R value of 16.2%. The number of positional parameters was reduced from 116 to five by assuming regular helices, with 10 atoms in three turns and the following coordinates for the helical axes (these coordinates pertain to the orthorhombic pseudo-cell):

Helix	y_0	z_0
A	0	$\frac{1}{4}$
B	0	$\frac{3}{4}$
C	y	z
D	y	$\frac{1}{2} + z$
E	$\frac{1}{4}$	$\frac{1}{4}$
F	$\frac{1}{4}$	$\frac{3}{4}$

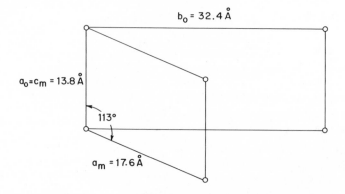

Fig. 16-27.   Relation between the true monoclinic cell (subscript m) and the pseudo orthorhombic cell (subscript o) of fibrous sulfur (according to Lind and Geller, 1969). The vertical axis (c_o and b_m) is 13.8 Å in both cases.

The coordinates of the 10 atoms in each helix, y and z relative to those of the axis, are given by the relations

$$x_n =\ \ x_0 + 0.1(n - 1)$$

$$y_n = - \frac{r}{b} \sin\{\theta_0 + p(n - 1)\ 108°\}$$

$$z_n =\ \ \frac{r}{c} \cos\{\theta_0 + p(n - 1)\ 108°\},\ n = 1 \text{ to } 10$$

where r is the radius of the helices, θ_0 is the angle of rotation at the first atom, and x_0 the translation along the axis of that atom; for a right-handed helix $p = 1$, for left-handed, $p = -1$. The following relations were assumed for the x_0 and θ_0:

Helix	x_0	θ_0
A	0	0°
B	$\frac{7}{20}$	18°
C	x	θ
D	$x + \frac{7}{20}$	$\theta - 18°$
E	$\frac{3}{5}$	18°
F	$\frac{1}{4}$	0°

Helices C and D are left handed, the others, right handed.   The final values of the refined parameters (and their estimated standard errors) are:

$$x\ = 0.4482\ (0.0012)$$
$$y = 0.1247\ (0.0004)$$
$$z = 0.0433\ (0.0008)$$
$$\theta\ = 17.4°\ (0.4°)$$
$$r\ = 0.95\ Å\ (0.003\ Å)*$$

*The order of magnitude difference between the significance on r and on its standard error was not explained.

These lead to the following values for the molecule:

S–S bond length	$= 2.066$ Å
S–S–S bond angle	$= 106.0°$
S–S–S–S torsion angle	$= 85.3°$

Because of the assumptions used in the refinement, the standard errors of these values are without significance. A projection of a portion of the structure along the helical axes is shown in Fig. 16-28. Atoms not included in this figure may be obtained by use of the equivalent positions (xyz), $(\bar{x}\ \bar{y}\ z)$, $(\frac{1}{2}+x\ \frac{1}{2}+y\ z)$, $(\frac{1}{2}-x\ \frac{1}{2}-y\ z)$.

For two other forms of sulfur thought to consist of infinite chains see Sections 1.2.a and 1.2.b, below, p. 355 ff.

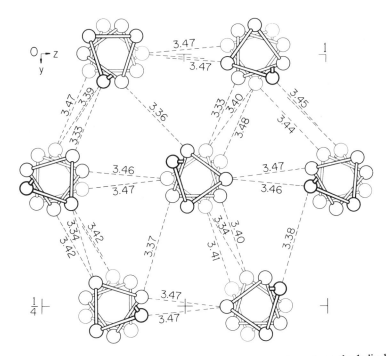

Fig. 16-28.   Projection of a portion of the fibrous sulfur structure along the helical axes, showing the shortest interatomic distances between chains (after Lind and Geller, 1969).

I. Provisional, Doubtful, and Insufficiently Characterized Forms of Sulfur

The total number of allotropes reported is close to 50, but very few of these save those discussed above have been satisfactorily characterized. A specified method of preparation, even if it leads to a material with a reproducible powder pattern, is insufficient evidence for the existence of a single species, for witness the cases of mercury, polonium, and fibrous sulfur, where structures were proposed on the basis of diffraction patterns which were later shown to be due to two different phases present in the preparations. Various samples believed to be sulfur allotropes are treated below, with the more credible ones coming first. Not included are molecular species observed only in mass spectrometry or in matrices at low temperatures after having been trapped from sulfur vapor at very high temperatures. Such noncrystalline allotropes lie outside the scope of this treatise.

1. Muthmann's Sulfur IV, δ-Sulfur

This form is obtained, in one experiment out of ten, from cold, saturated, alcoholic solutions of ammonium polysulfide (Muthmann, 1890). It has also been observed by Korinth (1928b) and Erämetsä (1963a). It appears in the form of hexagonal tablets which rapidly transform at room temperature. Thackray (1965) identified as δ-sulfur some regions observed during the crystallization of molten sulfur, but it is not clear why these regions were thought to consist of the δ-sulfur of Muthmann. If they indeed are, then this phase is very probably yet another form of cyclooctasulfur. No diffraction data have been reported.

2. Insoluble Sulfur

Sulfur which is insoluble in carbon disulfide may be prepared in a variety of ways, including the hydrolysis of sulfur monochloride (Das, 1938), quenching liquid sulfur (Das and Ghosh, 1939), sublimation of flowers of sulfur (Das and Ghosh; Boule, 1947), the reaction of sulfur dioxide with hydrogen sulfide at $0°C$ (Das, 1955), irradiation of a saturated solution of α-sulfur in carbon disulfide with sunlight (Erämetsä and Suonuuti, 1959), or by carrying out the irradiation in weak light or the dark (Erämetsä, 1963b). Insoluble sulfur is available commercially under the trade name "Crystex", also known as supersublimated sulfur.

The above are *not* the same as some preparations which consist of the residue after extraction of quenched liquid sulfur with carbon disulfide (e.g., Das 1938, Pinkus, Kim, McAtee and Concilio, 1957, 1959b), and which give powder patterns virtually identical to those of α-sulfur (Das. 1938; Donohue and Caron, 1961). It has been suggested by Das (1938) and by Das and Ghosh (1939) that the insolubility of this material, which is essentially α-sulfur, may in some way be due to the presence of sulfur dioxide. Although the powder patterns of the various preparations of authentic insoluble sulfurs are not always identical, as discussed below, they are entirely different from that of α-sulfur.

Erämetsä and Suonuuti (1959) showed that most of the specimens which had previously been called ω-sulfur (fibrous, insoluble) were actually mixtures of (at least) two independent allotropes which were termed ω-sulfur and μ-sulfur. This conclusion was confirmed by Erämetsä (1963b) and by Tuinstra (1967a), who at first termed these two components ω1 and ω2, respectively, but then suggested that the names "second fibrous sulfur" and "laminar sulfur" be used instead. These appellations seem unambiguous and thus are used below. The respective symbols suggested by Tuinstra (1967b) are Sψ, and Sχ.

(a) **Second Fibrous Sulfur.** Tuinstra (1967a, b) recorded the powder pattern of Crystex and noted that if the material was carefully heated to 110°C, the insoluble residue gave a pattern which lacked some of the lines of the original. On the other hand, the pattern of specimens, homemade by quenching, if carefully heated to 80°C, also lost some lines, the *remaining* lines in this case being exactly those which *disappeared* in the Crystex experiments. A powder pattern consisting of 29 observed lines was tabulated, but it is not entirely clear how the sample used to obtain this pattern was prepared.

Erämetsä and Suonuuti (1959) had previously published diffraction patterns of two preparations which they called ω-sulfur, one prepared by the hydrolysis of sulfur monochloride, the other from the washed (with carbon disulfide) precipitate obtained when a saturated solution of α-sulfur in carbon disulfide was exposed to sunlight for 6 to 8 hours. They published a pattern of eight lines, which they indexed on the hexagonal unit of Das (1955), which has $a = 8.24$ and $c = 9.15$ Å.

We now encounter yet another set of curious contradictions, not uncommon in the structural chemistry of sulfur. The above unit cell was supposedly derived by Das from data previously published by him, that is, a powder pattern of seven lines (1938). However, comparison of that pattern with that of Erämetsä and Suonuuti reveals that they are not the same. Furthermore, Tuinstra, although correct in stating that the patterns from his treated, homemade specimens were "similar" to those from the ω-sulfur of Erämetsä and Suonuuti, ignored two serious discrepancies,* and it is thus possible that the materials examined in these two investigations were not identical.

---

*These are a line at 3.56 Å observed by Erämetsä and Suonuuti but not by Tuinstra, and a line at 2.92 Å observed by Tuinstra but not by Erämetsä and Suonuuti. These discrepancies might not be so serious were it not for the relatively small number of observed lines.

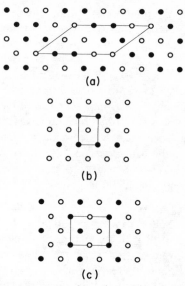

Fig. 16-29. Hexagonal planar structures of equal numbers of right-handed (open circles) and left-handed helices (filled circles) tested by Tuinstra (1967a, b). (a) 3-3 Coordination; (b) and (c) 4-2 coordination. Structure (c) was assigned to second fibrous sulfur.

Tuinstra presented a detailed interpretation of his data. It is based on the packing of parallel helices having the dimensions used in his model for fibrous sulfur. The observed spacings could be indexed on an orthorhombic unit cell with $a = 9.02$, $b = 8.33$, $c = 4.58$ Å. (This repeat in the c direction is one-third the fiber axis repeat of 13.7 Å observed in fibrous sulfur.) Keeping the helical parameters fixed, Tuinstra tested three different combinations of right- and left-handed helices, those shown in Fig. 16-29. The first two were found to be unsatisfactory, but acceptable agreement between observed and calculated intensities was obtained with the third arrangement, as presented in a table which, however, gave only 20 reflections as having observable intensity; quantitative data are given for 10 of these, which give an R value of 9.5% when calculated on the basis of $\sqrt{I}$. The arrangement of the helices in this structure is shown in Fig. 16-30.

(b) Laminar Sulfur. In the experiments of Tuinstra (1967a, b) described in the first paragraph of the preceding section, the other component, that is, the one which remains when Crystex is heated to 110°C and the one which disappears when homemade quenched sulfur is heated to 80°C, was named by him laminar sulfur. On the basis of diffraction patterns, this is the same form which had been previously observed by others, as summarized in Table 16-6, which also gives the synonymy. There is an important, inexplicable, and not

Table 16-6. Reports of Laminar Sulfur, as Identified by Similarity of the Powder Patterns

Reference	Name Used	Method of Preparation
Das, 1938	White sulfur, $S\omega$	Hydrolysis of S_2Cl_2
Das & Ghosh, 1939	$S\omega$	Sublimation; quenching
Schenk, 1955	$S\mu$	Quenching
Erämetsä & Suonuuti, 1959	μ-sulfur	Quenching; irradiation of carbon disulfide solutions of α-sulfur in weak light
Erämetsä, 1963b	μ-sulfur	Irradiation of ethanol, methanol, chloroform, etc, solutions of α-sulfur in sunlight
Geller, 1966	Pressure induced sulfur I	Pressure-temperature experiments: see Fig. 16-34
Tuinstra, 1967a, b	χ-sulfur, ω2-sulfur, laminar sulfur	Heating Crystex to $110°C$

357

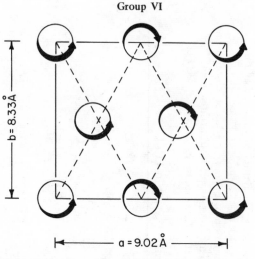

Fig. 16-30.   Schematic drawing of the structure of second fibrous sulfur according to Tuinstra (1967a, b), viewed along the helical axes. (Cf. Figs. 16-25 and 16-26.)

unexpected discrepancy regarding this form: Tuinstra prepared it by heating Crystex, which is presumably a mixture of second fibrous sulfur and laminar sulfur, to 110°C, while both Das (1938) and Das and Ghosh (1939) found that their S_ω (=white sulfur = laminar sulfur) transformed to α-sulfur when heated above 88°C.

The seven-line pattern of Das (1938) agrees well with those found in the other references of Table 16-6. In yet another singular sequence in sulfur structural chemistry, Das (1955) then stated that he had obtained a powder pattern of white sulfur consisting of 13 lines, did not list the spacings, but added that these could be indexed on a hexagonal cell having $a = 8.24$, $c = 9.15$ Å. This cell does not pertain to the patterns from the substances listed in Table 16-6, but to the second fibrous sulfur of the preceding section. The conclusion to be reached, however unreasonable it may seem, is that the preparation described by Das in 1938 was different from the one described by him in 1955, and that he was unaware of this situation.

Erämetsä (1963b) indexed his pattern of five lines on a hexagonal unit cell having $a = 14.12$, $c = 14.44$ Å, but made no structural conjectures.

Geller (1966) presented a powder pattern of 26 lines but stated that twinning and pseudo-symmetry prevented a clear determination of the unit cell, which appeared to be pseudo-tetragonal, or perhaps monoclinic. The unique axis was suggested to be of length 32.4 Å, with the two equal axes at 6.55 ($= 4.6\sqrt{2}$) Å. Other than pointing out that it was both unlikely that S_8 or S_6 molecules were present, and that the coordination of the sulfur atoms was different from two, Geller also made no structural conjectures but did point out the the seven strongest lines matched those of Das (1938).

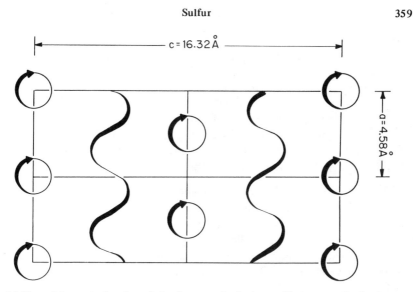

Fig. 16-31.   Schematic drawing of the "cross-grained plywood" structure for laminar sulfur proposed by Tuinstra (1967b).

The powder pattern of Tuinstra (1967b) consists of 13 lines (his Table I) or 12 lines (his Table III, unfortunately misprinted as pertaining to S_{ω_1} instead of S_{ω_2}). This was indexed on a body centered tetragonal unit cell having $a = b = 4.58$, $c = 16.32$ Å. This cell is closely related to the one tentatively suggested by Geller (see preceding paragraph), but there is no obvious relation between it and the hexagonal cell of Erämetsä. Because the fiber diagrams suggested an arrangement of parallel chain molecules, Tuinstra tried "all possible conformations of this type", but these were ruled out on the basis of improbably short nonbonded distances. A structure consisting of planes of molecules perpendicular to the c-axis, with the molecules in successive planes at right angles, was said to be satisfactory, with the best packing achieved when the molecules in neighboring planes are of opposite chirality. For the seven reflections for which quantitative intensity data were given, the R value, based on $\sqrt{I}$, has the remarkably small value of 3.8%

This "cross-grained plywood" structure is depicted schematically in Fig. 16-31. The feature of having helices oriented perpendicular to each other is unique in polymer structures, and is possible for sulfur because the pitch of the helix is the same as the distance between adjacent helix axes.

3. High Pressure Forms of Sulfur

Studies of sulfur under pressure have led to a maze of contradictions and inconsistencies. For this reason the results are considered chronologically. First, however, it is noted that there is serious disagreement among the six reported melting curves, as shown in Fig. 16-32. Not only are there major differences in the general slope of this curve, but one set of investigators found evidence of three polymorphic transitions, two sets found evidence for one transition, and three sets, no evidence of a transition. (These transitions are in addition to the α-orthorhombic–β-monoclinic transition which occurs at about 1 kbar and does not show in Fig. 16-32.) The discrepancies among the various results have been noted by Vezzoli, Dachille, and Roy (1969b). These disparities concerning the melting of sulfur at high pressures suggest that other results in this area might well be treated with circumspection.

   (a) Metallic Sulfur. David and Hamann (1958) subjected sulfur to shock pressures of about 230,000 atm and observed a reduction of the specific resistance from $> 10^{18}$ ohm-cm to under 0.03 ohm-cm, and were inclined to think that the sulfur was metallic. However, in an alternate interpretation, Berger, Joigneau, and Bothet (1960) suggested that the increase in conductivity may have been caused by impurities as the sulfur melted along the shock front.

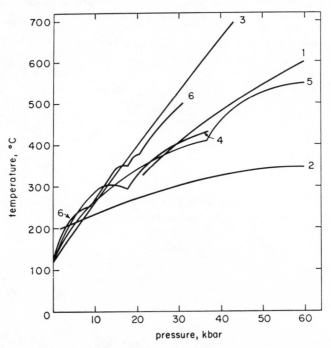

Fig. 16-32.   Reported melting curves of sulfur. (1) Susse, Epain, & Vodar, 1964; (2) Baak, 1965; (3) Deaton & Blum, 1965; (4) Paukov, Tonkov, & Mirinski, 1965; (5) Ward & Deaton, 1976; (6) Vezzoli, Dachille, & Roy, 1969a. The α-β transition at ca. 1 kbar is not shown.

Harris and Jura (1965) found no evidence for a metallic structure even at pressures as high as 400 kbar, finding specific resistances of over 10^7 ohm-cm, but did not consider their results at variance with those of David and Hamann because those measurements were made under adiabatic conditions.

(b) Cubic Sulfur of Bäak. Bäak (1965) studied sulfur in the temperature: pressure ranges of 20 to 400°C:6 to 85 kbar; his phase diagram is shown in Fig. 16-33. A diffraction pattern of 25 lines was obtained from a sample quenched from the "cubic" region and indexed on a cell having a = 13.66 Å. This value, together with the observed density of 2.18_5 g cm^{-3}, gives 104 atoms per unit cell (calc. 104.6). Cubic sulfur was described as being light yellow and insoluble in carbon disulfide. No structural conjectures were made. It must be pointed out, however, that the agreement between the calculated spacings (not given by Bäak) and the observed spacings is not entirely satisfactory.

(c) Pressure-Induced Phases of Geller. Geller (1966) obtained, reproducibly, three pressure-induced phases of sulfur, in addition to plastic and orthorhombic sulfur, in the ranges 10 to 65 kbar and annealing temperatures 100 to 350°C (other annealing temperatures above 350°C were stated to be probably above the melting point). The conditions under which the three forms appear are shown in Fig. 16-34. The forms were identified by their distinct powder patterns, recorded at room temperature from quenched specimens.

(i) Phase I (= Laminar Sulfur). The powder pattern of Phase I consisted of 26 lines, the seven strongest of which, as pointed out by Geller, agree well with the pattern of "white sulfur" presented by Das (1938). As discussed

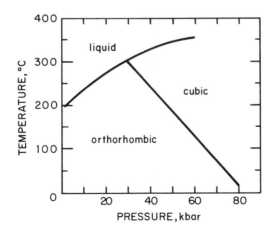

Fig. 16-33.   The phase diagram of sulfur according to Bäak (1965).

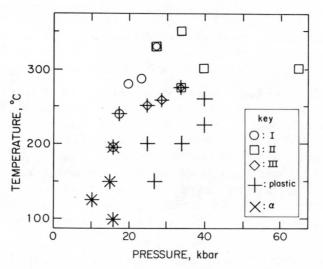

Fig. 16-34.   Conditions observed for the production of various phases of sulfur reported by Geller (1966); see text for identification of phases I and II.

above in Section I.2.b, this is the form designated by Tuinstra (1967b) as
laminar, or $\omega 2$ sulfur. The 13-line pattern of Tuinstra for this form is in excellent agreement with that of Geller's phase I. A structure for this form proposed by Tuinstra is described in the section called Laminar Sulfur.

Ward and Deaton (1967) also observed phase I in the range 20 to 29 kbar,
from about 240°C to the melting point. They did not present any diffraction
data but stated that their patterns agreed with those of Geller.

Roof (1972) indexed the powder pattern of Geller's phase I on a monoclinic unit cell having $a = 7.086$, $b = 6.215$, $c = 5.319$ Å, $\beta = 96.19°$, and
suggested that the space group is one of the five in the monoclinic system
which is primitive and without glide planes. No other structural details were
conjectured. (If the density is taken as 2.0 ± 0.1 g cm^{-3}, then this unit cell
contains 8.8 ± 0.4 atoms of sulfur.)

(ii) Phase II (= Fibrous Sulfur). Phase II of Geller is the same as the well-
characterized fibrous sulfur, described in detail previously. This phase was
also observed, almost simultaneously, by Sclar, Carrison, Gager, and Stewart
(1966). Their 19-line powder pattern for what they termed the 4.04 Å phase
is in excellent agreement with that given by Geller for his phase II. Sclar et
al. noted that their pattern most closely resembles that of fibrous sulfur reported by Prins and Tuinstra (1963a), and concluded that the 4.04 Å phase
may be structurally related. This work preceded, of course, the definitive
work on fibrous sulfur described in the section on S_∞. The conditions under
which various forms of sulfur appear are indicated in Fig. 16-35.

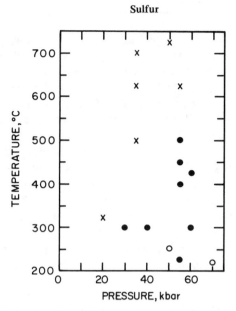

Fig. 16-35.   Products of the high pressure experiments on sulfur of Sclar et al. (1966).
X, liquid; open circles, orthorhombic; filled circles, "4.04 Å phase" (=fibrous sulfur,
phase II of Geller, 1966).

Shortly thereafter Ward and Deaton (1967) also observed pressure-induced
fibrous sulfur, noting that their diffraction patterns agreed with those of
Geller. They stated that their data indicated an orthorhombic-fibrous-liquid
triple point at 37 kbar and 410°C. However, this apparent triple point could
be changed to 29 kbar and 380°C by using samples having initial nucleations
of fibrous sulfur.

(iii) Phase III. The powder pattern of phase III was stated by Geller to be
"essentially the pattern of phase I" (unfortunately misprinted as "phase II"),
but with the lines at 3.53, 2.66, and 2.32 Å missing or reduced in intensity,
and that these differences were definitely not a result of preferred orientation.
Other than remarking that it was closely related to phase I, no structural conjectures were made by Geller.

Roof (1972) indexed the powder pattern of Geller's phase III on a monoclinic unit cell having $a = 6.299$, $b = 7.240$, $c = 5.673$ Å, $\beta = 95.51°$. The
possible space groups he suggested are the same as those he gave for phase I,
above. The close similarity of the two unit cells was said to indicate that relatively minor differences may exist between the structures of phases I and III.
(If the density of phase III is 2.0 ± 0.1 g cm^{-3}, then there are 9.7 ± 0.5 atoms
in the unit cell.)

(d) The Twelve Crystalline Fields of Vezzoli, Dachille, and Roy. In 1969a, Vezzoli, Dachille, and Roy reported the results of x-ray diffraction measurements on samples at -20°C quenched from various pressures and temperatures together with transitions detected by volumetric, optical, and electrical resistance measurements. They found triple points along the melting curve at 9 kbar–255°C, 18 kbar–344°C, and 20.6 kbar–375°C. Their proposed phase diagram is shown in Fig. 16-36. Phases I, II, and XII were identified as α-orthorhombic, β-monoclinic, and fibrous sulfur, respectively; phases V, VIII, and IX were said to revert completely to α-sulfur despite the rapid quenching. The remaining six phases were considered different, and different from any of the other forms of sulfur. Confirmation of this rather astonishing phase behavior is much to be desired. It is interesting that Bridgman (1935) observed no polymorphic transition in sulfur, presumably at room temperature, up to 50,000 kg cm^{-2} (= ca. 50 kbar). Furthermore, the specific volume data of Harris and Jura (1965) show no evidence of any transitions up to 100,000 kg cm^{-2}.

4. The Four Sulfurs of Korinth

(a) ζ-Sulfur. Colorless, or sometimes pale red-brown orthorhombic crystals of ζ-sulfur crystallize out of a chloroform solution of sulfur which contains a small amount of selenium. The tabular crystals are probably monoclinic. Their production also requires patience and luck. They transform quickly into α-sulfur, sometimes in less than a minute (Korinth, 1928a, b; Erämetsä, 1958b). On the other hand, Erämetsä (1958c) later expressed the suspicion that ζ-sulfur may be α-sulfur having an unusual face development. If that is so, its rapid transformation to something else is somewhat mysterious.

(b) η-Sulfur. Korinth (1928a, b) obtained what he termed η-sulfur by evaporation of carbon disulfide solutions of sulfur to which a small amount of selenium had been added. This modification was described as almost colorless hexagonal tablets. Erämetsä (1963a) stated that he succeeded only three times (in 1957) in obtaining the η-sulfur described by Korinth, after hundreds of attempts. The crystals transformed to α-sulfur within about 10 minutes and were said to be more stable than δ-sulfur but less stable than ζ-sulfur.

(c) θ-Sulfur. Korinth (1928a, cited at length in Erämetsä, 1958b) produced a form of sulfur by allowing a carbon disulfide solution of sulfur, to which rubber and nitrobenzene or benzonitrile had been added, to evaporate slowly on a glass slide. The crystals were stated to belong to the tetragonal system, with $a{:}c$ = 1:0.46801. After 20 to 30 seconds the crystals begin to round at the corners and dissolve completely within a short time. Erämetsä also obtained it by chilling hot solutions of sulfur in α-pinene, and by

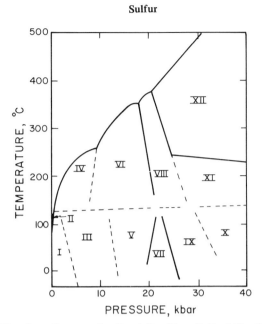

Fig. 16-36.   The phase diagram of sulfur (after Vezzoli, Dachille, & Roy, 1969a).

evaporation of cold, saturated solutions in nitrobenzene. The crystals were said to be easily recognizable under the microscope, but their occurrence in a given series of experiments was characterized as accidental.

(d) ξ-Sulfur. Triclinic crystals of sulfur were obtained by sublimation from the upper part of a sulfur bath. They were described as extremely unstable (Korinth, 1928a).

5. The π-Sulfur of Aten and Erämetsä

Aten (1914) explained the decrease of the melting point of sulfur from an initial value of 119 to 114.5°C by the formation of a new allotrope which he termed π-sulfur. Pure π-sulfur was obtained by cooling plastic sulfur and extracting with carbon disulfide at –80°C. Intense yellow crystals, said to be very unstable at room temperature, resulted on evaporation of the extract. Ebullioscopic measurements indicated the molecular formula S_4, but Braune (1954) and Poulsen-Nautrup (1954) both found, by the same method, the formula S_8 for a molecule assumed to be a chain, and relatively stable.

These conflicting results led Erämetsä (1959c) to investigate further material prepared by the method of Aten. Chromatography of the carbon disulfide extract on alumina gave *four* zones, with the properties summarized in Table 16-7.

Table 16-7.   The Four Components of Aten's π-Sulfur, According to
Erämetsä

Name	Color	Atoms/ Molecule	Remarks
π_1	White	6	Decomposes to μ, ω, and α in 12 hours
π_2	Intense yellow	~5.2	Gives S_8 when CS_2 solution is boiled; original molecule may have been S_4 or S_5
π_3	Lilac	?	Insufficient material obtained for x-ray diagram
π_4	Intense yellow	?	Decomposes to α-S_8 in solution, but was not originally S_8 because of adsorption properties

6. The Sixteen Sulfurs of Erämetsä (in chronological order)

(a) κ-Sulfur. Erämetsä (1958a) obtained black, pine-needle shaped crystals, presumably monoclinic, by cooling a nearly saturated solution of sulfur in o-xylene from the boiling point. They were said to have greater birefringence than β-sulfur and to be more stable than γ-sulfur.

(b) φ-Sulfur. During an unusually severe winter when the laboratory temperature fell to 8°C, Erämetsä observed that if an α-pinene solution of sulfur was allowed to stand overnight between glass plates, thin rectangular plates appeared. These plates, unlike the similarly appearing crystals of θ-sulfur, remained unchanged for several days. They exhibited unusually low birefringence.

(c) ψ-Sulfur. ψ-Sulfur arises under the same conditions as does φ-sulfur. The crystals, apparently orthorhombic, are rhombic plates or needles, with an unusually low refractive index. The production of both φ-and ψ-sulfur requires much patience and luck (Erämetsä, 1958b).

(d) τ-Sulfur. According to Erämetsä (1958b), τ-sulfur is one of the most difficult allotropes to obtain: in 5000 crystallizations it arose twice from o-xylene and once from menthenone. The crystals were said to be cubic tetartohedral with a strikingly high refractive index. After 12 hours they transform completely to α-sulfur.

(e) Red Sulfur. Erämetsä (1959a) produced what he termed red sulfur by reaction of ammonium polysulfide solution with sulfur monochloride at a pH of 8. A carbon disulfide extract of the red, puttylike precipitate was chromatographed on silica gel, giving three zones: A, orange-yellow, most strongly adsorbed; B, dark red; C, light yellow. Zone A could not be eluted, zone B was found by x-ray diffraction to be amorphous, and zone C decomposed in a paler material. The part not adsorbed on silica gel was a beautiful red color, and 100.0% sulfur. This was chromatographed on alumina, giving four zones: D, orange, not mentioned again; E, brownish yellow; F, pale violet; G, pale yellow. Pure α-sulfur was found in the eluate. According to control analyses, E, F, and G were pure sulfur.

(i) E-Sulfur. E-sulfur gave a powder pattern which contained maxima not present in the patterns of any known allotropes of sulfur. Cryoscopic determination of the molecular weight (in camphor) gave a value corresponding to 12.4 atoms per molecule.

(ii) F-Sulfur. The powder pattern of F-sulfur was likewise different from that of all known allotropes. The molecular weight corresponds to $S_{13.1}$. After two weeks, F-sulfur had transformed almost completely into α-sulfur.

(iii) G-Sulfur. The x-ray diagram of component G was quite similar to that of α-sulfur, with, however, additional maxima. Furthermore, the chromatographic behavior of G-sulfur deviates from that of α-sulfur. The molecular weight corresponds to S_9.

(f) Orange Sulfur. Erämetsä (1959b) produced orange sulfur by the same reaction that gave red sulfur, but at a different polysulfide concentration. The precipitate, after various preliminary treatments, was extracted with carbon disulfide and chromatographed, giving four zones: I, orange-yellow, most strongly adsorbed; K and L, both pale yellow; M, lilac.

(i) I-Sulfur. Zone I analyzed as 100.2% sulfur. The ebullioscopically determined molecular weight of 32.5 corresponds to a monatomic molecule. This component did not give an x-ray diffraction diagram and was accordingly stated to be amorphous.

(ii) K-Sulfur. This zone was 99.9% sulfur; the crystals consist of octaätomic molecules and have a melting point of 112°C. The x-ray diagram was similar to, but not identical with, that of α-sulfur. One strong maximum in it was attributed to E-sulfur.

(iii) L-Sulfur. L-Sulfur was 99.5% sulfur, with the same melting point and molecular formula as K-sulfur. The diffraction pattern contained maxima belonging to no previously known species of sulfur.

(iv) M-Sulfur. The melting point of crystals from this zone was also 112°C, and the molecular formula also S_8. In the powder pattern the lines of α-sulfur were almost completely absent. It was said to be very short lived.

(g) ν-Sulfur. The undissolved orange-yellow residue after red sulfur (see above) had been extracted with carbon disulfide was designated component H by Erämetsä (1959b). It contained about 49% sulfur and was insoluble in boiling xymol, toluene, glycerin, methylbenzoate, and hexane. On slowly heating to 300°C, the color changed to dark brown, and pale yellow crystals sublimed out. These consisted of about one-half of the original mass. The major fraction of the sublimate was soluble in carbon disulfide, and was found to be α-sulfur. The pale yellowish gray residue proved to be pure sulfur and gave a diffraction pattern which clearly differed from all previously known allotropes. This form was designated ν-sulfur. It was said to be cubic and had an experimental density of 2.025 g cm^{-3}. No lattice constant was given. The production of ν-sulfur was "entirely reproducible".

In replying to criticism (not cited) that ν-sulfur was a mixture of plastic sulfur and ammonium chloride Erämetsä (1962) described his preparation steps in great detail. After various washing and drying procedures, the residual ν-sulfur was shown, by chemical analysis, to be pure sulfur. The x-ray diffraction pattern, at first glance, was very similar to that of ammonium chloride, but with measurable differences. The lattice constant of ν-sulfur was given as 3.8656 Å, while that of ammonium chloride, determined under the same conditions, as 3.8650 Å. The calculated number of sulfur atoms in the unit cell, not calculated by Erämetsä, is thus 2.198. There is no apparent explanation for the significant departure from nonintegrality.

(h) i-Sulfur. Erämetsä (1959c) reported that when dry π_2-sulfur was allowed to stand overnight, part of it became insoluble in carbon disulfide. The x-ray diffraction pattern showed that it was a form of insoluble sulfur different from μ-, ω-, or ν-sulfur, which was termed i-sulfur. It is formed only from dry π_2-sulfur—from the solution one obtains in the twilight only mere α-sulfur. ("An der Lösung enthält man in der Dämmerung nur blossen α-Schwefel.")

(i) m-Sulfur. During the preparation of ν-sulfur, Erämetsä (1962) found that after ν crystals had formed the carbon disulfide solution, on standing, yielded a new allotrope insoluble in that solvent, which he termed m-sulfur. Its x-ray diagram was said to indicate a rhombohedral structure (on p. 155), but orthorhombic lattice constants of $a = 13.5$, $b = 20.1$, $c = 21.4$ Å were given on p. 157. Publication of a more comprehensive study was promised for the immediate future.

(j) o-Sulfur. If the solution from which η-sulfur crystals are obtained is allowed to evaporate freely, it often happens, according to Erämetsä (1963a), that an allotropic form appears which is different from all previously described modifications. When the crystal has grown for about 5 seconds it begins to change to γ-sulfur. The boundary line between the phases migrates over the whole crystal in 2 to 3 seconds. o-Sulfur is thus the shortest lived of all of the sulfur allotropes.

(k) f-Sulfur. In his investigations of μ-sulfur Erämetsä (1963b) found that if liquid sulfur was heated not above 140°C the fraction insoluble in carbon disulfide gave a powder pattern different from those of either μ- or ω-sulfur. He called it f-sulfur.

7. The Black Sulfur of Schenk

When gaseous S_2 is condensed, a black deposit is sometimes observed. It was presumed to be a form of sulfur (Schenk, 1953).

8. The Black Sulfur of Skjerven

Skjerven (1962) heated highly purified sulfur and obtained a brown-black residue after extraction with carbon disulfide. Irradiation caused a transformation to α-sulfur, so that it was not possible to obtain a diffraction pattern.

9. The Violet Sulfur of Meyer

Meyer (1960) slowly condensed a molecular beam of S_2 at 60°K or lower, and obtained a violet-purple solid. The presence of S_2 in this condensate was demonstrated by Meyer and Schumacher (1960a, b). Above 80°K the color changes to green.

10. The Green Sulfur of Rice and of Meyer

By condensing cyclo-S_8 vapor at liquid air temperature Rice and Ditter (1953) obtained a green product. It is also obtained by warming violet sulfur (see above). According to Meyer (1964) green sulfur is a solid with a complicated composition, with at least four different molecular species present. At 173°K the color changes to yellow.

11. The Purple Sulfur of Rice

Rice and co-workers (1953, 1959) condensed S_2 vapor in liquid nitrogen traps and obtained a purple deposit, which was attributed to the S_2 radical. Meyer (1964), however, stated that the color varied with the speed of condensation, with fast condensation giving purple sulfur and slow giving green sulfur.

Group VI

SELENIUM

Structures have been reported for six crystalline forms of selenium: α (normal, metallic, trigonal), which is the form stable under normal conditions; two monoclinic forms, α and β, obtained simultaneously on evaporation of carbon disulfide extracts of vitreous selenium, the latter being obtained by melting and quenching metallic selenium; and three cubic forms obtained as thin films by vacuum evaporation.

α-Selenium

The structure of the ordinary form of selenium was first determined by Bradley (1924b), following a determination of the unit cell by Slattery (1923). It is trigonal, space group $P3_1 21-D_3^4$ and the enantiomorphous $P3_2 21-D_3^6$, three atoms per unit cell at $(x0\frac{1}{3}\ 0x\frac{2}{3}\ \bar{x}\bar{x}0)$. The value of x, as determined by Bradley, is 0.217, but a more precise value, recently determined by Cherin and Unger (1967b), is 0.2254 ± 0.0010, as obtained in a least squares analysis of 12 reflections from a single crystal. Various determinations of the lattice

Table 34-1.   The Lattice Constants of α-Selenium at 25°C[a]

a, Å	c, Å	c/a	Reference
4.35[b]	4.96[b]	1.14	Slattery, 1923, 1925
4.36[b]	4.98[b]	1.14	Bradley, 1924
4.146[b]	5.514[b]	1.330	Olshausen, 1925
4.355[b]	4.962[b]	1.139	Parravano & Caglioti, 1930
4.369[b]	4.972[b]	1.138	Tanaka, 1934
4.35[b]	4.96[b]	1.14	Prins & Dekeyser, 1937
4.3656[c]	4.9590[c]	1.1359	Straumanis, 1940
4.3656[d]	4.9584[d]	1.1358	Krebs, 1949
4.3639[e]	4.9590[e]	1.1365	Straumanis, 1949
4.384[f]	4.96[f]	1.131	Grison, 1951
4.3659[g]	4.9537[g]	1.1346	Swanson, Gilfrich, & Ugrinic, 1955
4.3667[h]	4.9552[h]	1.1348	Deshpande & Pawar, 1964
4.3605[f]	4.9604	1.1376	Bonnier, Hicter, Aleonard, & Laugier, 1964
4.30[h]	4.89[h]	1.14	Avilov & Imamov, 1969
4.36[f]	4.94[f]	1.13	Bakinov, Ibaev, & Agaev, 1971
av. 4.3655	4.9576	1.1356	
±0.0010	±0.0024	±0.0007	

[a]Values corrected when necessary to 25°C with the thermal expansion coefficients given in the text.   [e]From kX and 20.1°C.
[b]From kX, omitted from the average.   [f]Omitted from the average.
[c]From kX.   [g]From 26°C.
[d]From kX and 20°C.   [h]From 30°C.

The structure consists of infinite helices parallel to the c-axis, three atoms per turn. The Se–Se bond distance is 2.374 ± 0.005 Å, the Se–Se–Se bond angle $103.1 \pm 0.2°$, and the Se–Se–Se–Se torsion angle $100.7 \pm 0.1°$. The interatomic distances between chains, four per atom, is 3.436 Å. The structure in projection down c is shown in Fig. 34-1.

This structure may also be described as a distortion of a simple cubic structure. This point of view is discussed in detail in the section Tellurium. This distortion is much greater in selenium than in tellurium: the interchain distance in the former is 45% longer than the bond distance, as opposed to 23% in the latter (1.062 versus 0.660 Å).

There is no change in structure up to at least $213°C$ (Bonnier, Hicter, Aleonard, and Laugier, 1964). (The melting point is $217°C$.) There is, however, a contraction along c. The average thermal expansion coefficients are $\alpha_\perp = 75 \times 10^{-6}$ deg^{-1} and $\alpha_\parallel = -17 \times 10^{-6}$ deg^{-1}. The variation of the lattice constants with temperature is shown in Fig. 34-2.

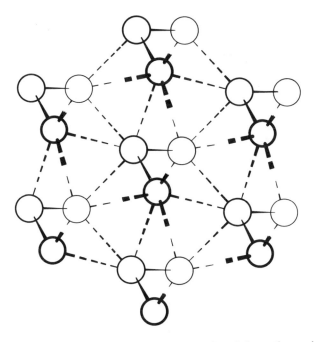

Fig. 34-1.   The structure of α (metallic)-selenium projected down the c-axis. The bond distance (solid lines) is 2.374 Å, the interchain distances (dashed lines) are 3.436 Å.

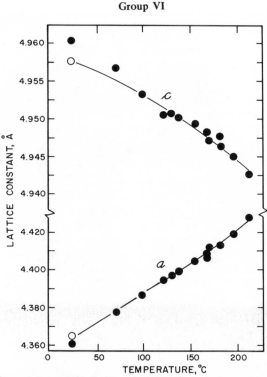

Fig. 34-2.   The variation of the lattice constants of α-selenium with temperature. Open circles, Table 34-1; filled circles, data of Bonnier et al., 1964.

α-Monoclinic Selenium

α-Monoclinic selenium is deposited as chunky, complicated crystals from carbon disulfide extracts of glassy selenium. β-Monoclinic selenium also appears at the same time under these conditions. The two are similar in color, deep red, but usually can be distinguished by their face development, as shown in Figs. 34-3 and 34-4. Both forms revert to α-selenium on heating.

The unit cell of α-monoclinic selenium was first determined by Halla, Mehl, and Bosch (1931). Various determinations of the lattice constants are presented in Table 34-2; the recent ones of Cherin and Unger (1972) are doubtless the most precise, and are used below.

The space group was incorrectly stated by Halla et al. and by Klug (1934) to be $P2_1/m\text{--}C_{2h}^2$. The correct space group, as determined by Burbank (1951), is $P2_1/n\text{--}C_{2h}^5$. The unit cell contains 32 selenium atoms in eight sets of position 4e, at $\pm(xyz)(\frac{1}{2}+x\ \frac{1}{2}-y\ \frac{1}{2}+z)$. The structure was solved and refined by the use of Fourier analysis of the three prism zones: 88 $h0\ell$, 86 $0k\ell$, and 85 $hk0$ reflections gave a final R value of 18% (unobserved reflections ex-

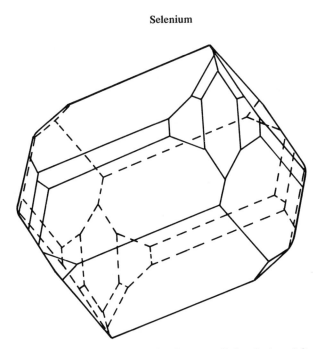

Fig. 34-3.   A common habit of crystals of α-monoclinic selenium (after Groth, 1906).

cluded). A more recent least squares refinement of 1108 reflections (Cherin and Unger, 1972) gave the positional parameters presented in Table 34-3. They differ slightly from those of Burbank, and correspond to an R value of 7.2%.

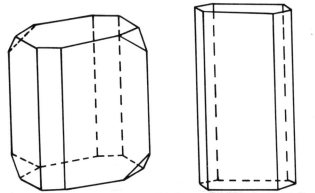

Fig. 34-4.   Two common habits of crystals of β-monoclinic selenium (after Groth, 1906).

Table 34-2. Lattice Constants of α-Monoclinic Selenium at Room Temperature

a, Å	b, Å	c, Å	β	Reference
9.00[a]	9.00[a]	11.52[a]	90°57'	Halla, Mehl, & Bosch, 1931
9.01[a]	8.99[a]	11.54[a]	91°34'	Klug, 1934
9.05	9.07	11.61	90°46'	Burbank, 1951
9.064[a]	9.072[a]	11.596[a]	90°52'	Newton & Colby, 1951
9.042	9.052	11.574	90°48'	Cherin & Unger, 1966
9.054	9.083	11.601	90°49'	Cherin & Unger, 1972

[a]From kX.

Table 34-3. Positional Parameters for α-Monoclinic Selenium

Atom	x	y	z
1	0.3209	0.4840	0.2362
2	0.4254	0.6625	0.3569
3	0.3178	0.6376	0.5378
4	0.1343	0.8186	0.5529
5	−0.0862	0.6904	0.5203
6	−0.1565	0.7322	0.3294
7	−0.0814	0.5217	0.2290
8	0.1301	0.5990	0.1337

The structure consists of eight-membered rings of selenium atoms packed together in a complicated fashion. The rings, within the limits of error of the experiment, have symmetry $\bar{8}2m$-D_{4d}. The molecular dimensions are shown in Figs. 34-5 and 34-6. The average Se–Se bond distance is 2.336 ± 0.007 Å, the average Se–Se–Se bond angle 105.7 ± 1.6°, and the average Se–Se–Se–Se torsion angle 101.3 ± 3.2°, as compared with the respective values of 2.374 ± 0.005 Å, 103.1 ± 0.2°, and 100.7 ± 0.1° in metallic selenium. In α-monoclinic system the bond distance is thus significantly shorter than that in metallic selenium.

The shortest distances between molecules do not appear to follow any regular pattern. The spectrum of all such distances less than 4.0 Å is shown in Fig. 34-7. This distance was chosen as the cutoff because it is twice the generally used value of the Pauling van der Waals radius. It is interesting in this regard that numerous intermolecular interatomic distances are much shorter than this value, as is the interchain distance of 3.44 Å in metallic (α)-selenium. Table 34-4 lists all of the 47 neighbors less than 4.0 Å distant.

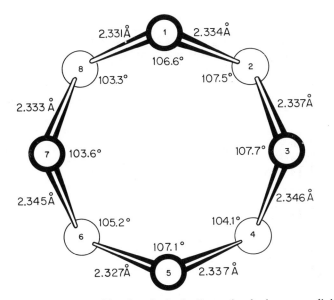

Fig. 34-5.   Bond distances and bond angles in the Se₈ molecules in α-monoclinic selenium.

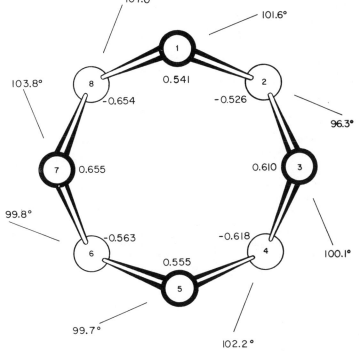

Fig. 34-6.   Torsion angles and distances from the mean plane (in Å) in the Se₈ molecules in α-monoclinic selenium.

Table 34-4. Interatomic Distances Less Than 4.0 Å between Molecules in α-Monoclinic Selenium[a]

Atom Number	Type	Distance, Å	Atom Number	Type	Distance, Å
From atom 1			From atom 5		
4	$4_{0\bar{1}0}$	3.70*	8	$2_{\bar{1}10}$	3.48
2	$4_{\bar{0}\bar{1}0}$	3.82*	3	3_{011}	3.70*
8	$4_{0\bar{1}0}$	3.83*	7	3_{011}	3.79*
6	$4_{0\bar{1}0}$	3.89*	5	3_{011}	3.83*
5	3_{011}	3.90	1	3_{011}	3.90*
4	$2_{01\bar{1}}$	3.99	1	$2_{\bar{1}10}$	3.97
5	$2_{0\bar{1}\bar{1}}$	3.97	From atom 6		
From atom 2			3	$2_{\bar{1}1\bar{1}}$	3.59
7	4_{000}	3.70	7	$4_{\bar{1}00}$	3.60
3	3_{111}	3.77	4	$2_{\bar{1}1\bar{1}}$	3.73
1	4_{000}	3.82	2	$1_{\bar{1}00}$	3.86
6	1_{100}	3.86	1	4_{000}	3.89
8	4_{000}	4.00	3	3_{011}	3.98
From atom 3			From atom 7		
6	2_{010}	3.59	4	$2_{\bar{1}1\bar{1}}$	3.57
5	3_{011}	3.70	6	$4_{\bar{1}\bar{1}0}$	3.60
7	3_{011}	3.76	2	$4_{0\bar{1}0}$	3.70
2	3_{111}	3.77	3	3_{011}	3.76
8	2_{010}	3.86	5	3_{011}	3.79
7	2_{010}	3.91	3	$2_{\bar{1}1\bar{1}}$	3.91
6	3_{011}	3.98	From atom 8		
From atom 4			5	$2_{01\bar{1}}$	3.48
7	2_{010}	3.57	1	4_{000}	3.83
1	4_{000}	3.70	3	$2_{\bar{1}1\bar{1}}$	3.86
6	2_{010}	3.73	4	$4_{0\bar{1}0}$	3.98
8	4_{000}	3.98	2	$4_{0\bar{1}0}$	4.00
1	$2_{\bar{1}10}$	3.99			

*See text.

[a]For each atom, the number of the neighbor, and its type, are given. The type numbers correspond to the equivalent positions: $1 = xyz$; $2 = \frac{1}{2}+x, \frac{1}{2}-y, \frac{1}{2}+z$; $3 = \bar{x}\bar{y}\bar{z}$; $4 = \frac{1}{2}-x, \frac{1}{2}+y, \frac{1}{2}-z$. Translations of a, b, and c are denoted by subscripts.

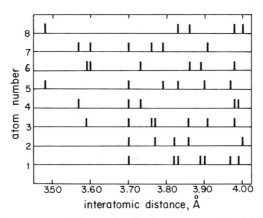

Fig. 34-7.   The spectrum of intermolecular distances less than 4.0 Å in α-monoclinic
selenium.

In two cases an atom of a neighboring molecule appears to be nestling above
the center of a square of intramolecular atoms; these are designated by an
asterisk in the table.

The structure, as viewed in projection along the three crystallographic
axes, is shown in Figs. 34-8, 34-9, and 43-10.

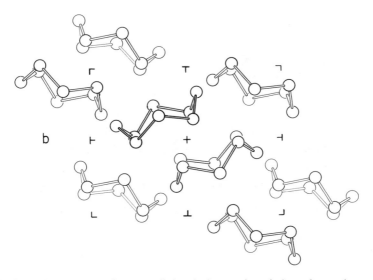

Fig. 34-8.   The structure of α-monoclinic selenium projected along the a-axis.

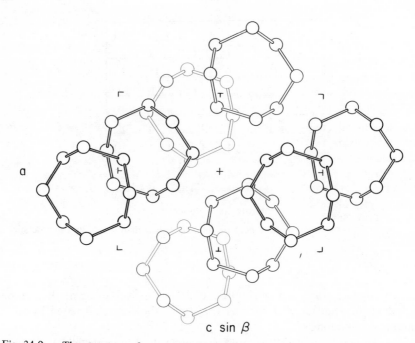

c sin β

Fig. 34-9.   The structure of α-monoclinic selenium projected along the *b*-axis.

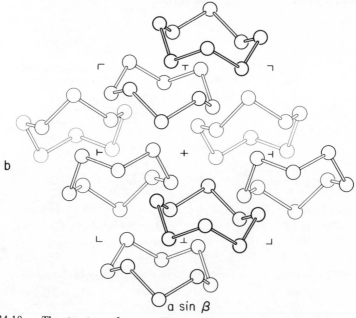

a sin β

Fig. 34-10.   The structure of α-monoclinic selenium projected along the *c*-axis.

β-Monoclinic Selenium

The unit cell and space group of β-monoclinic selenium were first determined by Klug (1934), who reported a = 12.77, b = 8.06, c = 9.27 Å (from kX), β = 93°4′, 32 atoms per cell in eight sets of position 4e, at $\pm(xyz)(\frac{1}{2}+x \ \frac{1}{2}-y \ z)$, space group P2$_1$/a–C$_{2h}^5$. Closely similar results were later reported by Burbank (1952), that is, a = 12.85, b = 8.07, c = 9.31 Å, β = 93°8′, same space group and number of atoms per unit cell. The more recent lattice constants are used below.

A structure for this allotrope was first reported by Burbank (1952), who used reflections from the three prism zones to calculate the corresponding projections along the a-, b-, and c-axes. That structure consisted of eight-atom chain molecules, quite different from the eight-membered rings found previously in α-monoclinic selenium. Furthermore, surprisingly short intermolecular interatomic packing distances of 2.85 and 2.86 Å were reported, whereas the shortest such distances in α-monoclinic selenium are 3.5 Å. These abnormal features suggested to Marsh, Pauling, and McCullough (1953) that the structure was erroneous. They found that if the y-coordinate of one of the atoms was given the value $\frac{1}{2}-y$ the expected eight-membered ring results, and if a shift of the origin by $\frac{1}{4}$ along x in the b-axis projection was made, the minimum intermolecular Se-Se distance increased to the satisfactory value of 3.50 Å. (This shift amounts to an interchange of the centers of symmetry and the twofold screw axes in the $h0\ell$ projection.) These two changes have no effect on the calculated magnitudes of the structure factors of the $h0\ell$ or $0k\ell$ reflections nor on the $hk0$ reflections if h is even, a situation which accounts for the satisfactory overall agreement attained by Burbank. Marsh et al. refined the corrected structure by Fourier methods, using the observed values of the $F_{h0\ell}$ reported by Burbank, and a set of observed F_{hk0} values (which had not been published by Burbank) obtained partly from an 18-year-old set of oscillation photographs and partly by estimation of the intensities on a reproduction of an $hk0$ Weissenberg photograph included in the paper of Burbank.

The final parameters of Marsh et al. are presented in Table 34-5. It should be noted that these values give an R value of 17.4% for the $hk0$ reflections having h odd, while the parameters published by Burbank give the value 40.8%. This result confirms the correctness of the structure of Marsh et al. (The R value for the remaining reflections is a little less than 20%.)

The structure, like that of α-monoclinic selenium, consists of eight-membered rings of selenium atoms packed together in a complicated fashion, but there is no obvious relation between the two structures. The rings in the β form also have symmetry $\bar{8}2m$–D$_{4d}$, within the limits of error. The molecular geometry, shown in Figs. 34-11 and 34-12, is virtually the same in the two allotropes (cf. Figs. 34-5 and 34-6). In the rings in the β form the average

Table 34-5.   Positional Parameters for β-Monoclinic Selenium

Atom	x	y	z
1	0.584	0.315	0.437
2	0.477	0.227	0.246
3	0.328	0.398	0.240
4	0.352	0.580	0.050
5	0.410	0.831	0.157
6	0.590	0.840	0.142
7	0.660	0.754	0.368
8	0.710	0.479	0.334

Se–Se bond distance is 2.337 ± 0.019 Å, the average Se–Se–Se bond angle 105.7 ± 1.0°, and the average Se–Se–Se–Se torsion angle, 101.4 ± 1.8°, none of which is significantly different from those in both metallic selenium and α-monoclinic selenium.

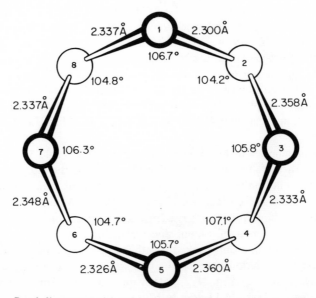

Fig. 34-11.   Bond distances and bond angles in the Se_8 molecules in β-monoclinic selenium.

Again, the shortest intermolecular interatomic distances do not appear to follow any regular pattern. The spectrum of all such distances less than 4.0 Å is seen in Fig. 34-13. As may be seen, many of them are smaller than the van der Waals distance of 4.0 Å. Table 34-6 lists all of the 39 neighbors less than 4.0 Å distant from one molecule. In one case an atom of a neighboring molecule appears to be nestling near the center of a square of intramolecular atoms; these are designated by an asterisk in the table.

The structure, as viewed in projection along the three crystallographic axes, is shown in Figs. 34-14, 34-15, and 34-16.

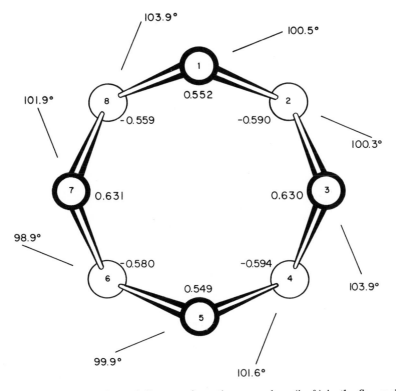

Fig. 34-12.   Torsion angles and distances from the mean plane (in Å) in the Se$_8$ molecules in β-monoclinic selenium.

Table 34-6.   Intermolecular Interatomic Distances Less Than 4.0 Å in β-Monoclinic Selenium[a]

Atom Number	Type	Distance, Å	Atom Number	Type	Distance, Å
From atom 1			From atom 5		
7	$2_{1\bar{1}1}$	3.70	2	1_{010}	3.40
7	3_{111}	3.75*	8	$4_{\bar{1}10}$	3.48
1	3_{111}	3.90*	7	$4_{\bar{1}10}$	3.92
3	3_{111}	3.91*	1	3_{111}	3.96
5	3_{111}	3.96*	From atom 6		
From atom 2			4	4_{010}	3.58
5	$1_{0\bar{1}0}$	3.40	2	1_{010}	3.60
6	$1_{0\bar{1}0}$	3.60	2	3_{110}	3.71
6	3_{110}	3.71	3	4_{010}	3.79
4	3_{110}	3.94	4	3_{110}	3.92
8	$4_{\bar{1}00}$	3.94	From atom 7		
From atom 3			8	2_{101}	3.64
8	$4_{\bar{1}00}$	3.53	1	2_{101}	3.70
7	$4_{\bar{1}\bar{1}0}$	3.77	1	3_{111}	3.75
6	$4_{\bar{1}\bar{1}0}$	3.79	3	4_{010}	3.77
7	3_{111}	3.85	3	3_{111}	3.85
1	3_{111}	3.91	5	4_{010}	3.92
From atom 4			From atom 8		
6	$4_{\bar{1}10}$	3.58	5	4_{010}	3.48
8	3_{110}	3.65	3	4_{000}	3.53
6	3_{110}	3.92	7	$2_{1\bar{1}1}$	3.64
2	3_{110}	3.94	4	3_{110}	3.65
			2	4_{000}	3.94

*See text.

[a]For each atom the number of the neighbor and its type are given. The type numbers correspond to the equivalent position: $1 = xyz$; $2 = \frac{1}{2}-x, \frac{1}{2}+y, \bar{z}$; $3 = \bar{x}\bar{y}\bar{z}$; $4 = \frac{1}{2}+x, \frac{1}{2}-y, z$. Translations of a, b, and c are denoted by subscripts.

The volume per Se_8 molecule is 241.0 Å^3 in β-monoclinic selenium, as opposed to 238.5 Å^3 in the α form. This small difference, which may not be significant, may result from the larger number of intermolecular interatomic distances less than 4.0 Å in α-monoclinic selenium (cf. Tables 34-4 and 34-6). The volume per eight selenium atoms in metallic selenium is much smaller, 218.2 Å^3. The corresponding calculated densities are $\alpha = 4.400$, $\beta = 4.352$; metallic, 4.807 g cm^{-3}.

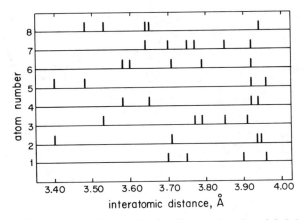

Fig. 34-13.   The spectrum of intermolecular distances less than 4.0 Å in β-monoclinic selenium.

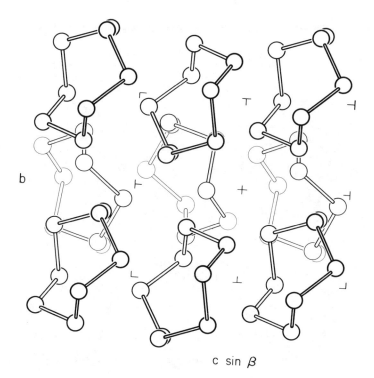

c sin β

Fig. 34-14.   The structure of β-monoclinic selenium viewed along the a-axis.

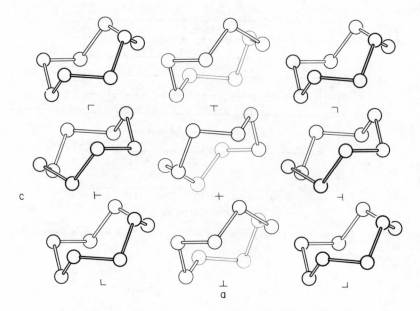

Fig. 34-15.   The structure of β-monoclinic selenium viewed along the *b*-axis.

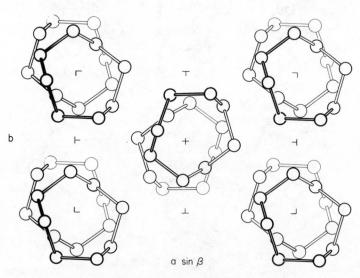

Fig. 34-16.   The structure of β-monoclinic selenium projected down the *c*-axis.

Cubic Selenium

Andrievski, Nabitovich, and Kripjakevich (1959) examined thin films of selenium evaporated onto films of "Zapon" (apparently some sort of plastic material) supported on copper at various temperatures. At 35 to 40°C they found α-monoclinic selenium, at 65°C, the β-monoclinic form, and at 150 to 160°C, a new modification which they termed β-cubic selenium. It was said to be simple cubic, one atom in a unit cell having $a = 2.970 \pm 0.004$ Å. Heating the β-cubic form gave a second cubic form, termed α-cubic selenium. It was said to be face centered cubic, four atoms in a unit cell having $a = 5.755 \pm 0.007$ Å. The possibility that β-cubic selenium was really cupric selenide (which is face centered cubic, $a = 5.84$ Å), was considered unlikely.

Semiletov (1960), on the other hand, pointed out that the films used as backings are stable only to 120 to 130°C, and that β-cubic selenium is in fact cupric selenide.

Andrievski and Nabitovich (1960) replied that they now had eliminated the possibility of any reaction with copper and obtained a third cubic form with the diamond structure, space group $Fd3m-O_h^7$, eight atoms per cell, $a = 6.04 \pm 0.01$ Å.

Some of the properties of these three cubic seleniums are summarized in Table 34-7.

It is remarkable that the variation of nearest neighbor distances with number of nearest neighbors is very nearly linear, as shown in Fig. 34-17.

Confirmation of the existence of these three cubic allotropes is much to be desired.

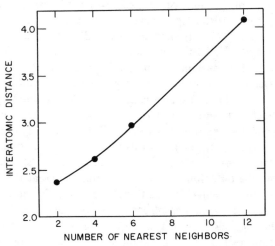

Fig. 34-17.   Variation of interatomic distance with number of nearest neighbors in selenium; from left to right, the structures are metallic, diamond, simple cubic, and face centered cubic.

Table 34-7.   Some Properties of the Cubic Seleniums

	a, Å	Shortest Interatomic Distance, Å	Number of Shortest Distances	Calculated Density, g cm^{-3}
Diamond	6.04	2.615	4	4.76
Simple	2.970	2.970	6	5.00
Face centered	5.755	4.069	12	2.75

TELLURIUM

Three forms of tellurium have been observed: α-(trigonal, metallic, normal), and two high pressure forms: β, stable between about 40 and 70 kbar at room temperature, of unknown structure, and γ (rhombohedral), stable over ca. 70 kbar at room temperature. A fourth form, said to be isotypic with α-arsenic, was reported by Kabalkina, Vereshchagin and Shulenin (1964), but has not been observed by others, as described below.

α-Tellurium

The unit cell of tellurium was first determined by Slattery (1923), and the structure was first determined by Bradley (1924b). It is isostructural with α-selenium, that is, trigonal, space group $P3_121$–D_3^4, and the enantiomorphous $P3_221$–D_3^6, three atoms per unit cell at ($x0\frac{1}{3}$, $0x\frac{2}{3}$ $\bar{x}\bar{x}0$). Bradley gave a rough value of 0.27 for the positional parameter, and recent, precise determination of it by Cherin and Unger (1967a) yielded $x = 0.2633 \pm 0.0005$ from a least squares refinement of 22 $hk0$ reflections from a single crystal. Various determinations of the lattice constants are presented in Table 52-1.

The structure consists of infinite helices parallel to c, with three atoms per turn of the helix. The bond distance is 2.834 ± 0.002 Å, the bond angle, $103.2 \pm 0.1°$. Each tellurium atom has four neighbors in adjacent chains at 3.494 Å. A view of the structure down the c-axis is presented in Fig. 52-1.

This arrangement may be thought of as a distortion of a simple cubic lattice, taking place in two parts, as shown in Fig. 52-2: a simple cubic lattice is elongated along the direction of a body diagonal, while at the same time displacing the atoms from their ideal positions, as shown. In the resulting distorted structure, Fig. 52-3, each atom instead of having six equal nearest neighbors now has two closer, and four 23% more distant. A different way of expressing this concept is that a tellurium-type structure having an axial ratio $c/a = \sqrt{3/2}$ (= 1.225) and a positional parameter of $\frac{1}{3}$, instead of 1.3301 and 0.2633, respectively, would, in fact, be simple cubic.

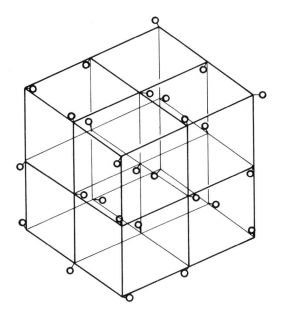

Fig. 52-2. The structure of α-tellurium, shown as a distorted simple cubic structure. The displacements shown give rise to the two shorter (covalent) bonds per atom shown in Fig. 52-3.

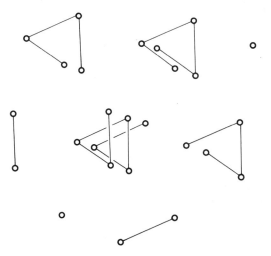

Fig. 52-3. The α-tellurium structure, showing the same atoms as in view of Fig. 52-2, and only the covalent bonds formed by those atoms. Part of one helical chain is apparent in the center of the figure.

Table 52-1.   The Lattice Constants of α-Tellurium at 25°C

a, Å	c, Å	c/a	Reference
4.45[a]	5.92[a]	1.33	Slattery, 1923, 1925
4.454[a]	5.924[a]	1.33	Bradley, 1924b
4.41[a]	5.87[a]	1.330	Olshausen, 1925
4.454[a]	5.87[a]	1.321	Rollier, Hendricks, & Maxwell, 1936
4.4565[b]	5.9268[b]	1.3299	Straumanis, 1940
4.456[c]	5.93[c]	1.331	Grison, 1951
4.4570	5.9290	1.3303	Swanson & Tatge, 1953
4.464[a]	5.892[a]	1.320	Zorll, 1954a
4.4543	5.9261	1.3304	Bonnier, Hicter, & Aléonard, 1964
4.4566	5.9264	1.3298	Deshpande, & Pawar, 1965
av. 4.4561	5.9271	1.3301	
±0.0012	±0.0013	±0.0005	

[a]From kX, omitted from the average.
[b]From kX.
[c]Omitted from the average.

Although there is no change in structure up to at least 428°C, the behavior at higher temperature is interesting (Bonnier, Hicter, and Aléonard,
1964). In the range 25 to 428°C the expansion along a is constant, $\alpha_\perp = 32 \times$
10^{-6} deg^{-1}, while in the range 25 to 375°C there is a contraction along c,
$\alpha_\parallel = -10 \times 10^{-6}$ deg^{-1}, the value then decreasing to about zero at 428°C.
The experimental data are shown in Fig. 52-4.

Tellurium at High Pressure

Reports of transitions in tellurium under pressure are in incomplete agreement.
Bridgman (1935, 1941) found two transitions at room temperature, one at
about 40 kbar, the other at about 70 kbar. Kennedy and Newton (1963)
found one transition on the melting curve, at 30 kbar, but they made studies
only to about 50 kbar. However, Kabalkina, Vereshchagin, and Shulenin
(1964) reported a transition at about 15 to 20 kbar. From a sample at 30
kbar they observed 12 powder lines, which they indexed as an α-arsenic type
structure (see above), having $a = 4.208$, $c = 12.036$ Å (hexagonal unit cell)
or $a = 4.69$ Å, $\alpha = 53°18'$ (primitive rhombohedral unit cell). The positional
parameter was stated to be 0.230. In this structure each tellurium atom has
three nearest neighbors at 2.87 Å and three next-nearest neighbors at 3.48 Å.
Kabalkina et al. also observed a powder pattern of nine lines from a sample
at greater than 45 kbar, but they could not index it. Jamieson and McWhan

(1965), on the other hand, found no evidence of a transition below 40 kbar, but they did observe those at 40 kbar and 70 kbar, in agreement with Bridgman. The pattern obtained at 53 kbar consisted of 17 lines and agreed with that of Kabalkina et al. Jamieson and McWhan were also unable to index this pattern satisfactorily, so the structure of the 40 to 70 kbar (β) phase remains unsolved, as does the question of what was observed by Kabalkina et al.

γ-Tellurium

The structure of the form stable above 70 kbar was solved by Jamieson and McWhan (1965), who observed 10 powder lines from a sample at 115 kbar. It is isostructural with β-polonium (see Fig. 84-2), space group $R\overline{3}m$–D_{3d}^5, one atom per unit cell at (000). The lattice constants are $a = 3.002 \pm 0.015$ Å, $\alpha = 103.3 \pm 3°$. Each tellurium atom is thus equidistant from six others at 3.002 Å. The structure may be thought of as a change of the normal α-tellurium structure to one of higher symmetry—compare Fig. 52-3 with Fig. 84-2—brought about by a shift of the positional parameter to exactly $\frac{1}{3}$ (and accompanied by a change of the rhombohedral angle from the value $86.77°$ calculated for the hexagonal cell of α-tellurium).

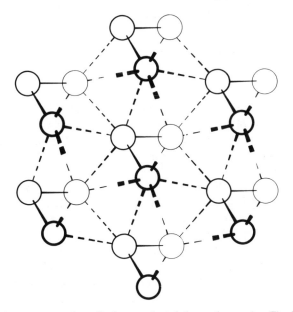

Fig. 52-1.   The structure of α-tellurium projected down the c-axis. The bond distance (solid lines) is 2.834 Å, the interchain distances (dashed lines) are 3.494 Å.

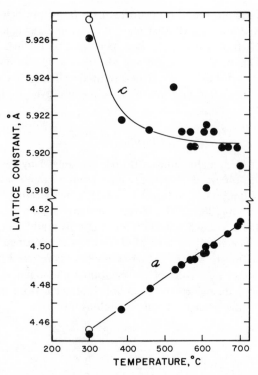

Fig. 52-4.   Variation of the lattice constants of α-tellurium with temperature. Open circles, Table 52-1; filled circles, data of Bonnier et al., 1964.

POLONIUM

A structure for polonium was first proposed by Rollier, Hendricks, and Maxwell (1936), who observed 21 powder lines in an electron diffraction experiment at room temperature. They indexed the first 12 of these lines on a monoclinic unit cell containing 12 atoms, with lattice constants a = 7.43, b = 4.30, c = 14.13 Å (from kX), β = 92°. A structure based on space group C2–C$_2^3$ with the atoms in three sets of the general position gave moderate agreement between the observed and calculated intensities.

This was the accepted structure for polonium—as Strukturbericht type A19— for 10 years, when Beamer and Maxwell (1946) discovered that at room temperature polonium is actually a mixture of a low temperature (α) form and a high temperature (β) form. The diffraction pattern of Rollier et al., which had been indexed as a characteristic of a single phase, was shown to consist of six lines of the α form, seven lines of the β form, and eight lines coincident from the two forms.

The phase transformation was first thought to occur at 75°C (Beamer and

Maxwell, 1946, 1949), but in later work Goode (1957) showed that the change $\alpha \rightarrow \beta$ takes place at 54°C, and the change $\beta \rightarrow \alpha$ at 18°C. Between these two temperatures the two phases are believed to coexist.

α-Polonium

The structure of α-polonium was first determined by Beamer and Maxwell (1946, 1949). It is cubic, space group Pm3m-O_h^1, one atom per unit cell. The atoms thus lie at the lattice points of a primitive cubic lattice.

There are conflicting reports concerning the value of the lattice constant. Beamer and Maxwell (1946) give a = 3.35 Å (from kX), later revised by them (1949) to the values (both from kX) a = 3.352 Å for samples at 10 ± 10°C, and a = 3.295 Å for samples at 75 ± 15°C. They stated that they were "not able to offer a suitable explanation for this temperature effect," which corresponds to large negative thermal expansion coefficient, α = -300 × 10^{-6} deg^{-1}. Goode (1953) found a much higher value, a = 3.366 Å (from kX) at 39 ± 15°C. This was obtained by measurements of four different patterns from two different samples, and is probably to be preferred over the earlier values. Brocklehurst, Goode, and Vassamillet (1957) remeasured the thermal expansion coefficient in the range -94 to 24°C and found that it is positive, as expected: α = 23 × 10^{-6} deg^{-1}. They did not, however, give any of the values of the lattice constant which they had determined.

In this structure, which is shown in Fig. 84-1, each polonium atom has six nearest neighbors lying on the vertices of a regular octahedron, at 3.366 Å. The atomic volume is 38.14 Å^3 (at 39°C).

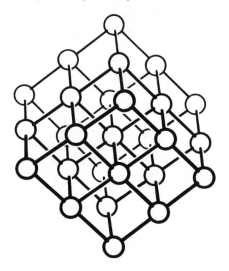

Fig. 84-1.   The structure of α-polonium.

β-Polonium

The structure of β-polonium was first determined by Beamer and Maxwell (1946, 1949). It is rhombohedral, space group $R\bar{3}m$–D_{3d}^{5}, one atom per unit cell. The atoms thus lie at the lattice points of a primitive rhombohedral lattice. Beamer and Maxwell's (1946) preliminary value for the lattice constants are a = 3.37 Å (from kX) and α = 98°13′; the value of a was subsequently revised by them (1949) to 3.366 ± 0.002 Å (from kX), at 75 ± 10°C. The values reported by Goode (1953) are a = 3.373 ± 0.002 Å (from kX), α = 98°5′, at 39 ± 15°C. The two sets combined again lead to a negative thermal expansion coefficient, but this property has not been directly investigated.

In the β-polonium structure (Fig. 84-2) each atom is surrounded by six others at the vertices of a somewhat flattened trigonal antiprism at 3.373 Å. The atomic volume, if Goode's lattice constants are accepted, is 37.11 Å^3, or 2.7% smaller than that of α-polonium, both at 39°C.

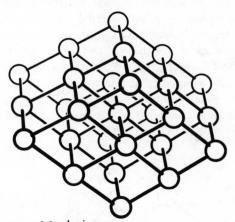

Fig. 84-2.   The structure of β-polonium.

Chapter 10

Group VII

FLUORINE

There are two forms of crystalline fluorine: α, stable below $45.6°K$, and β, stable between $45.6°K$ and the melting point of $53.5°K$.

α-Fluorine

A structure for α-fluorine at $23°K$ was reported by Meyer, Barrett and Greer (1968). It is based on 22 observed powder lines which were indexed on a monoclinic unit cell having $a = 5.50$, $b = 3.28$, $c = 10.01$ Å, $\beta = 134.66°$ (sic). With four molecules per unit cell, the calculated density is 1.97 g cm^{-3}, as compared with an observed value of 1.78 g cm^{-3} at $45°K$, and the molecular volume is 32.11 Å^3. A structure based on space group C2/m was described in detail, but the agreement between the observed and calculated intensities is so poor, the average discrepancy being over 75%, with the corresponding conventional R value of over 50%, that it cannot be accepted and will not be discussed further here.

Pauling, Keaveny, and Robinson (1970) reinterpreted the data of Meyer et al. They used a less obtuse unit cell of $a = 5.50$, $b = 3.28$, $c = 7.28$ Å, $\beta = 102.17°$, and noted that the absences were consistent with space group C2/c. (This space had not been completely ruled out by Meyer et al.) The eight fluorine atoms in the unit cell were placed in position 8(f) of C2/c, at $(000 \frac{1}{2}\frac{1}{2}0) \pm (xyz) (\bar{x} y \frac{1}{2}+z)$. Approximate values of the positional parameters were estimated on the basis of packing considerations, and these were refined by least squares, using observed F values obtained from data in the paper of

Meyer et al., to an R value of 19.9%. The final values of these parameters
are $x = 0.287, y = 0.319, z = 0.100$. (These are the values given in the text
by Pauling et al. They differ slightly from the values $x = 0.285, y = 0.317$,
$z = 0.0997$ in their abstract.)

The structure may be described as nearly close packed hexagonal layers of
molecules in planes parallel to the *ab* plane. The observed ratio *a:b* of 1.677
differs by 3% from the ideal value of $\sqrt{3}$. The molecules are tipped ±18° in
the *b* direction in order nearly to equalize the contacts between layers. Each
atom has three nearest neighbors in an adjacent layer, two at 2.82 and one at
2.87 Å. Within a given layer, each atom has four neighbors at 3.20 and two
at 3.28 Å; the neighbors in the other plane of molecular layer are one at
1.49 Å (the F–F bond) plus one each at 3.17, 3.26, and 3.38 Å.

A view of the structure along the *b*-axis is shown in Fig. 9-1.

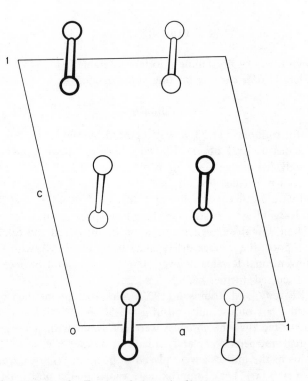

Fig. 9-1.   The structure of α-F$_2$ viewed along the *b*-axis.

β-Fluorine

The structure of β-fluorine was first described by Jordan, Streib, Smith, and
Lipscomb (1964), with full details being given by Jordan, Streib, and Lipscomb
(1964). It is isostructural with γ-oxygen, that is cubic, a = 6.67 Å at 50°K,
space group Pm3n- O_h^3, eight molecules per unit cell, the centers lying at 000,
$\frac{1}{2}\frac{1}{2}\frac{1}{2}$, $\frac{1}{4}\frac{1}{2}0$, $\frac{3}{4}\frac{1}{2}0$, $0\frac{1}{4}\frac{1}{2}$, $0\frac{3}{4}\frac{1}{2}$, $\frac{1}{2}0\frac{1}{4}$, $\frac{1}{2}0\frac{3}{4}$. The molecular volume is 37.09 Å^3. The
structure is highly disordered, as shown in Fig. 9-2. The molecules at 000
and $\frac{1}{2}\frac{1}{2}\frac{1}{2}$ each have 12 nearest neighboring molecules, the distance between
molecular centers being 3.73 Å. Each of the other molecules has two mole-
cules at 3.34 Å, four at 3.73 Å, and eight at 4.08 Å. These distances are
short enough to cause the disorder to be strongly hindered.

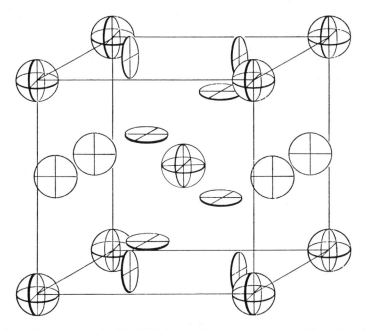

Fig. 9-2.   The structure of β-F$_2$. This is also the structure of γ-0$_2$. For a description
of the disorder, see legend to Fig. 8-7.

CHLORINE

A structure for solid chlorine was first proposed by Keesom and Taconis (1936b, e, f). On the basis of 18 powder lines observed by Köhler (1934), they determined the structure as tetragonal, with a = 8.58, c = 6.13 Å (from kX), eight molecules per unit cell, space group P4/ncm-D_{4h}^{16}, 16 atoms in the general position with x = 0.125, y = 0.167, z = 0.107. In this structure the bond distance is 1.82 Å, and the shortest intermolecular approach 2.53 Å. Because this bond distance is much shorter than the value 2.01 Å found in gaseous chlorine (Pauling and Brockway, 1934), and because the intermolecular contact is considerably shorter than the value 3.6 Å predicted on the basis of the van der Waals radius, the structure was reinvestigated by Collin (1952, 1956), who used single crystals.

Collin found that the crystals were really orthorhombic, with a = 6.24, b = 4.48, c = 8.26 Å at –160°C. (The accidental coincidences that $c_{tetr} \approx a_{orth}$ and $a_{tetr}^2 \approx \frac{1}{2}[(2b)_{orth}^2 + c_{orth}^2]$ enabled Keesom and Taconis to to obtain good agreement with an incorrect cell for a majority of powder lines.) The space group is Cmca-D_{2h}^{18}, and there are eight atoms in the unit cell, at $(000\ \frac{1}{2}\frac{1}{2}0) \pm (0yz, \frac{1}{2}\ y\ \frac{1}{2}-z)$. Collin determined values for y and z by trial and error, obtaining y = 0.130 and z = 0.100. The data of Collin were later subjected to a refinement by the method of least squares by Donohue and Goodman (1965), who obtained y = 0.1173 ± 0.0038 and z = 0.1016 ± 0.0009. The corresponding intramolecular Cl–Cl bond distance is 1.980 ± 0.022 Å, in excellent agreement with that found in gaseous chlorine, 1.986 Å by electron diffraction (Shibata, 1963) and 1.988 Å by infrared spectroscopy (Herzberg, 1950).

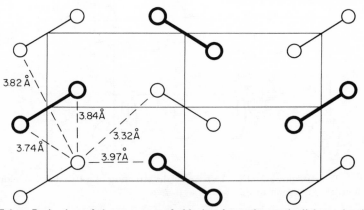

Fig. 17-1.   Projection of the structure of chlorine down the a-axis, light molecules at x = 0, heavy molecules at $x = \frac{1}{2}$. The intermolecular distances shown are those at –160°C.

The structure consists of Cl_2 molecules very weakly bonded together to form layers parallel to the *bc* plane. Some intermolecular interatomic distances are indicated in Fig. 17-1. The weak intralayer interactions may be seen in Fig. 17-2, which is a packing drawing of a portion of the structure.

The thermal expansion behavior of solid chlorine has been reported by Hawes and Cheesman (1959), but since they indexed their powder photographs with the incorrect tetragonal cell, these results are of no value.

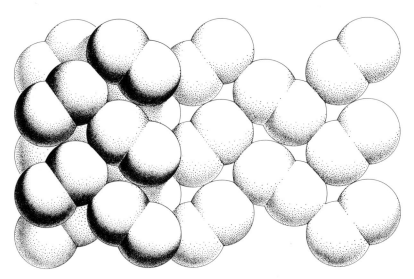

Fig. 17-2.   The structure of chlorine viewed down the *a*-axis.

BROMINE

The structure of solid bromine was determined by Vonnegut and Warren (1936). It is isostructural with chlorine: orthorhombic, space group Cmca-D_{2h}^{18}, a = 6.68, b = 4.49, c = 8.74 Å (from kX) at -150°C, eight atoms per unit cell at $(000 \frac{1}{2}\frac{1}{2}0) \pm (0yz, \frac{1}{2}y\frac{1}{2}-z)$, with y = 0.135 and z = 0.110. The positional parameters were determined by trial and error, and are not very precise. The calculated Br–Br bond distance is 2.27 ± 0.10 Å, in fortuitously good agreement with the value 2.2836 Å found in the gas (Herzberg, 1950).

The weak intermolecular "bonds" are somewhat stronger than those in chlorine, as may be seen from the packing drawing, Fig. 35-1. The closest intermolecular interatomic distances are indicated in Fig. 35-2.

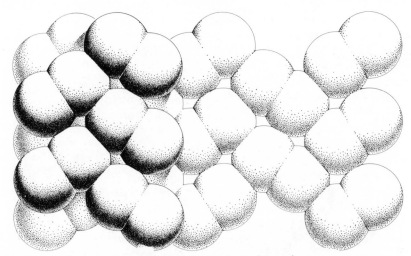

Fig. 35-1.   The structure of bromine viewed down the *a*-axis.

The thermal expansion of bromine between –106.0 and –23.5°C has been reported by Hawes (1959). The lattice constants were found to vary linearly with the temperature:

$$a = 6.752 + 5.84 \times 10^{-4}t$$
$$b = 4.564 + 6.07 \times 10^{-4}t$$
$$c = 8.765 + 2.59 \times 10^{-4}t$$

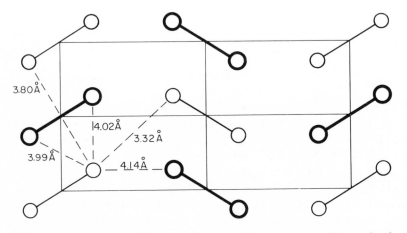

Fig. 35-2.   Projection of the structure of bromine down the *a*-axis, light molecules at $x = 0$, heavy molecules at $x = \frac{1}{2}$. The intermolecular distances are those found at –150°C.

IODINE

The first determination of the unit cell of iodine was by Ferrari (1927), and the first determination of the structure is that of Harris, Mack, and Blake (1928). Various determinations of the lattice constants are presented in Table 53-1. The structure is the same as that of crystalline chlorine and bromine: the space group Cmca-D_{2h}^{18}, $a = 7.268$, $b = 4.797$, $c = 9.797$ Å at room temperature, eight atoms per unit cell at $(000\frac{1}{2}\frac{1}{2}0) \pm (0yz, \frac{1}{2}y\frac{1}{2}-z)$, with $y = 0.149$ and $z = 0.116$. These values for the positional parameters, which were determined at room temperature by Kitaigorodski et al. (1953), are not as precise as a set determined at $-163°C$ by van Bolhuis et al. (1967); both sets appear in Table 53-1. Various interatomic distances calculated with the two sets of data are shown in Fig. 53-1.

The covalent I–I bond distance is calculated to be 2.69 Å from the room temperature data and 2.715 ± 0.006 Å from the low temperature data. Both of these are longer, the latter significantly so, than the value 2.662 Å observed in the gas by Karle (1955). This lengthening is doubtless a consequence of the fact that the weak intralayer bonds, which are evident in Fig. 53-2, are much stronger than those in either chlorine or bromine. More precise determinations of the parameters in chlorine and bromine and in iodine at room temperature are needed to clarify this situation.

The thermal expansion of iodine is remarkable (Table 53-1). In the c direction the observed value is about equal to that found in many metals, while in the a and b directions it is over 10 times as large.

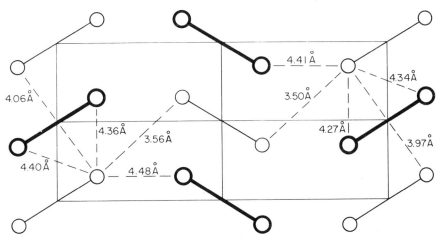

Fig. 53-1.   Projection of the structure of iodine down the a-axis, light molecules at $x = 0$, heavy molecules at $x = \frac{1}{2}$. Intermolecular distances at room temperature given on the left, those at $-163°C$ on the right.

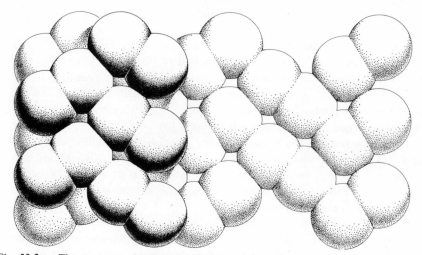

Fig. 53-2.   The structure of iodine viewed down the a-axis.

ASTATINE

The structure of astatine has not been determined. The longest lived known isotope has a half-life of $7\frac{1}{2}$ hours.

Table 53-1. Some Crystallographic Data for Iodine

a, Å	b, Å	c, Å	t, °C	y	z	Reference
7.178[a]	4.770[a]	9.803[a]	room			Ferrari, 1927
7.270[b]	4.805[b]	9.800[b]	room	0.150	0.117	Harris, Mack, & Blake, 1928
7.263[a]	4.801[b]	9.791[b]	room			Neuberger and Schiebold, 1936
7.2697[b]	4.7903[b]	9.7942[b]	25			Straumanis & Sauka, 1943
7.26[c]	4.80[c]	9.80[c]	room	0.149	0.116	Kitaigorodskii, Khotsyanova, & Struchkov, 1953
7.271	4.792	9.803	26			Swanson, Fuyat, & Ugrinic, 1954
7.136[c]	4.686[c]	9.784[c]	-163	0.1543	0.1174	van Bolhuis, Koster, & Migchelson, 1967
av. 7.270	4.797	9.797	room			
±0.001	±0.007	±0.005				
97	113	7	room			Average thermal expansion coefficient, -163° to 25°, $deg^{-1} \times 10^6$

[a]From kX, omitted from the average.
[b]From kX.
[c]Omitted from the average.

References

The boldface numbers in parentheses that follow each reference are the atomic numbers of the elements treated in the reference. Russian literature is cited in their English translations, where these exist. Readers who wish to consult the originals will find references to them in the translations.

1865

W. Hittorf, *Ann. Phys. Chem.* **126,** 193. (*15*)

1866

W. Hittorf, *Phil. Mag.* **31,** 311. (*15*)

1890

W. Muthmann, *Z. Krist.* **17,** 336. (*16*)

1891

M. R. Engel, *Compt. rend. Acad. Sci. Paris* **112,** 866. (*16*)

1904

A. Stock & O. Guttmann, *Chem. Ber.* **37,** 885. (*51*)

1906

P. Groth, *Chemische Kristallographie,* Englemann, Leipzig, p. 33. (*34*)

1907

T. W. Richards & T. N. Brink, *J. Am. Chem. Soc.* **29,** 117. (*19, 37, 55*)

1913

W. H. Bragg & W. L. Bragg, (a) *Nature,* 91, 557 (*6*); (b) *Proc. Roy. Soc.* A89, 277. (*6*)

1914

A. H. W. Aten, *Z. physik. Chem.* 88, 321. (*16*)
W. H. Bragg, *Proc. Roy. Soc.* A89, 575. (*16*)
W. L. Bragg, *Phil. Mag.* 28, 355. (*29*)
P. W. Bridgman, *J. Am. Chem. Soc.* 36, 1344. (*15*)
E. Cohen & W. D. Halderman, *Z. physik. Chem.* 87, 409. (*48*)

1915

E. Jänecke, *Z. physik. Chem.* 90, 313. (*48*)

1916

P. Debye & P. Scherrer, *Physik. Z.* 17, 277. (*6, 14*)
L. Vegard, (a) *Phil. Mag.* 31, 83 (*47*); (b) *Phil. Mag.* 32, 65. (*79, 82*)

1917

P. Debye, *Physik, Z.* 18, 483. (*74*)
P. Debye & P. Scherrer, *Physik. Z.* 18, 291. (*6*)
A. W. Hull, (a) *Phys. Rev.* 9, 84 (*26*); (b) *Phys. Rev.* 9, 564 (*13, 14*); (c) *Phys. Rev.* 10, 661. (*6, 12, 26, 28*)

1918

A. J. Bijl & N. H. Kolkmeijer, (a) *Chem. Weekblad* 15, 1077 (*50*); (b) *Chem. Weekblad* 15, 1264. (*50*)

1919

A. J. Bijl & N. H. Kolkmeijer, (a) *Proc. Acad. Sci. Amsterdam* 21, 494 (*50*); (b) *Proc. Acad. Sci. Amsterdam* 21, 501. (*50*)
A. W. Hull, *Phys. Rev.* 14, 540. (*24, 26, 27, 28*)

1920

H. Bohlin, *Ann. Physik* 61, 421. (*28,90*)
A. W. Hull, *Science* 52, 227. (*20, 22, 30, 44, 46, 48, 49, 73, 77*)
R. W. James & N. Tunstall, *Phil. Mag.* 40, 233. (*51*)
S. Nishikawa & G. Asahara, *Phys. Rev.* 15, 38. (*81*)

1921

A. W. Hull, (a) *Phys. Rev.* 17, 571. (*24, 27, 28, 30, 42, 44, 45, 46, 48, 49, 73, 77, 78*); (b) *Phys. Rev.* 18, 88. (*22, 40, 58, 76, 90*)
A. W. Hull & W. P. Davey, (a) *Phys. Rev.* 17, 266 (*30, 48, 49*); (b) *Phys. Rev.* 17, 549. (*30, 48, 49*)

W. Gerlach, *Physik. Z.* **22**, 557. (*14*)

R. W. James, *Phil. Mag.* **42**, 193. (*83*)

H. Kahler, *Phys. Rev.* **18**, 210. (*47, 78, 79, 83*)

G. Meier, Dissertation, Göttingen. (*4*)

A. Ogg, *Phil. Mag.* **42**, 163. (*83*)

A. Westgren, (a) *J. Iron Steel Inst.* **103**, 303 (*26*); (b) *Engineering* **111**, 727 (*26*); (c) *Engineering* **111**, 757. (*26*)

A. Westgren & A. E. Lindh, *Z. physik. Chem.* **98**, 181. (*26*)

1922

N. Alsen & G. Aminoff, *Geol. Foren. Stockholm Forh.* **44**, 124. (*80*)

W. Gerlach, *Physik. Z.* **23**, 114. (*14, 29*)

A. W. Hull, *Phys. Rev.* **20**, 113. (*6, 23, 32*)

F. Kirchner, *Ann. Physik* **69**, 59. (*29, 79*)

N. H. Kolkmeijer, *Proc. Acad. Sci, Amsterdam* **25**, 125. (*32*)

H. Mark, M. Polanyi, & E. Schmid, *Z. Physik* **12**, 58. (*30*)

L. W. McKeehan, (a) *Proc. Nat. Acad. Sci.* **8**, 254 (*19*); (b) *Phys. Rev.* **20**, 424. (*46, 47, 79*)

L. W. McKeehan & P. P. Cioffi, *Phys. Rev.* **19**, 444. (*80*)

P. Stoll, *Arch. Sci. Phys. Mat.* **3**, 546. (*42*)

A. Westgren & G. Phragmen, (a) *J. Iron Steel Inst.* **105**, 241 (*26*); (b) *Engineering* **113**, 630 (*26*); (c) *Z. physik. Chem.* **102**, 1. (*26*)

1923

K. Becker & F. Ebert, *Z. Physik* **16**, 165. (*25, 73, 81*)

H. Küstner & H. Remy, *Physik. Z.* **24**, 25. (*14*)

H. Mark & M. Polanyi, *Z. Physik* **18**, 75. (*50*)

H. Mark, M. Polanyi, & E. Schmid, *Naturwiss.* **11**, 256. (*50*)

H. Mark, K. Weissenberg, & K. W. Gonell, *Z. Elektrochem.* **29**, 364. (*16*)

L. W. McKeehan, (a) *Phys. Rev.* **21**, 334 (*46*); (b) *Phys. Rev.* **21**, 402 (*26, 28*); (c) *J. Franklin Inst.* **195**, 59. (*83*)

E. A. Owen & G. D. Preston, (a) *Proc. Phys. Soc.* **35**, 101 (*26, 29, 82*); (b) *Proc. Phys. Soc.* **36**, 49. (*29, 30*)

M. K. Slattery, *Phys. Rev.* **21**, 378. (*34, 52*)

N. Uspenski & S. Konobejewski, *Z. Physik* **16**, 215. (*78*)

R. B. Wilsey, *Phil. Mag.* **50**, 487. (*47*)

R. W. G. Wyckoff, *Z. Krist.* **59**, 55. (*77*)

J. F. T. Young, *Phil. Mag.* **46**, 291. (*25, 29*)

1924

A. E. van Arkel, *Proc. Acad. Sci. Amsterdam* **27**, 97. (*50*)

J. D. Bernal, *Proc. Roy. Soc.* **106A**, 749. (*6*)

A. J. Bradley, (a) *Phil. Mag.* **47**, 657 (*33*); (b) *Phil. Mag.* **48**, 477. (*34, 52*)

W. P. Davey, *Phys. Rev.* **23**, 292. (*26, 28, 29, 42, 47, 74, 78, 79, 82, 83*)

O. Hassel & H. Mark, (a) *Z. Physik* **23**, 269 (*83*); (b) *Z. Physik* **25**, 317. (*6*)

K. Heindlhofer, *Phys. Rev.* **24**, 426. (*26*)

A. Huber, *Physik. Z.* **25**, 45. (*79*)

E. R. Jette, G. Phragmen, & A. Westgren, *J. Inst. Metals* **31**, 193. (*29*)

W. M. Lehmann, Z. Krist. 60, 379. (14)
G. R. Levi, Nuovo Cimento 1, 137. (81, 82)
H. Mark & M. Polanyi, Z. Physik 22, 200. (50)
H. Mark & E. Wigner, Z. physik. Chem. 111, 398. (16)
R. A. Patterson, Ind. Eng. Chem. 16, 689. (29)
F. Simon & C. von Simson, Z. Physik 25, 160. (18)
C. von Simson, Z. physik. Chem. 109, 183. (30, 48)
F. Wever, (a) Z. Elektrochem. 30, 376 (28); (b) Z. Physik 28, 69. (26)

      1925

T. Barth & G. Lunde, Z. physik. Chem. 117, 478. (44, 46, 76)
F. C. Blake, Phys. Rev. 26, 60. (26)
A. J. Bradley, Phil. Mag. 50, 1018. (25)
G. L. Clark, W. C. Asbury, & R. M. Wick, J. Am. Chem. Soc. 47, 2661. (28)
W. P. Davey, Phys. Rev. 25, 753. (26, 28, 29, 42, 46, 47, 74, 78, 79, 82, 83)
J. E. Jones & A. E. Ingham, Proc. Roy. Soc. 107A, 636. (2, 10, 18, 36, 54, 86)
H. Lange, Ann. Physik 76, 476. (28, 29, 79)
G. R. Levi, Z. Krist. 61, 559. (81, 82)
G. R. Levi & G. Tacchini, Gazz. Chim. Ital. 55, 28. (28)
G. Linck & H. Jung, Z. anorg. allg. Chem. 147, 288. (15)
C. Mauguin, J. Physiol. 6, 38. (6)
W. Noethling & S. Tolksdorf, Z. Krist. 62, 255. (40, 72)
S. Olshausen, Z. Krist. 61, 463. (15, 25, 33, 34, 52)
R. A. Patterson, (a) Phys. Rev. 25, 581 (22, 24); (b) Phys. Rev. 26, 56. (22, 24)
W. M. Peirce, E. A. Anderson, & P. van Dyck, J. Franklin Inst. 200, 349. (30)
W. C. Phebus & F. C. Blake, Phys. Rev. 25, 107. (24, 28, 30, 50, 82)
A. Sacklowski, Ann. Physik 77, 241. (29, 47)
M. K. Slattery, Phys. Rev. 25, 333. (34, 52)
J. de Smedt & W. H. Keesom, (a) Physica 5, 344 (7, 8, 18); (b) Comm. Phys. Lab. Univ.
      Leiden 178b, 19. (7, 8, 18)
A. Westgren & G. Phragmen, (a) Z. Physik 33, 777 (25); (b) Phil. Mag. 50, 311. (29,
      47, 79)
R. W. G. Wyckoff & E. D. Crittenden, J. Am. Chem. Soc. 47, 2866. (26)

      1926

A. E. van Arkel, Physica 6, 64. (42, 74)
G. Asahara & T. Sasahara, Sci. Papers Inst. Phys. Chem. Res. Tokyo 5, 79. (81)
T. Barth & G. Lunde, Z. physik. Chem. 121, 78. (44, 45, 47, 77, 78, 79)
K. Becker, Z. Physik 40, 37. (14, 74)
A. J. Bradley & E. F. Olland, Nature 117, 122. (24)
W. P. Davey, Z. Krist. 63, 316. (26, 28, 29, 42, 46, 47, 74, 78, 79, 82)
W. P. Davey & T. A. Wilson, Phys. Rev. 27, 105. (29, 74)
W. Ehrenburg, Z. Krist. 63, 320. (6)
A. Erdal, Z. Krist. 65, 69. (29, 47)
J. R. Freeman, F. Sillers, & P. F. Brandt, Sci. Papers U. S. Bur. Stand. 20, 611. (30)
V. M. Goldschmidt, Geochemische Verteilungsgesetze der Elemente, Skrifter Norske
      Videnskaps-Acad. Oslo, I. Mat.-Naturv. Kl. (32)
K. A. Hoffmann, U. Hoffmann, & K. Herrmann, Chem. Ber. 59, 2433. (6)
S. Holgersson, Ann. Physik 79, 35. (28, 29, 47, 79)

F. M. Jaeger, P. Terpstra, & H. G. K. Westenbrink, *Proc. Acad. Sci. Amsterdam* **29**, 1193. (*31*)

H. Jung, (a) *Z. Krist.* **64**, 413 (*29, 47, 79*); (b) *Centralblatt Min. Geol.* 1926, 107. (*15, 33*)

W. H. Keesom & H. Kammerlingh-Onnes, *Comm. Phys. Lab. Univ. Leiden* **174**, 43. (*82*)

G. R. Levi, *Nuovo Cimento* **3**, 297. (*81*)

G. R. Levi & R. Haardt, *Gazz. Chim. Ital.* **56**, 369. (*44, 76*)

C. Mauguin, *Bull. Soc. Franc. Mineral.* **48**, 32. (*6*)

P. Terpstra, *Z. Krist.* **63**, 318. (*81*)

1927

A. E. van Arkel, *Z. physik. Chem.* **130**, 100. (*40, 72*)

K. Becker, *Z. Physik* **42**, 479. (*81*)

A. J. Bradley & J. Thewlis, *Proc. Roy. Soc.* **115A**, 456. (*25*)

G. Bredig & R. Allolio, *Z. physik. Chem.* **126**, 41. (*28, 29, 78*)

A. Ferrari, *Rend. Acad. Lincei.* **5**, 582. (*53*)

F. M. Jaeger, P. Terpstra, & H. G. K. Westenbrink, *Z. Krist.* **66**, 195. (*31*)

J. C. McLennan & J. O. Wilhelm, *Phil. Mag.* **3**, 383. (*8*)

S. Sekito, *Sci. Reports Tohoku Imp. Univ.* **16**, 545. (*27*)

F. Sillers, *Trans. Am. Electrochem. Soc.* **52**, 301. (*24*)

S. Umino, *Sci. Rep. Tohoku Imp. Univ.* **16**, 593. (*27*)

L. Vegard & H. Dale, *Skrifter Norske Videnskaps-Akad.*, No. 14. (*27, 28*)

1928

G. Allard, *Compt. rend. Acad. Sci. Paris* **187**, 222. (*47*)

A. E. van Arkel, *Z. Krist.* **67**, 235. (*14, 29, 42, 45, 47, 74, 78, 79*)

G. L. Clark, A. J. King, & J. F. Hyde, *Proc. Nat. Acad. Sci.* **14**, 617. (*56*)

F. Halla & H. Staufer, *Z. Krist.* **67**, 440. (*82*)

P. M. Harris, E. Mack, & F. C. Blake, *J. Am. Chem. Soc.* **50**, 1583. (*53*)

E. Korinth, (a) Dissertation, Jena (*16*); (b) *Z. anorg. allg. Chem.* **174**, 57. (*16*)

A. Osawa & Y. Ogawa, *Z. Krist.* **68**, 177. (*26, 30*)

H. Ott, *Ann. Physik* **85**, 81. (*6*)

H. Perlitz, *Z. Physik* **50**, 433. (*81*)

E. Persson & A. Westgren, *Z. physik. Chem.* **136**, 208. (*51, 81*)

E. Posnjak, *J. Phys. Chem.* **32**, 254. (*19*)

G. D. Preston, (a) *Phil. Mag.* **5**, 1198 (*25*); (b) *Phil. Mag.* **5**, 1207. (*25*)

F. Simon & E. Vohsen, (a) *Z. physik. Chem.* **133**, 165 (*3, 11, 19, 37, 55*); (b) *Z. physik. Chem.* **133**, 181. (*38*)

C. S. Smith, *Min. Met.* **9**, 458. (*29, 79*)

H. Terrey & C. M. Wright, *Phil. Mag.* **6**, 1055. (*29, 80*)

M. Wolf, *Nature* **122**, 314. (*80*)

1929

R. Bach, *Helv. Phys. Acta* **2**, 95. (*26*)

F. Ebert & H. Hartmann, *Z. anorg. allg. Chem.* **179**, 418. (*38, 56*)

K. Frolich, R. L. Davidson, & M. R. Fenske, *Ind. Eng. Chem.* **21**, 109. (*29*)

V. M. Goldschmidt, (a) *Naturwiss.* **17**, 134 (*75*); (b) *Z. physik. Chem.* **B2**, 244. (*75*)

F. von Göler & G. Sachs, *Z. Physik* **55**, 581. (*29*)

G. Greenwood, *Z. Krist.* **72**, 309. (*28*)

G. Hägg & G. Funke, *Z. physik. Chem.* **B6**, 272. (*28*)

K. Horovitz, *Phys. Rev.* **33**, 121. (*80*)

N. Katoh, *Z. physik. Chem.* **B6**, 27. (*29*)

A. J. King, *Proc. Nat. Acad. Sci.* **15**, 337. (*38*)

A. J. King & G. L. Clark, *J. Am. Chem. Soc.* **51**, 1709. (*56*)

E. J. Lewis, *Phys. Rev.* **34**, 1575. (*4*)

G. Mayer, *Z. Krist.* **70**, 383. (*26*)

L. Mazza & A. G. Nasini, *Phil. Mc̣.* **7**, 301. (*28*)

J. C. McLennan & R. J. Monkman, *Trans. Roy. Soc. Canada* **23**, 255. (*30, 41, 48, 68, 73*)

A. Osawa, *J. Inst. Metals* **41**, 511. (*47*)

A. Osawa & S. Oya, *Sci. Reports Tohoku Imp. Univ.* **18**, 727. (*23*)

E. Persson & E. Ohman, *Nature* **124**, 333. (*25*)

W. Schmidt, *Arch. Eisenhüttenw.* **3**, 293. (*26*)

S. Sekito, (a) *Z. Krist.* **72**, 406 (*25*); (b) *Sci. Reports Tohoku Imp. Univ.* **18**, 59. (*29*)

J. de Smedt, W. H. Keesom, & H. H. Mooy, (a) *Comm. Phys. Lab. Univ. Leiden,* 202a (*7*); (b) *Proc. Acad. Sci. Amsterdam* **32**, 745. (*7*)

G. P. Thomson, *Nature* **123**, 912. (*28*)

S. Valentiner & G. Becker, *Naturwiss.* **17**, 639. (*28*)

L. Vegard, (a) *Nature* **124**, 267 (*7*); (b) *Nature* **124**, 337 (*7*); (c) *Z. Physik* **58**, 497. (*7*)

O. Weinbaum, *Z. Metallk.* **21**, 397. (*29*)

A. Westgren & A. Almin, *Z. physik. Chem.* **B5**, 14. (*47*)

F. Wever & V. Hashimoto, *Mitl. Kais.-Wilh.-Inst. Eisenforsch.* **11**, 293. (*24*)

M. Wolf, *Z. Physik* **53**, 72. (*80*)

1930

R. H. Aborn & R. L. Davidson, *J. Phys. Chem.* **34**, 522. (*29*)

N. Ageew, M. Hansen, & G. Sachs, *Metallwirt.* **9**, 91. (*29*)

N. Ageew & G. Sachs, *Z. Physik* **63**, 293. (*47*)

C. Agte & K. Becker, *Z. techn. Physik* **11**, 107. (*73, 74*)

W. G. Burgers & J. C. M. Basart, *Z. Krist.* **75**, 155. (*28, 29*)

W. G. Burgers & J. van Liempt, *Z. anorg. allg. Chem.* **193**, 144. (*90*)

W. F. Ehret & R. D. Fine, *Phil. Mag.* **10**, 551. (*83*)

A. W. Frost, *J. Russ. Phys.-Chem. Ges.* **62**, 2235. (*15*)

G. Hägg, (a) *Z. physik. Chem.* **B7**, 339 (*42, 74*); (b) *Z. phys. Chem.* **B11**, 433. (*22, 23, 40, 73*)

C. Johansson & J. O. Linde, (a) *Ann. Physik* **5**, 762 (*79*); (b) *Ann. Physik* **6**, 458. (*78*)

N. Katoh, *Z. Krist.* **76**, 228. (*29*)

W. H. Keesom & H. H. Mooy, (a) *Comm. Phys. Lab. Univ. Leiden,* 209b, 13 (*36*); (b) *Nature* **125**, 889 (*36*); (c) *Proc. Acad. Sci. Amsterdam* **33**, 447. (*36*)

W. H. Keesom, J. de Smedt, & H. H. Mooy, (a) *Nature* **126**, 757 (*1*); (b) *Proc. Acad. Sci. Amsterdam* **33**, 814 (*1*); (c) *Comm. Phys. Lab. Univ. Leiden,* 209d, 35. (*1*)

J. C. McLennan & R. W. McKay, (a) *Trans. Roy. Soc. Canada* **24**, 1 (*92*); (b) *Trans. Roy. Soc. Canada* **24**, 33. (*57*)

R. F. Mehl & C. S. Barrett, *Trans. AIME* **89**, 575. (*80*)

K. Meisel, *Z. anorg. allg. Chem.* **190**, 237. (*41*)

G. Natta & A. G. Nasini, (a) *Nature* **125**, 457 (*54*); (b) *Nature* **125**, 889. (*36*)

G. Natta & L. Passerini, *Nature* **125**, 707. (*15*)

E. Öhman, (a) *Svensk. Kem. Tidskr.* 42, 210 (*25*); (b) *Metallwirt.* 9, 825. (*25*)

A. Osawa, *Sci. Reports Tohoku Imp. Univ.* 19, 109. (*27, 28*)

A. Osawa & S. Oya, *Sci. Reports Tohoku Imp. Univ.* 19, 95. (*23*)

N. Parravano & V. Caglioti, *Gazz. Chim. Ital.* 60, 923. (*34, 83*)

E. Persson, *Z. physik. Chem.* 9, 25. (*25, 29*)

O. L. Roberts & W. P. Davey, *Met. Alloys* 1, 648. (*26*)

G. Sachs & J. Weerts, *Z. Physik* 60, 481. (*47, 79*)

K. Sasaki & S. Sekito, *J. Soc. Chem. Ind. Japan* 33, 482B. (*24*)

S. Sekito, *Z. Krist.* 74, 189. (*81*)

J. de Smedt, W. H. Keesom, & H. H. Mooy, *Comm. Phys. Lab. Univ. Leiden,* 203e. (*10*)

D. Solomon & W. Morris-Jones, *Phil. Mag.* 10, 470. (*82*)

G. P. Thomson, *Proc. Roy. Soc.* A128, 649. (*78*)

1931

C. Agte, H. Alterthum, K. Becker, G. Heyne, & K. Moers, *Z. anorg. allg. Chem.* 196, 129. (*75*)

A. E. van Arkel & W. G. Burgers, *Z. Metallk.* 23, 149. (*26*)

S. Arrhenius & A. Westgren, *Z. physik. Chem.* 14B, 66. (*29*)

E. G. Bowen & W. Morris-Jones, *Phil. Mag.* 12, 441. (*50*)

G. Bredig & E. S. von Bergkampf, *Z. physik. Chem., Bodenstein Festband,* 172. (*28*)

W. G. Burgers & J. A. M. van Liempt, *Rec. Trav. Chim. Pay-Bas* 50, 1050. (*74*)

P. P. Ewald & C. Hermann, (a) *Strukturbericht* 1, 42 (*80*); (b) *Strukturbericht* 1, 737 (*80*); (c) *Strukturbericht* 1, 738. (*31*)

G. Greenwood, *Z. Krist.* 78, 242. (*78*)

F. Halls, E. Mehl, & F. X. Bosch, *Z. physik. Chem.* 11B, 455. (*34*)

H. Hartmann, F. Ebert, & O. Bretschneider, *Z. anorg. allg. Chem.* 198, 116. (*74*)

F. M. Jaeger & E. Rosenbohm, (a) *Proc. Acad. Sci. Amsterdam* 34, 85 (*45*); (b) *Proc. Acad. Sci. Amsterdam* 34, 308. (*44*)

F. M. Jaeger & J. E. Zanstra, *Proc. Acad. Sci. Amsterdam* 34, 14. (*45, 46*)

C. H. M. Jenkins & G. D. Preston, *J. Inst. Metals* 45, 307. (*48*)

S. Kaya & A. Kussmann, *Z. Physik* 72, 293. (*28*)

C. H. Mathewson, E. Spire, & C. H. Samans, *Trans. Am. Soc. Steel Treatment* 19, 357. (*23*)

K. Moeller, *Naturwiss.* 19, 575. (*75*)

M. C. Neuberger, (a) *Z. Krist.* 78, 164 (*41*); (b) *Z. anorg. allg. Chem.* 197, 219. (*41*)

O. Nial, A. Almin, & A. Westgren, *Z. physik. Chem.* B14, 81. (*47*)

G. Phragmén, *J. Iron Steel Inst.* 123, 405. (*26, 28*)

B. Ruhemann & F. Simon, *Z. physik. Chem.* B15, 389. (*36, 54*)

D. Solomon & W. Morris-Jones, *Phil. Mag.* 11, 1090. (*50, 82, 83*)

W. S. Stenzel & J. Weerts, *Festschrift der Platinschmelze,* G. Siebert, Hanau, 288. (*46, 47, 78, 79*)

J.-J. Trillat & J. Forestier, *Compt. rend. Acad. Sci. Paris* 192, 559. (*16*)

S. Zeidenfeld, *Proc. Phys. Soc.* 43, 512. (*42, 74, 82*)

E. Zintl & A. Harder, *Z. physik. Chem.* 154A, 47. (*82*)

1932

F. A. Bannister & M. H. Hey, *Min. Mag.* 23, 188. (*78*)

W. Boas, *Metallwirt.* 11, 603. (*30*)

A. J. Bradley & A. H. Jay, *Proc. Phys. Soc.* 44, 563. (*26, 28*)

W. G. Burgers, (a) *Nature* **129**, 281 (*40*); (b) *Z. anorg. allg. Chem.* **205**, 81. (*40*)

J. A. Darbyshire, *J. Chem. Soc.* 211. (*82*)

F. P. J. Dwyer & D. P. Mellor, *Proc. Roy. Soc. N. S. Wales* **66**, 234. (*49*)

A. Goetz & R. C. Hergenrother, (a) *Phys. Rev.* **40**, 137 (*83*); (b) *Phys. Rev.* **40**, 643. (*83*)

C. Hermann & M. Ruhemann, *Z. Krist.* **83**, 136. (*80*)

K. Iwasé and N. Nasu, *Bull. Chem. Soc. Japan* **7**, 305. (*26, 28*)

E. R. Jette & F. Foote, *Phys. Rev.* **39**, 1018. (*83*)

H. Kersten, *Physics* **2**, 276. (*51*)

F. Laves, *Naturwiss.* **20**, 472. (*31*)

M. LeBlanc & G. Wehner, *Ann. Physik* **14**, 481. (*29, 79*)

J. O. Linde, *Ann. Physik* **15**, 249. (*29*)

H. D. Megaw, *Phil. Mag.* **14**, 130. (*29, 47*)

H. H. Mooy, *Comm. Phys. Lab. Univ. Leiden*, 223a. (*8*)

Z. Nishiyama, *Sci. Reports Tohoku Imp. Univ.* **12**, 364. (*26*)

E. A. Owen & J. Iball, *Phil. Mag.* **13**, 1020. (*13, 28, 29, 42, 45, 46, 47, 73, 74, 77, 79, 82*)

R. T. Phelps & W. P. Davey, *Trans. AIME* **99**, 234. (*47*)

G. D. Preston, *Phil. Mag.* **13**, 419. (*24, 26*)

L. L. Quill, (a) *Z. anorg. allg. Chem.* **208**, 59 (*39*); (b) *Z. anorg. allg. Chem.* **208**, 257 (*41, 73*); (c) *Z. anorg. allg. Chem.* **208**, 273. (*57, 58, 60*)

A. Rossi, *Rend. Acad. Lincei* **15**, 298. (*59*)

M. Ruhemann, *Z. Physik* **76**, 368. (*7, 8*)

W. Stenzel & J. Weerts, *Z. Krist.* **84**, 20. (*12, 30, 48, 50, 75*)

O, E. Swjaginzeff & B. K. Brunowski, *Z. Krist.* **83**, 187. (*76*)

N. W. Taylor, *J. Am. Chem. Soc.* **54**, 2713. (*48*)

J.-J. Trillat & J. Forestier, *Bull. soc. chim. France* **51**, 248. (*16*)

L. Vegard, *Z. Physik* **79**, 471. (*7*)

T. Yuching, *Phys. Rev.* **40**, 662. (*6*)

1933

N. A. Ageew & D. Shoyket, *J. Inst. Met.* **52**, 119. (*47*)

R. W. Drier & H. L. Walker, *Phil. Mag.* **16**, 294. (*45*)

F. Ebert, H. Hartmann, & H. Peisker, *Z. anorg. allg. Chem.* **213**, 126. (*20*)

H. Esser & G. Mueller, *Arch. Eisenhüttenw.* **7**, 265. (*26*)

G. I. Finch & A. G. Quarrell, *Proc. Roy. Soc.* **141A**, 398. (*12, 13, 30*)

L. Graf, *Metallwirt.* **12**, 649. (*20*)

P. Hidnert & H. S. Krider, *J. Res. Nat. Bur. Stand.* **11**, 279. (*41*)

B. Jacobsen & A. Westgren, *Z. physik. Chem.* **B20**, 361. (*28*)

F. M. Jaeger & J. E. Zanstra, *Proc. Acad. Sci. Amsterdam* **36**, 636. (*4*)

E. R. Jette & E. B. Gebert, *J. Chem. Phys.* **1**, 753. (*14, 30, 47, 48, 50, 51, 82, 83*)

F. Laves, *Z. Krist.* **84**, 256. (*31*)

M. C. Neuberger, (a) *Z. Krist.* **85**, 232 (*28, 74*); (b) *Z. anorg. allg. Chem.* **212**, 40. (*80*)

I. Obinata, *Metallwirt.* **12**, 101. (*82*)

I. Obinata & F. Wasserman, *Naturwiss.* **21**, 382. (*29*)

E. A. Owen & J. Iball, *Phil. Mag.* **16**, 479. (*30*)

E. A. Owen & L. Pickup, (a) *Proc. Roy. Soc.* **139A**, 526 (*29*); (b) *Proc. Roy. Soc.* **140A**, 179. (*30*)

E. A. Owen & L. Yates, (a) *Phil. Mag.* **15**, 472 (*13, 26, 29, 45, 46, 47, 77, 78, 79, 82*); (b) *Phil. Mag.* **16**, 606. (*45, 46, 78, 79*)

H. Saini, *Helv. Phys. Acta* **6**, 597. (*47*)

G. Shinoda, *Mem. Coll. Sci., Kyoto Imp. Univ.* **16A**, 193. (*49*)

S. Stenbeck, *Z. anorg. allg. Chem.* **214**, 16. (*47*)

J. G. Thompson, *Met. Alloys* **4**, 114. (*90*)

J. Weigle, *Helv. Phys. Acta* **7**, 51. (*30*)

P. 'Viest, *Z. Physik* **81**, 121. (*79*)

T. A. Wilson, (a) *Physics* **4**, 148 (*92*); (b) *Phys. Rev.* **43**, 781. (*92*)

E. Zintl & S. Neumayr, (a) *Z. Elektrochem.* **39**, 81 (*49*); (b) *Z. Elektrochem.* **39**, 84. (*57*)

1934

W. G. Burgers & J. C. M. Basart, *Z. anorg. allg. Chem.* **216**, 223. (*41, 73*)

W. Büssem & F. Gross, *Z. Physik* **87**, 778. (*28*)

L. Graf, *Physik. Z.* **35**, 551. (*20*)

W. P. Jesse, *Physics* **5**, 147. (*28*)

E. R. Jette, W. L. Brunner, & F. Foote, *Am. I. M. M. E. Tech. Pub. 526*, 66. (*79*)

E. R. Jette, V. H. Nordstrom B. Queneau, & F. Foote, *Trans. AIME* **111**, 361. (*24*)

H. P. Klug, *Z. Krist.* **88**, 128. (*34*)

J. W. L. Köhler, Dissertation, Leiden. (*17*)

K. H. Meyer & Y. Go, *Helv. Chim. Acta* **17**, 1081. (*16*)

M. C. Neuberger, *Z. anorg. allg. Chem.* **217**, 154. (*74*)

A. Ölander, *Z. physik. Chem.* **168A**, 274. (*82*)

E. A. Owen & L. Pickup, *Z. Krist.* **88**, 116. (*28, 29*)

E. A. Owen & E. L. Yates, *Phil. Mag.* **17**, 113. (*30, 47, 78*)

L. Pauling & L. O. Brockway, *J. Chem. Phys.* **2**, 867. (*17*)

A. Rossi, *Nature* **133**, 174. (*57, 59*)

G. Shinoda, (a) *Mem. Coll. Sci. Kyoto Imp. Univ.* **17**, 27 (*40, 48*); (b) *Proc. Phys. Math. Soc. Japan* **16**, 436. (*40, 48*)

K. Tanaka, *Mem. Coll. Sci. Kyoto Imp. Univ.* **17**, 59. (*34*)

L. Vegard, *Z. Physik* **88**, 235. (*7*)

L. Vegard & A. Kloster, *Z. Krist.* **89**, 560. (*29, 79*)

F. Weibke, *Z. anorg. allg. Chem.* **220**, 293. (*29*)

W. H. Willott & E. J. Evans, *Phi.. . g.* **18**, 114. (*33*)

1935

F. Birch, in Hultgren, Gingrich, & Warren (1935). (*15*)

A. J. Bradley, *Z. Krist.* **91**, 302. (*31*)

P. W. Bridgman, *Phys. Rev.* **48**, 841. (*16, 31, 52, 80, 81, 83*)

L. K. Frevel & E. Ott, *J. Am. Chem. Soc.* **57**, 228. (*49*)

G. Hägg & A. G. Hybinette, *Phil. Mag.* **20**, 913. (*33, 50, 51*)

R. Hultgren, N. S. Gingrich, & B. E. Warren, *J. Chem. Phys.* **3**, 351. (*15*)

R. Hultgren & B. E. Warren, (a) *Bull. Am. Phys. Soc.* **10**, 30 (*15*); (b) *Phys. Rev.* **47**, 808. (*15*)

E. R. Jette & F. Foote, *J. Chem. Phys.* **3**, 605. (*12, 13, 14, 26, 28, 30, 42, 47, 48, 50, 51, 74, 79, 83*)

G. F. Kossolapow & A. K. Trapesnikow, *Z. Krist.* **91**, 410. (*48*)

M. C. Neuberger, (a) *Z. Krist.* **92**, 313 (*14*); (b) *Z. Krist.* **92**, 474. (*4*)

E. A. Owen & L. Pickup, *Phil. Mag.* **20**, 1155. (*4*)

E. A. Owen, L. Pickup, & J. O. Roberts, *Z. Krist.* **91**, 70. (*4, 12, 30, 44, 76*)

E. A. Owen & J. Rogers, *J. Inst. Metals* **57**, 257. (*29, 47*)

M. Straumanis & O. Mellis, *Z. Physik* **94**, 184. (*29, 47*)

L. Vegard, (a) *Z. Physik* **98**, 1 (*8*); (b) *Nature* **136**, 720. (*8*)

B. E. Warren & J. T. Burwell, *J. Chem. Phys.* **3**, 6. (*16*)

F. Weibke & H. Eggers, *Z. anorg. allg. Chem.* **222**, 146. (*47*)

T. A. Wilson, *Phys. Rev.* **47**, 332. (*25*)

L. Wright, H. Hirst, & J. Riley, *Trans. Faraday Soc.* **31**, 1253. (*24*)

1936

N. W. Ageev & V. Ageeva, *J. Inst. Metals* **59**, 311. (*49*)

W. G. Burgers & F. M. Jacobs, *Z. Krist.* **94**, 299. (*22, 40*)

M. V. Cohen, *Z. Krist.* **94**, 288. (*74*)

W. Guertler, *Handbuch der Metallographie,* Berlin. (*53*)

U. Hoffmann & D. Wilm, *Z. Elektrochem.* **42**, 504. (*6*)

W. Hume-Rothery, *The Structure of Metals and Alloys,* The Institute of Metals, London. (*24*)

W. Hume-Rothery, G. F. Lewin, & P. W. Reynolds, *Proc. Roy. Soc.* **157A**, 167. (*29, 47)*

A. Ievinš & M. Straumanis, (a) *Z. physik. Chem.* **33B**, 265 (*13*); (b) *Z. physik. Chem.* **34B**, 402. (*13*)

E. R. Jette & F. Foote, *Metals Tech.* **3**, 1. (*26, 28*)

W. H. Keesom & K. W. Taconis, (a) *Comm. Phys. Lab. Univ. Leiden,* No. 240d (*8*); (b) *Comm. Phys. Lab. Univ. Leiden,* No. 240e (*17*); (c) *Proc. Roy. Soc. Amsterdam* **39**, 149 (*8*); (d) *Physica* **3**, 141 (*8*); (e) *Physica* **3**, 237 (*17*); (f) *Proc. Roy. Soc. Amsterdam* **39**, 314. (*17*)

G. F. Kossolapow & A. K. Trapeznikow, *Z. Krist.* **94**, 53. (*4, 50*)

W. Köster & W. Dannöhl, *Z. Metallk.* **28**, 248. (*79*)

D. P. Mellor, S. B. Cohen, & E. B. Underwood, *J. Proc. Austral. Chem. Inst.* **3**, 329. (*5*)

G. Natta & L. Passerini, *Atti accad. naz. Lincei, Classe sci. fis., mat. nat.* **24**, 464. (*15*)

M. C. Neuberger, (a) *Z. Krist.* **93**, 158 (*41*); (b) *Z. Krist.* **93**, 312 (*73*); (c) *Z. Krist.* **93**, 314. (*23*)

M. C. Neuberger & E. Schiebold, in Guertler (1936). (*53*)

E. A. Owen & T. L. Richards, *Phil. Mag.* **22**, 310. (*4*)

E. A. Owen & E. W. Roberts, *Phil. Mag.* **22**, 290. (*48*)

E. A. Owen & E. L. Yates, *Phil. Mag.* **21**, 809. (*28*)

M. A. Rollier, S. B. Hendricks, & L. R. Maxwell, *J. Chem. Phys.* **4**, 648. (*52, 84*)

M. Straumanis & A. Ievinš, *Z. Physik* **98**, 461. (*14, 26, 74, 79, 82*)

B. Vonnegut & B. E. Warren, *J. Am. Chem. Soc.* **58**, 2459. (*35*)

1937

H. van Bergen, *Naturwiss.* **25**, 415. (*29*)

A. J. Bradley, A. H. Jay, & A. Taylor, *Phil. Mag.* **23**, 545. (*26*)

J. T. Burwell, *Z. Krist.* **97**, 123. (*16*)

J. E. Dorn & G. Glockler, *J. Phys. Chem.* **41**, 499. (*51*)

H. Esser, W. Eilender, & K. Bungardt, *Arch. Eisenhüttenw.* **12**, 157. (*29*)

C. Gottfried & F. Schossberger, *Strukturbericht* **3**, 3. (*31*)

J. D. Hanawalt & L. K. Frevel, *Z. Krist.* **98**, 84. (*12*)

C. W. Jacob & B. E. Warren, *J. Am. Chem. Soc.* **59**, 2588. (*92*)

W. Klemm & H. Bommer, *Z. anorg. allg. Chem.* **231**, 138. (*57, 58, 59, 60, 63, 64, 65, 66, 68, 69, 70, 71*)

K. Moeller, *Z. Krist.* **97**, 170. (*47, 74, 78*)

H. Nitka, *Physik. Z.* **38**, 896. (*32*)

E. A. Owen & J. I. Jones, *Proc. Phys. Soc.* **49**, 587. (*46*)

E. A. Owen & E. W. Roberts, *Z. Krist.* **96**, 497. (*44, 76*)

E. A. Owen & E. L. Yates, *Proc. Phys. Soc.* **49**, 307. (*26*)

J. A. Prins & W. Dekeyser, *Physica* **4**, 900. (*34*)

A. G. Quarrell, *Proc. Phys. Soc.* **49**, 279. (*47*)

M. Renninger, *Z. Physik.* **106**, 141. (*6*)

J.-J. Trillat & S. Oketani, *Z. Krist.* **98**, 334. (*16*)

W. Trzebiatowski, (a) *Roczn. Chem.* **17**, 73 (*6*); (b) *Z. anorg. allg. Chem.* **233**, 376. (*75*)

W. A. Wood, *Phil. Mag.* **23**, 984. (*24*)

1938

E. Aruja & H. Perlitz, *Z. Krist.* **100**, 195. (*11*)

H. van Bergen, *Ann. Physik* **33**, 737. (*29*)

W. Betteridge, *Proc. Phys. Soc.* **50A**, 519. (*49*)

P. W. Bridgman, *Proc. Am. Acad. Arts Sci.* **72**, 207. (*55*)

S. R. Das, *Ind. J. Phys.* **12**, 163. (*16*)

H. Esser, W. Eilender, & K. Bungardt, *Arch. Eisenhüttenw.* **12**, 157. (*26, 78, 79*)

W. Hume-Rothery & P. W. Reynolds, *Proc. Roy. Soc.* **167A**, 25. (*47*)

A. Ievinš, M. Straumanis, & K. Karlsons, *Z. physik. Chem.* **B40**, 347. (*12, 50, 83*)

G. Johannsen & H. Nitka, *Physik. Z.* **39**, 440. (*25*)

W. H. Keesom & K. W. Taconis, *Physica* **5**, 161. (*2*)

W. Trzebiatowski & E. Bryjak, *Z. anorg. allg. Chem.* **238**, 255. (*33, 51*)

1939

A. E. van Arkel, *Reine Metalle*, J. Springer, Berlin. (*23, 24*)

B. Böhm & W. Klemm, *Z. anorg. allg. Chem.* **243**, 69. (*19, 37, 55*)

H. Bommer, (a) *Z. anorg. allg. Chem.* **242**, 277 (*67*); (b) *Z. Elektrochem.* **45**, 357. (*21, 39*)

S. R. Das & K. Ghosh, *Ind. J. Phys.* **13**, 91. (*16*)

J. D. Fast, *Rec. Trav. Chim.* **58**, 972. (*22*)

G. Grube, E. Oestreicher, & O. Winkler, *Z. Elektrochem.* **45**, 776. (*25*)

G. Grube & O. Winkler, *Z. Elektrochem.* **45**, 784. (*25*)

F. Halla & R. Weil, *Z. Krist.* **101**, 435. (*5*)

W. Klemm & H. Bommer, *Z. anorg. allg. Chem.* **241**, 264. (*59, 60*)

W. Köster & W. Rauscher, *Z. Metallk.* **39**, 178. (*25*)

G. LeClerc & A. Michel, *Compt. rend. Acad. Sci. Paris* **208**, 1583. (*28*)

C. W. Mason & G. E. Pelissier, *Am. Inst. Min. Met., Tech. Pub. 1043*. (*50*)

K. Meisel, *Naturwiss.* **27**, 230. (*21*)

E. A. Owen & E. W. Roberts, *Phil. Mag.* **27**, 294. (*29, 47*)

E. A. Owen, J. Rogers, & J. C. Guthrie, *J. Inst. Metals* **65**, 457. (*48*)
G. V. Raynor & W. Hume-Rothery, *J. Inst. Metals* **65**, 477. (*12*)
H. Stöhr, *Z. anorg. allg. Chem.* **242**, 138. (*33*)

1940

H. H. Chiswik & R. Hultgren, *Metals Tech.* **7**, TP 1169. (*83*)
J. C. Felipe, *Rev. Acad. Madrid* **34**, 180. (*29, 47*)
F. Foote & E. R. Jette, *Phys. Rev.* **58**, 81. (*29, 47*)
W. Hume-Rothery & G. V. Raynor, *Proc. Roy. Soc.* **A174**, 457. (*48*)
P. H. Miller, Jr., & J. W. M. DuMond, *Phys. Rev.* **57**, 198. (*13, 47*)
A. E. Newkirk, Dissertation, Cornell University. (*5*)
E. A. Owen & V. W. Rowlands, *J. Inst. Met.* **66**, 361. (*29, 47, 79*)
H. Perlitz & E. Aruja, *Phil. Mag.* **30**, 55. (*3*)
G. V. Raynor, *Proc. Roy. Soc.* **A174**, 457. (*12*)
H. Söchtig, *Ann. Physik* **38**, 97. (*24*)
M. Straumanis, *Z. Krist.* **102**, 432. (*34, 52*)
A. Taylor & D. Laidler, *Nature* **146**, 130. (*6*)

1941

H. van Bergen, *Ann. Physik* **39**, 553. (*13, 26*)
P. W. Bridgman, *Phys. Rev.* **60**, 351. (*52, 83*)
L. Carapella & R. Hultgren, *Metals Tech.* **8**, No. 1405. (*29, 50*)
J. Fitzwilliam, A. P. Kaufmann, & C. F. Squire, *J. Chem. Phys.* **9**, 678. (*40*)
F. Foote & E. R. Jette, (a) *Trans. AIME* **143**, 124 (*82*); (b) *Trans. AIME* **143**, 151. (*47*)
R. Fricke, *Naturwiss.* **29**, 365. (*24, 28, 29, 82*)
F. M. Jaeger & E. Rosenbohm, *Proc. Acad. Sci. Amsterdam* **44**, 144. (*44*)
W. Klemm & G. Mika, *Z. anorg. allg. Chem.* **248**, 155. (*20, 38, 56*)
H. Lipson, N. J. Petch, & D. Stockdale, *J. Inst. Met.* **67**, 79. (*47*)
H. Lipson & A. K. Stokes, *Nature* **148**, 437. (*81*)
S. S. Lu & Y. L. Chang, *Proc. Phys. Soc.* **53**, 517. (*13, 26, 28, 29, 42, 48, 51, 74, 82*)
A. R. Stokes & A. J. C. Wilson, *Proc. Phys. Soc.* **53**, 658. (*82*)

1942

P. W. Bridgman, *Proc. Am. Acad. Arts Sci.* **74**, 425. (*83*)
L. Carapella & R. Hultgren, *Trans. AIME* **147**, 232. (*25, 29, 50*)
A. Colombani & J. Wyart, *Compt. rend. Acad. Sci. Paris* **215**, 129. (*28*)
G. Hass, *Kolloidzeit.* **100**, 230. (*47*)
W. Hume-Rothery & K. W. Andrews, *J. Inst. Metals* **68**, 19. (*29*)
A. J. King, *J. Am. Chem. Soc.* **64**, 1226. (*38*)
O. Kubaschewski & A. Schneider, *Z. Elektrochem.* **48**, 671. (*24, 42, 74*)
H. Lipson & A. R. Stokes, (a) *Nature* **149**, 328 (*6*); (b) *Proc. Roy. Soc.* **181A**, 101. (*6*)
K. H. Meyer, *Natural and Synthetic High Polymers,* Interscience, New York, p. 52. (*16*)
H. P. Rooksby, *J. Roy. Soc. Arts* **90**, 673. (*28*)
A. J. C. Wilson, *Proc. Phys. Soc.* **54**, 487. (*13*)

1943

A. G. H. Anderson & A. W. Kingsbury, *Trans. AIME* **152**, 38. (*29*)
C. S. Barrett, *Structure of Metals,* McGraw-Hill, New York. (*28*)

A. W. Laubengayer, D. T. Hurd, A. E. Newkirk, & J. L. Hoard, *J. Am. Chem. Soc.* **65**, 1924. (*5*)

C.-S. Lu & E. W. Malmberg, *Rev. Sci. Instr.* **14**, 271. (*79*)

M. M. Popov, V. P. Simanov, S. M. Skuratov, & M. N. Suzdalceva, *Izv. Sek. Fiz. Khim. Anal.* **16**, 111. (*78, 79*)

J. A. Robertson, Dissertation, Cornell University. (*5*)

G. Siebel, *Z. Elektrochem.* **49**, 2i8. (*13*)

M. Straumanis & J. Sauka, *Z. physik. Chem.* **53B**, 320. (*53*)

A. R. Troiano & G. I. Williams, *Trans. Am. Soc. Met.* **31**, 340. (*26*)

F. Trombe & M. Foëx, *Compt. rend. Acad. Sci. Paris* **217**, 501. (*57, 58*)

O. Winkler, *Z. Elektrochem.* **49**, 221. (*75, 76*)

S. Yoshida, *Proc. Phys. Math. Soc. Japan* **17**, 535. (*24*)

1944

A. Colombani, *Ann. Physik* **19**, 272. (*28*)

L. D. Ellsworth & F. C. Blake, *J. Appl. Phys.* **15**, 507. (*25*)

U. Esch & A. Schneider, *Z. Elektrochem.* **50**, 268. (*28, 78*)

H. Lipson & L. E. R. Rogers, *Phil. Mag.* **35**, 544. (*14*)

K. Lonsdale, *Nature* **153**, 22. (*6*)

C. Palache, H. Berman, & C. Frondel, *The System of Mineralogy,* Wiley, New York. (*33*)

N. J. Petch, *Nature* **154**, 337. (*74*)

D. P. Riley, *Nature* **153**, 587. (*6*)

H. P. Rooksby, *Nature* **154**, 337. (*74*)

F. Trombe & M. Foëx, *Ann. Chim.* (xi) **19**, 417. (*58*)

J. G. White, B. Sc. Thesis, Glasgow University. (*16*)

1945

C. G. Fink, E. R. Jette, S. Katz, & F. J. Schnettler, *Trans. Electrochem. Soc.* **88**, 229. (*49, 50*)

M. L. Huggins, *J. Chem. Phys.* **13**, 37. (*16*)

W. Hume-Rothery & K. Lonsdale, *Phil. Mag.* **36**, 842. (*37*)

R. I. Jaffee, E. M. Smith, & B. W. Gonser, *Metals Tech.* **12**, 1. (*79*)

K. Lonsdale & W. Hume-Rothery, *Phil. Mag.* **36**, 799. (*3*)

A. Michel, J. Bénard, & G. Chaudron, *Bull. Soc. Chim. France* **12**, 336. (*46*)

J. B. Nelson & D. P. Riley, *Proc. Phys. Soc.* **57**, 477. (*6*)

E. A. Owen & E. A. Roberts, *J. Inst. Met.* **71**, 213. (*79*)

E. V. Potter & R. W. Huber, *Phys. Rev.* **68**, 24. (*25*)

1946

W. H. Beamer & C. R. Maxwell, *J. Chem. Phys.* **14**, 569. (*84*)

H. Bückle, *Metallforsch.* **1**, 53. (*41, 42, 73, 74*)

J. Gibson, *Nature* **158**, 752. (*6*)

H. P. Klug, *J. Am. Chem. Soc.* **68**, 1493. (*82*)

V. G. Kuznecov, *Izv. Sek. Platiny Akad. Nauk SSSR* (20), 5. (*46*)

H. Nowotny, K. Schubert, & U. Dettinger, *Z. Metallk.* **37**, 137. (*46*)

A. J. Rose, *Compt. rend. Acad. Sci. Paris* **222**, 805. (*29*)

A. von Wiedebach-Nostiz, *Metallforsch.* **1**, 56. (*30*)

1947

C. J. Barrett, *Phys. Rev.* **72**, 245. (*3*)

A. Boule, *Ann. Agron.* **217**, 575. (*16*)

G. Brauer, *Z. anorg. allg. Chem.* **255**, 101. (*55*)

G. I. Finch, H. Wilman, & L. Yang, *Disc. Faraday Soc.* **A43**, 144. (*28*)

H. J. Goldschmidt & T. Land, *J. Iron Steel Inst.* **155**, 221. (*45, 78*)

F. Heyd, F. Khol, & A. Kochanovska, *Coll. Czech. Chem. Comm.* **12**, 502. (*14*)

L. J. E. Hofer & W. C. Peebles, *J. Am. Chem. Soc.* **69**, 893. (*27*)

F. H. Horn & W. T. Ziegler, *J. Am. Chem. Soc.* **69**, 2762. (*41, 73*)

G. E. Klein, *Am. Miner.* **32**, 691. (*15*)

R. C. L. Mooney, *Phys. Rev.* **72**, 1269. (*43*)

J. B. Nelson & D. P. Riley, *Nature* **159**, 637. (*6*)

O. Nial, *Svensk Kem. Tidskr.* **59**, 165. (*50*)

L. Pauling, *J. Am. Chem. Soc.* **69**, 542. (*24*)

H. P. Rooksby & E. G. Stewart, *Nature* **159**, 638. (*6*)

W. L. Roth, T. W. DeWitt, & A. J. Smith, *J. Am. Chem. Soc.* **69**, 2881. (*15*)

B. M. Rovinski & T. V. Tagunova, *Ž. Tekh. Fiz.* **17**, 1137. (*26, 29*)

C. A. Snavely, *Trans. Electrochem. Soc.* **92**, 537. (*24*)

F. Trombe & M. Foëx, *Rev. Mét.* **44**, 349. (*58*)

1948

N. K. Andrushchenko, V. V. Tjapkina, & P. O. Dankov, *Dokl. Akad. Nauk* **59**, 8, 1113. (*47*)

H. J. Axon & W. Hume-Rothery, *Proc. Roy. Soc.* **A193**, 1. (*13*)

C. S. Barrett, *Am. Miner.* **33**, 749. (*11*)

C. S. Barrett & O. R. Trautz, *Trans. AIME* **175**, 579. (*3*)

P. W. Bridgman, *Proc. Am. Acad. Arts Sci.* **76**, 55. (*15*)

H. S. M. Coxeter, *Regular Polytopes,* Metheun, London. (*5*)

R. S. Dean, E. V. Potter, & R. W. Huber, *Trans. Am. Soc. Met.* **40**, 381. (*25*)

E. C. Ellwood & J. M. Silcock, *J. Inst. Metals* **74**, 457. (*13*)

J. D. Fast, *Chem. Weekblad.* **44**, 621. (*72*)

R. Goniche & R. Graf, *La Recherche Aéronautique,* **3**, 55. *(13)*

G. Hass, *Z. anorg. allg. Chem.* **257**, 166. (*14*)

R. I. Jaffee & H. P. Nielsen, *Metals Tech.* **15**, TP 2420. (*74*)

R. C. L. Mooney, (a) *Phys. Rev.* **73**, 653 (*43*); (b) *Acta Cryst.* **1**, 161. (*43*)

S. A. Nemnonov, *Ž. Tekh. Fiz.* **18**, 239. (*24*)

E. A. Owen, Y. H. Liu, & D. Morris, *Phil. Mag.* **39**, 831. (*13*)

D. E. Thomas, *J. Sci. Instr.* **25**, 440. (*26*)

A. R. Troiano & J. L. Tokich, *Trans. AIME* **175**, 728. (*27*)

A. S. Wilson & R. E. Rundle, U. S. Atomic Energy Comm., AECD-2046. (*92*)

R. W. G. Wyckoff, *Crystal Structures,* Interscience, New York. (*28*)

1949

W. H. Beamer & C. R. Maxwell, *J. Chem. Phys.* **17**, 1293. (*84*)

C. R. Berry, *Acta Cryst.* **2**, 393. (*47*)

S. J. Carlile, J. W. Christian, & W. Hume-Rothery, *J. Inst. Met.* **76**, 169. (*24, 25*)

H. T. Clark, (a) *J. Metals* **1**, 588 (*22*); (b) *Trans. AIME* **185**, 588. (*22*)

C. Crussard & F. Aubertin, *Rev. Mét.* **46**, 354. (*29*)

D. D. Cubicciotti & C. D. Thurmond, *J. Am. Chem. Soc.* **71**, 2149. (*38*)

P. Gordon, *J. Appl. Phys.* **20**, 908. (*4*)

B. W. Gosner, *Metals Progress* **55**, 346. (*22*)

E. S. Greiner & W. C. Ellis, *Trans. AIME* **180**, 657. (*22*)

D. D. van Horn, *Phys. Rev.* **75**, 1630. (*26*)

W. Hume-Rothery & T. H. Boultbee, *Phil. Mag.* **40**, 71. (*12, 13*)

R. Kiessling, (a) *Acta Chem. Scand.* **3**, 90 (*40*); (b) *Acta Chem. Scand.* **3**, 603. (*73*)

W. Klemm, *Anorganische Chemie* **1**, Dieterich, Wiesbaden. (*21*)

A. Kochanovska, *Physica* **15**, 191. (*13, 26*)

H. Krebs, *Z. Physik* **126**, 769. (*34*)

A. W. Lawson & T. Tang, *Phys. Rev.* **76**, 301. (*58*)

J. S. Lukesh, *Acta Cryst.* **2**, 420. (*92*)

S. V. Náray-Szabó & C. W. Tobias, *J. Am. Chem. Soc.* **71**, 1882. (*5*)

L. Pauling, *Proc. Nat. Acad. Sci.* **35**, 495. (*16*)

R. W. Powell, *Nature* **164**, 153. (*31*)

C. A. Snavely & D. A. Vaughan, *J. Am. Chem. Soc.* **71**, 313. (*24*)

M. E. Straumanis, *J. Appl. Phys.* **20**, 726. (*13, 34, 50, 82, 83*)

T. Sugawara & E. Kanda, *Sci. Rep. Res. Inst. Tohoku Univ.* **1**, 153. (*15*)

T. Sugawara, Y. Sakamoto, & E. Kanda, *Sci. Rep. Res. Inst. Tohoku Univ.* **1**, 29. (*15*)

A. H. Sully & H. K. Hardy, *J. Inst. Metals* **76**, 269. (*13*)

A. S. Wilson & R. E. Rundle, *Acta Cryst.* **2**, 126. (*92*)

A. Z. Zhmudski, *Zavod. Labor.* **15**, 1055. (*13, 26*)

U. Zwicker, *Z. Metallk.* **40**, 377. (*24*)

U. Zwicker, E. Jahn, & K. Schubert, *Z. Metallk.* **40**, 433. (*25*)

1950

G. E. Bacon, (a) *Acta Cryst.* **3**, 137 (*6*); (b) *Acta Cryst.* **3**, 320. (*6*)

C. S. Barrett, *Phase Transformations in Solids,* Wiley, New York. (*3*)

O. Beeck, *Disc. Faraday Soc.* **8**, 118. (*74*)

L. D. Brownlee, *Nature* **166**, 482. (*50*)

R. S. Busk, *J. Metals* **2**, 1460. (*12*)

P. Chiotti, U. S. Atomic Energy Comm., AECD-3072. (*90*)

J. E. Dorn, P. Pietrokowsky, & T. E. Tietz, *J. Metals,* **2**, 933. (*13*)

D. S. Eppelsheimer & R. R. Penman, (a) *Nature* **166**, 960 (*22*); (b) *Physica* **16**, 792. (*29*)

C. Frondel & R. E. Whitfield, *Acta Cryst.* **3**, 242. (*16*)

T. N. Godfrey & B. E. Warren, *J. Chem. Phys.* **18**, 1121. (*5*)

L. Guttman, *J. Metals* **2**, 1472. (*49*)

G. Herzberg, *Molecular Spectra and Molecular Structure. Infrared Spectra of Diatomic Molecules,* 2nd ed., Van Nostrand, New York. (*17*)

F. H. Horn, in Beeck (1950). (*74*)

K. H. Jack, *Acta Cryst.* **3**, 392. (*28*)

A. R. Kaufman, P. Gordon, & D. W. Lillie, *Trans. Am. Soc. Mech. Engrs.* **42**, 785. (*4*)

J. S. Lukesh, *Phys. Rev.* **80**, 226. (*6*)

K. H. Meyer, *Natural and Synthetic Polymers,* 2nd ed., Interscience, New York. (*16*)

R. W. Powell, *Nature* **166**, 1110. (*31*)

C. H. Schramm, P. Gordon, & A. R. Kaufmann, *J. Metals* **2**, 195. (*73, 74, 92*)

A. F. Schuch & J. H. Sturdivant, *J. Chem. Phys.* **18**, 145. (*58*)

S. S. Sidhu & C. O. Henry, *J. Appl. Phys.* **21**, 1036. (*4*)

M. E. Straumanis & E. Z. Aka, *Anal. Chem.* **22**, 1580. (*32*)

A. Taylor, *J. Inst. Metals* **77**, 585. (*27, 28*)

A. Taylor & R. W. Floyd, *Acta Cryst.* **3**, 285. (*27*)
C. W. Tucker, Jr., (a) *Trans. Am. Soc. Met.* **42**, 762 (*92*); (b) U. S. Atomic Energy
        Comm., AECD-2957 (*92*); (c) *Science* **112**, 448. (*92*)
L. Yang, *J. Electrochem. Soc.* **97**, 241. (*28*)

1951

Anonymous, *Structure Reports* **11**, 187. (*6*)
Z. S. Basinski & J. W. Christian, *J. Inst. Met.* **80**, 659. (*25*)
B. G. Bergman & D. P. Shoemaker, *J. Chem. Phys.* **19**, 515. (*92*)
D. S. Bloom & N. J. Grant, *J. Metals* **3**, 1009. (*24*)
R. D. Burbank, *Acta Cryst.* **4**, 140. (*34*)
G. J. Dickins, A. M. B. Douglas, & W. H. Taylor, (a) *Nature* **167**, 192 (*92*); (b) *J.
        Iron Steel Res.* **167**, 27. (*92*)
P. Duwez, *J. Appl. Phys.* **22**, 1174. (*72*)
J. W. Edwards, R. Speiser, & H. L. Johnston, *J. Appl. Phys.* **22**, 424. (*41, 42, 73, 78*)
M. E. Fine, E. S. Greiner, & W. C. Ellis, *J. Metals* **3**, 56. (*24*)
R. E. Franklin, *Acta Cryst.* **4**, 253. (*6*)
G. A. Geach & D. Summers-Smith, *J. Inst. Met.* **80**, 143. (*42, 73*)
E. Grison, *J. Chem. Phys.* **19**, 1109. (*34, 52*)
G. Grube, A. Schneider, & U. Esch, *Heraeus Festschrift* **20**. (*78, 79*)
J. L. Hoard, S. Geller, & R. E. Hughes, *J. Am. Chem. Soc.* **73**, 1892. (*5*)
J. S. Kasper, B. F. Decker, & J. R. Belanger, *J. Appl. Phys.* **22**, 361. (*92*)
N. Kato, *J. Phys. Soc. Japan* **6**, 502. (*79*)
F. B. Litton, *J. Electrochem. Soc.* **98**, 488. (*72*)
J. S. Lukesh, (a) *J. Chem. Phys.* **19**, 383 (*6*); (b) *J. Chem. Phys.* **19**, 1203 (*6*);
        (c) *Phys. Rev.* **84**, 1068. (*6*)
C. J. Newton & M. Y. Colby, *Acta Cryst.* **4**, 477. (*34*)
A. V. Seybolt, J. S. Lukesh, & D. W. White, *J. Appl. Phys.* **22**, 986. (*4*)
M. E. Straumanis & E. Z. Aka, (a) *Rev. Sci. Instr.* **22**, 843 (*6*); (b) *J. Am. Chem. Soc.*
        73, 5643. (*6*)
J. Thewlis, *Nature* **168**, 198. (*92*)
C. W. Tucker, Jr., *Acta Cryst.* **4**, 425. (*92*)
U. Ventriglia, *Periodico Miner.* **20**, 237. (*16*)
U. Zwicker, *Z. Metallk.* **42**, 327. (*25*)

1952

Z. S. Basinski & J. W. Christian, *J. Inst. Metals* **80**, 659. (*25*)
S. Beatty, *J. Metals* **4**, 987. (*23*)
D. S. Bloom, J. W. Putnam, & N. J. Grant, *J. Metals* **4**, 626. (*24*)
A. I. Bublik & B. Ya. Pines, *Dokl. Akad. Nauk SSSR* **87**, 215. (*28*)
R. D. Burbank, *Acta Cryst.* **5**, 236. (*34*)
R. S. Busk, *J. Metals* **4**, 207. (*12*)
M. Černohorský, *Spisy Vydávané Prirodovedeckou Fakultou Masarykovy University*
        7, 327. (*13*)
M. G. Charlton, *Nature* **169**, 109. (*74*)
R. L. Collin, *Acta Cryst.* **5**, 431. (*17*)
D. E. C. Corbridge & E. J. Low, *Nature* **170**, 629. (*15*)
D. A. Edwards, W. E. Wallace, & R. S. Craig, *J. Am. Chem. Soc.* **74**, 5256. (*48*)
D. J. Evans & M. R. Hopkins, *J. Electrodepositors' Tech. Soc.* **28**, 229. (*28*)

J. D. Fast, *J. Appl. Phys.* **23**, 350. (*40, 72*)

G. W. Fox, J. F. Carlson, G. C. Danielson, D. E. Hudson, E. N. Jensen, J. K. Knipp, J. M. Keller, L. J. Laslett, S. Legvold, G. Miller, & D. Zaffarano, U. S. Atomic Energy Comm. Report ISC-224, p. 32. (*57, 58, 59, 60, 64*)

E. S. Greiner, *J. Metals* **4**, 1044. (*32*)

N. Karlsson, *Acta Chem. Scand.* **6**, 1424. (*47*)

R. Kieffer & E. Cerwenka, *Z. Metallk.* **43**, 101. (*42*)

W. E. Pearson & W. Hume-Rothery, *J. Inst. Metals* **80**, 641. (*28*)

D. M. Poole & H. J. Axon, *J. Inst. Metals* **80**, 599. (*13*)

E. Rinck, *Compt. rend. Acad. Sci. Paris* **234**, 845. (*38*)

S. S. Sidhu & J. C. McGuire, *J. Appl. Phys.* **23**, 1257. (*72*)

R. Speiser, J. W. Spretnak, W. E. Few, & R. M. Parke, *J. Metals* **4**, 275. (*42*)

M. E. Straumanis & E. Z. Aka, *J. Appl. Phys.* **23**, 330. (*14, 32*)

D. Summers-Smith, *J. Inst. Metals* **81**, 73. (*73*)

N. Takahashi, *Compt. rend. Acad. Sci. Paris* **234**, 1619. (*29*)

A. Taylor & R. W. Floyd, *J. Inst. Metals* **80**, 577. (*24*)

J. Thewlis, *Acta Cryst.* **5**, 790. (*92*)

F. Trombe & M. Foëx, *Compt. rend. reunion ann. comm. thermodynam., Union intern. phys. Paris*, p. 308. (*58*)

C. W. Tucker, Jr., (a) *Acta Cryst.* **5**, 389 (*92*); (b) *Acta Cryst.* **5**, 395. (*92*)

R. A. Young & W. T. Ziegler, *J. Am. Chem. Soc.* **74**, 5251. (*57*)

W. H. Zachariasen, (a) *Acta Cryst.* **5**, 19 (*91*); (b) *Acta Cryst.* **5**, 660 (*93*); (c) *Acta Cryst.* **5**, 664. (*93*)

1953

B. Ancker, *Ann. Physik* **12**, 121. (*30*)

Z. S. Basinski & J. W. Christian, *Acta Met.* **1**, 754. (*25*)

R. L. Berry & G. V. Raynor, *Research* **6**, 21S. (*22*)

A. H. Daane, D. H. Dennison, & F. H. Spedding, *J. Am. Chem. Soc.* **75**, 2272. (*62, 70*)

R. F. Domagala & D. J. McPherson, U. S. Atomic Energy Comm., Publ. COO-181. (*40*)

J. S. Dugdale & F. E. Simon, *Proc. Roy. Soc.* **A218**, 291. (*2*)

F. H. Ellinger & W. H. Zachariasen, *J. Am. Chem. Soc.* **75**, 5650. (*62*)

G. Frohnmeyer & R. Glocker, *Acta Cryst.* **6**, 19. (*29*)

G. A. Geach & D. Summers-Smith, *J. Inst. Metals* **82**, 471. (*79*)

F. W. Glaser, D. Moskowitz, & B. Post, *J. Metals* **5**, 1119. (*72*)

J. M. Goode, U. S. Atomic Energy Comm., Report MLM-808. (*84*)

R. B. Hill & H. J. Axon, *Research* **6**, 23S. (*13*)

J. L. Hoard, NYO 3945: see *Structure Reports* **22**, 213 (1958). (*5*)

M. Hoch & H. L. Johnston, *J. Am. Chem. Soc.* **75**, 5224. (*14*)

A. I. Kitaigorodski, T. L. Khotsyanova, & Y. T. Struchkov, *Ž. Fiz. Khim.* **27**, 780. (*53*)

A. Kochanovska, *Czech. J. Phys.* **3**, 193. (*6*)

J. Lagrenaudie, *J. Chim. phys.* **50**, 629. (*5*)

B. W. Levinger, *J. Metals* **5**, 195. (*22*)

R. E. Marsh, L. Pauling, & J. D. McCullough, *Acta Cryst.* **6**, 71. (*34*)

E. R. Morgan, *Acta Met.* **1**, 377. (*25*)

A. B. Newkirk & A. H. Geisler, *Acta Met.* **1**, 456. (*27*)

I. Pakulla, Dissertation, Bonn. (*15*)

W. B. Pearson & W. Hume-Rothery, *J. Inst. Metals* **81**, 311. (*24*)

F. O. Rice & J. J. Ditter, *J. Am. Chem. Soc.* **75**, 6066. (*16*)

F. O. Rice, R. Potocki, & K. Gosselin, *J. Am. Chem. Soc.* **75**, 2003. (*15*)

F. O. Rice & C. J. Sparrow, *J. Am. Chem. Soc.* **75**, 848. *(16)*

M. A. Rollier, *Proc. 11th Int. Cong. Pure Appl. Chem., London, 1947* **5**, 935. *(5)*

R. B. Russell, *J. Appl. Phys.* **24**, 232. *(40, 72)*

P. W. Schenk, *Ang. Chem.* **65**, 325. *(16)*

A. U. Seybolt & H. T. Sumsion, *J. Metals* **5**, 292. *(23)*

E. A. Sheldon & A. J. King, *Acta Cryst.* **6**, 100. *(38)*

G. R. Skinner & H. L. Johnston, *J. Chem. Phys.* **21**, 1383. *(40)*

M. E. Straumanis, *Anal. Chem.* **25**, 700. *(6)*

A. H. Sully, E. A. Brandes, & K. W. Mitchell, *J. Inst. Metals* **81**, 585. *(24)*

H. E. Swanson & R. K. Fuyat, *Nat. Bur. Stand. Circular 539*, 2. *(6, 14, 31, 40, 50, 75)*

H. E. Swanson & E. Tatge, *Nat. Bur. Stand. Circular 539*, 1. *(12, 13, 28, 29, 30, 32, 42, 46, 47, 50, 52, 73, 74, 78, 79, 82)*

N. Takahashi, *J. Chim. phys.* **50**, 624. *(29)*

R. M. Treco, *J. Metals* **5**, 344. *(40)*

C. W. Tucker, Jr., & P. Senio, *Acta Cryst.* **6**, 753. *(92)*

P. L. Walker, Jr., H. A. McKinstry, & C. C. Wright, *Ind. Eng. Chem.* **45**, 1711. *(6)*

H. A. Wilhelm, O. N. Carlson, & H. E. Lunt, U. S. Atomic Energy Comm. Report AECD-3603. *(90)*

H. W. Worner, *J. Inst. Metals* **82**, 222. *(22)*

W. T. Ziegler, R. A. Young, & A. L. Floyd, *J. Am. Chem. Soc.* **75**, 1215. *(57)*

1954

J. R. Banister, S. Legvold, & F. H. Spedding, *Phys. Rev.* **94**, 1140. *(64, 66, 68)*

Z. S. Basinski & J. W. Christian, *Proc. Roy. Soc.* A223, 554. *(25)*

F. W. von Batchelder & R. F. Raeuchle, *Acta. Cryst.* **7**, 464. *(26, 28)*

G. Becherer & R. Ifland, *Naturwiss.* **41**, 471. *(47)*

G. Bergman & D. P. Shoemaker, *Acta Cryst.* **7**, 857. *(92)*

H. Braune, Dissertation, Hannover; see also Colloq. Sec. Inorg. Chem., IUPAC, Münster (Westfahlen), Sept. 1954, Verlag für Chem., Weinheim, p. 113. *(16)*

J. R. Brown, *J. Inst. Metals* **83**, 49. *(30)*

A. I. Bublik, *Dokl. Akad. Nauk SSSR* **95**, 521. *(47)*

M. G. Charlton, *Nature* **174**, 703. *(74)*

M. G. Charlton & G. L. Davis, *Nature* **175**, 131. *(74)*

P. Chiotti, *J. Electrochem. Soc.* **101**, 567. *(90)*

A. H. Daane, R. E. Rundle, H. G. Smith, & F. H. Spedding, *Acta Cryst.* **7**, 532. *(62)*

J. Drain, R. Bridelle, & A. Michel, *Bull. Soc. Chim. France* **1954**, 828. *(27)*

V. N. Eremenko, *Ž. Ukrain. Khem.* **20**, 227. *(41, 42)*

G. A. Geach & D. Summers-Smith, *J. Inst. Metals* **82**, 471. *(42)*

F. Grønvold, H. Haraldsen, & J. Vihovde, *Acta Chem. Scand.* **8**, 1927. *(26)*

W. Gruhl & E. -G. Nickel, *Giesserei.* **41**, 453. *(6)*

G. Hägg & N. Schönberg, *Acta Cryst.* **7**, 351. *(74)*

A. Hellawell & W. Hume-Rothery, (a) *Phil. Mag.* **45**, 797 *(44)*; (b) in *The Structure of Metals and Alloys*, W. Hume-Rothery & G. V. Raynor, The Institute of Metals, London, p. 94. *(46)*

J. A. Lee & G. V. Raynor, *Proc. Phys. Soc.* B67, 737. *(50)*

F. Lihl, *Arch. Eisenhüttenw.* **25**, 475. *(26)*

E. A. Owen & D. M. Jones, *Proc. Phys. Soc.* B67, 456. *(27)*

E. A. Owen & G. I. Williams, (a) *J. Sci. Inst.* **31**, 49 *(26, 47)*; (b) *Proc. Phys. Soc.* A67, 895. *(3)*

W. B. Pearson, *Can. J. Phys.* **32**, 708. *(3, 13)*

E. Poulsen-Nautrup, Dissertation, Hannover; see also *Colloq. Sec. Inorg. Chem., IUPAC, Münster* (Westfahlen), Sept. 1954, Verlag für Chem., Weinheim, p. 113. *(16)*

R. B. Russell, *J. Metals* **6**, 1045. *(40)*

N. Schönberg, *Acta Chem. Scand.* **8**, 240. *(73)*

A. U. Seybolt, *J. Metals* **6**, 774. *(41)*

S. S. Sidhu, *Acta Cryst.* **7**, 447. *(72)*

B. P. Stoicheff, *Can. J. Phys.* **32**, 630. *(7)*

H. E. Swanson, R. K. Fuyat, & G. M. Ugrinic, *Nat. Bur. Stand. Circular 539*, 3. *(22, 33, 45, 48, 49, 51, 53, 72, 83)*

J. Thewlis & A. R. Davey, *Nature* **174**, 1011. *(50)*

J. Thewlis & H. Steeple, *Acta Cryst.* **7**, 323. *(92)*

W. Trzebiatowski & J. Berak, *Bull. Acad. Polon. Sci. III* **2**, 37. *(75)*

C. W. Tucker, Jr., *Acta Cryst.* **7**, 752. *(92)*

C. Tyzack & G. V. Raynor, (a) *Acta Cryst.* **7**, 505 *(82)*; (b) *Trans. Faraday Soc.* **50**, 675. *(49)*

U. Zorll, (a) *Z. Physik* **138**, 167 *(52)*; (b) *Z. Physik* **139**, 654. *(49)*

1955

S. C. Abrahams, *Acta Cryst.* **8**, 661. *(16)*

K. Alexopoulos, *Acta Cryst.* **8**, 235. *(81)*

T. H. K. Barren & C. Domb, *Proc. Roy. Soc.* **A227**, 447. *(2)*

C. S. Barrett, *J. Inst. Metals* **84**, 43. *(11)*

Z. S. Basinski, W. Hume-Rothery, & A. L. Sutton, *Proc. Roy. Soc.* **A229**, 1179. *(26)*

Y. Baskin & L. Meyer, *Phys. Rev.* **100**, 544. *(6)*

S. R. Das, *Silicon-Sulfur-Phosphates*, Colloq. Sec. Inorg. Chem., IUPAC, Verlag Chemie, GmbH, Weinheim, p. 103. *(16)*

K. Dialer & W. Rothe, *Z. Elektrochem.* **39**, 84. *(58)*

E. R. Dobbs & K. Luszczynski, *Proc. Inter. Conf. Low Temp. Phys., Paris*, p. 349. *(36)*

J. D. H. Donnay, *Acta Cryst.* **8**, 245. *(16)*

F. H. Ellinger, *J. Metals* **7**, 411. *(60)*

G. I. Finch, K. P. Sinha, & A. Goswami, *J. Appl. Phys.* **26**, 250. *(28)*

J. Graham, A. Moore, & G. V. Raynor, *J. Inst. Metals* **84**, 86. *(49)*

G. Herzberg, *Spectra of Diatomic Molecules*, London. *(7)*

E. R. Jette, *J. Chem. Phys.* **23**, 365. *(94)*

I. L. Karle, *J. Chem. Phys.* **23**, 1739. *(53)*

F. M. Kelly & W. B. Pearson, *Can. J. Phys.* **33**, 17. *(37)*

H. Krebs, K. H. Müller, I. Pakulla, & G. Zürn, *Ang. Chem.* **67**, 524. *(15)*

H. Krebs, F. Schultze-Gebhardt, & R. Thees, *Z. anorg. allg. Chem.* **282**, 177. *(51)*

H. Krebs, H. Weitz, & K. H. Worms, *Z. anorg. allg. Chem.* **280**, 119. *(15)*

W. B. Pearson, *Can. J. Phys.* **33**, 473. *(49)*

E. Pipitz & R. Kieffer, *Z. Metallk.* **46**, 187. *(42)*

E. Raub & G. Wörwag, *Z. Metallk.* **46**, 513. *(46, 78)*

W. Rostoker & A. Yamamoto, *Trans. Am. Soc. Met.* **47**, 1002. *(23)*

P. W. Schenk, *Ang. Chem.* **67**, 344. *(16)*

C. T. Sims, C. M. Craighead, & R. I. Jaffee, *J. Metals* **7**, 168, *(75)*

A. Smakula & J. Kalnajs, *Phys. Rev.* **99**, 1737. *(13, 14, 32, 47)*

B. Staliński, *Bull. Acad. Polon. Sci.* **C13**, III, 613. *(57)*

C. Stein & N. J. Grant, *J. Metals* **7**, 127. *(24)*

M. E. Straumanis & C. C. Weng, *Acta Cryst.* **8**, 367. *(24)*

A. L. Sutton & W. Hume-Rothery, *Phil. Mag.* **46**, 1295. *(26)*

H. E. Swanson, R. K. Fuyat, & G. M. Ugrinic, *Nat. Bur. Stand. Circular 539,* **4.** (*26, 44, 56, 76, 77*)

H. E. Swanson, N. T. Gilfrich, & G. M. Ugrinic, *Nat. Bur. Stand. Circular 539,* **5.** (*24, 34*)

C. A. Swenson, (a) *J. Chem. Phys.* **23,** 1963 (*7*); (b) *Phys. Rev.* **100,** 1607. (*49, 81*)

I. Szántó, *Acta Tech. Acad. Sci. Hung.* **13,** 363. (*22*)

L. I. Tatarinova, *Trans. Inst. Kristall. Akad. Nauk SSSR* **1955,** 101. (*51*)

W. H. Zachariasen & F. H. Ellinger, *Acta Cryst.* **8,** 431. (*94*)

1956

E. P. Abrahamson & N. J. Grant, *J. Metals* **8,** 975. (*24*)

C. S. Barrett, (a) *Acta Met.* **4,** 528 (*3*); (b) *Acta Cryst.* **9,** 671 (*3, 11, 19, 37, 55*); (c) *J. Chem. Phys.* **25,** 1123. (*56, 63*)

P. Blum & A. Durif, *Acta Cryst.* **9,** 829. (*32*)

J. P. Bridge, C. M. Schwartz, & D. A. Vaughan, *J. Metals* **8,** 1282. (*92*)

V. P. Butuzov & E. G. Ponyatovski, *Sov. Phys.-Cryst.* **1,** 453. (*83*)

B. R. Coles, *J. Inst. Metals* **84,** 346. (*46*)

R. L. Collin, *Acta Cryst.* **9,** 537. (*17*)

B. D. Cullity, *Elements of X-Ray Diffraction,* Addison-Wesley, New York. (*28*)

R. Diament, *Métaux Corros.* **31,** 167. (*28*)

G. J. Dickins, A. M. B. Douglas, & W. H. Taylor, *Acta Cryst.* **9,** 297. (*92*)

E. R. Dobbs, B. F. Figgins, G. O. Jones, D. C. Piercey, & D. P. Riley, *Nature* **178,** 483. (*18*)

H. S. Dunsmore, L. D. Calvert, & W. A. Alexander, *Proc. Roy. Soc. Canada* **50,** Appendix C, 20. (*20*)

F. H. Ellinger, *Trans. Am. Inst. Metals Pet. Eng.* **206,** 1256. (*94*)

B. F. Figgins, G. O. Jones, & D. P. Riley, *Phil. Mag.* **1,** 747. (*13*)

P. Graf, B. B. Cunningham, C. H. Dauben, J. C. Wallmann, D. H. Templeton, & H. Ruben, *J. Am. Chem. Soc.* **78,** 2340. (*95*)

J. Güntert & A. Faessler, *Z. Krist.* **107,** 361. (*79*)

R. B. Hill & H. J. Axon, *J. Inst. Metals* **85,** 109. (*47*)

R. G. Hirst, A. J. King, & F. A. Kanda, *J. Phys. Chem.* **60,** 302. (*38, 56*)

W. J. James & M. E. Straumanis, *Acta Cryst.* **9,** 376. (*90*)

A. J. King, in Hirst, King, & Kanda (1956). (*56*)

H. Krebs, W. Holz, W. Lippert, & K. H. Worms, *Osterr. Chem. Ztg.* **57,** 137. (*33*)

H. Krebs & F. Schultze-Gebhardt, *Z. anorg. allg. Chem.* **283,** 263. (*33*)

K. Kuo & W. G. Burgers, *Proc. Acad. Sci. Amsterdam* B**59,** 288. (*50*)

F. Laves & Y. Baskin, *Z. Krist.* **107,** 337. (*6*)

G. Mannella & L. O. Hougen, *J. Phys. Chem.* **60,** 1148. (*74*)

H. Melsert, T. J. Tiedma, & W. G. Burgers, *Acta Cryst.* **9,** 525. (*20*)

H. Neff, *Z. Ang. Phys.* **8,** 505. (*14, 47, 79*)

J. A. Prins, J. Schenk, & P. A. M. Hospel, *Physica* **22,** 770. (*16*)

A. Schneider & G. Heymer, *Z. anorg. allg. Chem.* **286,** 118. (*48, 49, 81*)

V. Smirnova & B. F. Ormont, *Ž. Fiz. Khim.* **30,** 1327. (*73*)

J. F. Smith, O. N. Carlson, & R. W. Vest, *J. Electrochem. Soc.* **103,** 409. (*20*)

F. H. Spedding & A. H. Daane, *Progress in Nuclear Energy,* ser. 5, vol. 1, H. M. Finniston & J. P. Howe, eds., McGraw-Hill, New York, p. 413. (*62*)

F. H. Spedding, A. H. Daane, & K. W. Herrmann, *Acta Cryst.* **9,** 559. (*21, 39, 57, 58, 59, 60, 63, 64, 65, 66, 67, 68, 69, 70, 71*)

M. E. Straumanis & C. C. Weng, *Am. Miner.* **41,** 437. (*24*)

H. Thiel, *Ann. Physik* **17,** 122. (*15*)

J.-J. Trillat, L. Tertian, & N. Terao, *Compt. rend. Acad. Sci. Paris* **243**, 666. (*28*)

C. W. Tucker, Jr., P. Senio, J. Thewlis, & H. Steeple, *Acta Cryst.* **9**, 472. (*92*)

T. R. Waite, W. E. Wallace, & R. S. Craig, *J. Chem. Phys.* **24**, 634. (*73*)

H. Weyerer, (a) *Z. Ang. Phys.* **8**, 135 (*13*); (b) *Z. Ang. Phys.* **8**, 202 (*79*); (c) *Z. Ang. Phys.* **8**, 297 (*79*); (d) *Z. Ang. Phys.* **8**, 553. (*79*)

E. A. Wood, *J. Phys. Chem.* **60**, 508. (*14*)

1957

P. S. Aggarwal & A. Goswami, *Proc. Phys. Soc.* **70B**, 708. (*42*)

C. S. Barrett, *Acta Cryst.* **10**, 58. (*80*)

F. W. von Batchelder & R. F. Raeuchle, *Phys. Rev.* **105**, 59. (*12*)

R. E. Brocklehurst, J. M. Goode, & L. F. Vassamillet, *J. Chem. Phys.* **27**, 985. (*84*)

S. N. Chatterjee, *Ind. J. Phys.* **31**, 110. (*79*)

G. H. Cheesman & C. M. Soane, *Proc. Phys. Soc.* **B70**, 700. (*36, 54*)

E. J. Covington & D. J. Montgomery, *J. Chem. Phys.* **27**, 1030. (*3*)

J. M. Goode, *J. Chem. Phys.* **27**, 1269. (*84*)

M. A. Gurevic & B. F. Ormont, *Fiz. Metall. Metalloved.* **4**, 112. (*23*)

E. O. Hall & J. Crangle, *Acta Cryst.* **10**, 240. (*44*)

V. S. Kogan, B. G. Lazarev, & R. F. Bulatova, *Sov. Phys.-JETP* **4**, 593. (*1*)

H. Krebs, W. Holz, W. Lippert, & K. H. Worms, *Chem. Ber.* **90**, 1031. (*33*)

G. Mack, *Phys. Verh. Mosbach* **8**, 91. (*32*)

T. Millner, A. J. Hegedüs, K. Sasvari, & J. Neugebauer, *Z. anorg. allg. Chem.* **289**, 288. (*74*)

S. Nagakura, *J. Phys. Soc. Japan* **12**, 482. (*28*)

A. G. Pinkus, J. S. Kim, J. L. McAtee, Jr., & C. B. Concilio, *J. Am. Chem. Soc.* **79**, 4566. (*16*)

J. A. Prins, J. Schenk, & L. H. J. Wachters, *Physica* **23**, 746. (*16*)

A. A. Rudnicki & R. S. Polyakova, *Ž. Neorgan. Khim.* **2**, 2758. (*44*)

D. E. Sands & J. L. Hoard, *J. Am. Chem. Soc.* **79**, 5582. (*5*)

R. O. Simmons & R. W. Balluffi, *Phys. Rev.* **108**, 278. (*29*)

B. J. Skinner, *Am. Miner.* **42**, 39. (*6*)

A. Smakula & J. Kalnajs, *Nuovo Cim.* **6**, 214. (*32*)

F. H. Spedding, A. H. Daane, & K. W. Herrmann, *J. Metals* **9**, 895. (*57, 58, 59, 60*)

P. L. Walker, Jr., & G. Imperial, *Nature* **180**, 1184. (*6*)

H. Weyerer, *Z. Krist.* **109**, 338. (*79*)

W. H. Zachariasen & F. H. Ellinger, *J. Chem. Phys.* **27**, 811. (*94*)

1958

T. R. Anantharaman, *Curr. Sci.* **27**, 51. (*27*)

E. S. Bale, *Platinum Met. Rev.* **2**, 61. (*45*)

C. S. Barrett, *Phys. Rev.* **110**, 1071. (*64, 81*)

I. A. Black, L. H. Bolz, F. P. Brooks, F. A. Mauer, & H. S. Peiser, *J. Res. Nat. Bur. Stand.* **61**, 367. (*8*)

F. P. Bundy, *Phys. Rev.* **110**, 314. (*83*)

S. N. Chatterjee, *Acta Cryst.* **11**, 679. (*92*)

J. Cuthbert & J. W. Linnett, *Trans. Faraday Soc.* **54**, 617. (*2, 10, 18, 36, 54, 86*)

H. G. David & S. D. Hamann, *J. Chem. Phys.* **28**, 1006. (*16*)

J. Donohue, A. Caron, & E. Goldish, *Nature* **182**, 518. (*16*)

O. Erämetsä, (a) *Suom. Kemi.* **B31**, 237 (*16*); (b) *Suom. Kemi.* **B31**, 241 (*16*); (c) *Suom. Kemi.* **B31**, 246. (*16*)

Y. M. de Haan, *Physica* **24**, 855. (*16*)

D. G. Henshaw, (a) *Phys. Rev.* **109**, 328 (*2*); (b) *Phys. Rev.* **111**, 1470. (*10, 18*)

R. Herman & D. A. Swenson, *J. Chem. Phys.* **29**, 398. (*58*)

J. L. Hoard, R. E. Hughes, & D. E. Sands, *J. Am. Chem. Soc.* **80**, 4507. (*5*)

V. S. Kogan, B. G. Lazarev, & R. F. Bulatova, *Sov. Phys.-JETP* **7**, 165, (*1*)

J. A. Kohn, G. Katz, & A. A. Giardini, *Z. Krist.* **111**, 53. (*5*)

H. König, *Ang. Chem.* **70**, 110. (*47*)

S. T. Konobeevski, K. P. Bubrovin, M. Levitski, L. D. Panteleev, & N. F. Pravdynk, *Proc. 2nd U. N. Int. Conf. Peaceful Uses of Atomic Energy, Geneva* **5**, 574. (*92*)

G. Mack, *Z. Physik* **152**, 19. (*32*)

E. Matuyama, *Tanso* **7**, 12, (*6*)

L. V. McCarty, J. S. Kasper, F. H. Horn, B. F. Decker, & A. E. Newkirk, *J. Am. Chem. Soc.* **80**, 2592. (*5*)

R. L. Mills & E. R. Grilly, *Proc. 5th Int. Conf. Low-Temp. Phys. Chem.*, Univ. of Wisconsin, Aug. 26–31, 1957, J. R. Dillinger, ed., Univ. Wisconsin Press, Madison, 1958. (*2*)

L. Muldawer, D. M. Hoffman, & J. Riseman, *Bull. Am. Phys. Soc.* **3**, 111. (*24*)

S. Nagakura, *J. Phys. Soc. Japan* **13**, 1005. (*28*)

E. Parthé & J. T. Norton, *Z. Krist.* **110**, 166. (*5*)

J. C. Schottmiller, A. J. King, & F. A. Kanda, *J. Phys. Chem.* **62**, 1446. (*20*)

A. F. Schuch, *Proc. 5th Int. Conf. Low-Temp. Phys. Chem.*, Univ. of Wisconsin, Aug. 26–31, 1957, J. R. Dillinger, ed., University of Wisconsin Press, Madison, 1958, p. 79. (*2*)

A. F. Schuch, E. R. Grilly, & R. L. Mills, *Phys. Rev.* **110**, 775. (*2*)

M. Seal, *Nature* **182**, 1264. (*6*)

V. V. Sofina, Z. M. Azarkh, & N. N. Orlova, *Sov. Phys.-Cryst.* **3**, 544. (*22, 40*)

F. H. Spedding, J. J. Hanak, & A. H. Daane, *J. Metals* **10**, 97. (*63*)

D. R. Stern & L. Lynds, *J. Electrochem. Soc.* **105**, 676. (*5*)

C. A. Swenson, *Phys. Rev.* **111**, 82. (*80*)

A. Taylor & R. M. Jones, (a) *J. Appl. Phys.* **29**, 522 (*26*); (b) *J. Phys. Chem. Solids* **6**, 16. (*26*)

R. Uno, *J. Phys. Soc. Japan* **13**, 667. (*5*)

P. M. de Wolff, ASTM card no. 8-247. (*16*)

1959

A. I. Andrievski, I. D. Nabitovich, & P. I. Kripiakevich, *Sov. Phys.-Dokl.* **4**, 16. (*34*)

M. Atoji, J. E. Schirber, & C. A. Swenson, *J. Chem. Phys.* **31**, 1628. (*80*)

B. C. Banerjee & A. Goswami, (a) *J. Electrochem. Soc.* **106**, 20 (*28*); (b) *J. Electrochem. Soc.* **106**, 590. (*28*)

C. S. Barrett, *Acta Met.* **7**, 810. (*83*)

L. H. Bolz, M. E. Boyd, F. A. Mauer, & H. S. Peiser, *Acta Cryst.* **12**, 247. (*7*)

E. P. Bundy, *Phys. Rev.* **115**, 274. (*37*)

P. Chiotti, H. H. Klepfer, & R. W. White, *Trans. Am. Soc. Met.* **51**, 772. (*92*)

J. J. Couderc, G. Garigue, L. Lafourcade, & Q. T. Nguyen, *Compt. rend. Acad. Sci. Paris* **249**, 2037. (*29, 79*)

B. F. Decker & J. S. Kasper, *Acta Cryst.* **12**, 503. (*5*)

A. Defrain & I. Epelboin, *Compt. rend. Acad. Sci. Paris* **249**, 50. (*31*)

A. Defrain, I. Epelboin, & M. Erny, *Compt. rend. Acad. Sci. Paris* **248**, 1486. (*31*)

J. Donohue, (a) *Phys. Rev.* **114**, 1009 (*2*); (b) *Acta Cryst.* **12**, 697. (*91*)

O. Erämetsä, (a) *Suom. Kemi.* **B32**, 15 (*16*); (b) *Suom. Kemi.* **B32**, 97. (*16*)

O. Erämetsä & H. Suonuuti, *Suom. Kemi.* **B32**, 47. (*16*)

D. S. Evans & G. V. Raynor, *J. Nucl. Mater.* **1**, 281. (*90*)

B. Gale, *Acta Met.* **7**, 420. (*26*)

A. T. Grigoryev, L. N. Guseva, E. Sokolovskaya, & M. V. Maksimova, *Russ. J. Inorg. Chem.* **4**, 984. (*24*)

E. R. Grilly & R. L. Mills, *Ann. Phys.* **8**, 1. (*2*)

D. Hardie & R. N. Parkins, *Phil. Mag.* **4** (8), 815. (*12*)

L. L. Hawes, *Acta Cryst.* **12**, 34. (*35*)

L. L. Hawes & G. H. Cheesman, *Acta Cryst.* **12**, 477. (*17*)

F. H. Horn, (a) *J. Appl. Phys.* **30**, 1612 (*5*); (b) *J. Electrochem. Soc.* **106**, 905. (*5*)

P. C. Jamieson, A. W. Lawson, & N. D. Nachtrieb, *Rev. Sci. Inst.* **30**, 1016. (*83*)

Z. Johan, *Chem. Erde* **20**, 71. (*33*)

W. Kaiser, W. L. Bond, & M. Tannenbaum, *Bull. Am. Phys. Soc.* **4**, 27. (*6*)

J. A. Lee, P. G. Mardon, J. H. Pearce, & R. O. A. Hall, *J. Phys. Chem. Solids* **11**, 177. (*93*)

B. D. Lichter, *J. Metals* **11**, 581. (*40*)

A. M. Liquori & A. Ripamonte, *Ric. Sci.* **29**, 2186. (*16*)

P. G. Mardon & J. H. Pearce, *J. Less-Common Metals* **1**, 467. (*93*)

A. J. Martin & A. J. Moore, *J. Less-Common Metals* **1**, 85. (*4*)

R. L. Moss & I. Woodward, *Acta Cryst.* **12**, 255. (*74*)

H. Mu, *Proc. Int. Conf. Symp. High Temp. Tech.*, McGraw-Hill, New York. (*6*)

M. R. Nadler & C. P. Kempter, (a) *Anal. Chem.* **31**, 1922 (*41*); (b) *Anal. Chem.* **31**, 2109. (*3*)

T. Noda, in Mu (1959). (*6*)

A. G. Pinkus, J. S. Kim, J. L. McAtee, Jr., & C. B. Concilio, (a) *J. Polymer Sci.* **40**, 581 (*16*); (b) *J. Am. Chem. Soc.* **81**, 2652. (*16*)

E. Raub, H. Beeskow, & D. Menzel, *Z. Metallk.* **50**, 428. (*45*)

F. O. Rice & R. B. Ingalls, *J. Am. Chem. Soc.* **81**, 1856. (*16*)

W. L. Robb & L. C. Landauer, Abstracts of ACS Meeting, Boston. (*5*)

B. M. Rovinski, A. I. Samilov, & G. M. Rovenski, *Phys. Metals Metallogr.* **7**, 73. (*28*)

A. A. Rudnicki & R. S. Polyakova, *Russ. J. Inorg. Chem.* **4**, 631. (*46*)

D. R. Schwarzenberger, *Phil. Mag.* **4**, 1242. (*4*)

J. F. Smith & B. T. Bernstein, (a) *J. Electrochem. Soc.* **106**, 448 (*20*); (b) *Acta Cryst.* **12**, 419. (*20*)

J. Spreadborough & J. W. Christian, (a) *J. Sci. Instr.* **36**, 116 (*47*); (b) *Proc. Phys. Soc.* **74**, 609. (*22*)

M. E. Straumanis, *J. Appl. Phys.* **30**, 1965. (*13, 23*)

A. G. Streng & A. V. Grosse, *J. Am. Chem. Soc.* **81**, 805. (*8*)

R. E. Vogel & C. P. Kempter, U. S. Atomic Energy Comm., LA-2317. (*3, 6, 14, 50*)

H. Warlimont, *Z. Metallk.* **50**, 708. (*79*)

R. T. Weiner & G. V. Raynor, *J. Less-Common Metals* **1**, 309. (*58*)

W. H. Zachariasen, *Acta Cryst.* **12**, 698. (*91*)

W. H. Zachariasen & F. H. Ellinger, *Acta Cryst.* **12**, 175. (*94*)

1960

A. I. Andrievski & I. D. Nabitovich, *Sov. Phys.-Cryst.* **5**, 442. (*34*)

C. S. Barrett, *Austral. J. Phys.* **13A**, 209. (*83*)

H. J. Becher, *Z. anorg. allg. Chem.* **306**, 266. (*5*)

T. Berger, S. Joigneau, & G. Bothet, *Compt. rend. Acad. Sci. Paris* **250**, 4331. (*16*)

K. E. Beu, in Parrish (1960). (*6*)

W. L. Bond, *Acta Cryst.* **13**, 814. (*14*)

W. L. Bond & W. Kaiser, *J. Phys. Chem. Solids* **16**, 44. (*14*)

L. Bosio, A. Defrain, & I. Epelboin, *Compt. rend. Acad. Sci. Paris* **250**, 2553. (*31*)

A. Caron & J. Donohue, *J. Phys. Chem.* **64**, 1767. (*16*)

M. Černohorsky, *Acta Cryst.* **13**, 823. (*44, 45*)

N. T. Chebotarev & A. V. Beznosikova, *J. Nuc. Energy* **12**, 127. (*94*)

B. F. Decker & J. S. Kasper, *Acta Cryst.* **13**, 1030. (*5*)

A. Defrain, *Compt. rend. Acad. Sci. Paris* **250**, 483. (*31*)

A. Defrain & I. Epelboin, *J. phys. radium* **21**, 76. (*31*)

J. D. Dudley & H. T. Hall, *Phys. Rev.* **118**, 1211. (*49, 50*)

S. S. Dukhin, *Sov. Phys.-JETP* **10**, 1054. (*1*)

B. F. Figgins & B. L. Smith, *Phil. Mag.* **5**, 186. (*36*)

H. J. Goldschmidt, *J. Iron Steel Inst.* **194**, 169. (*41*)

A. T. Grigoryev, E. M. Sokolovskaya, Y. P. Simanov, I. G. Sokolova, M. V. Maksimova,
     & L. I. Pyatigorskaya, *Russ. J. Inorg. Chem.* **5**, 1036. (*24*)

H. T. Hall, *Rev. Sci. Instr.* **31**, 125. (*55*)

J. L. Hoard & A. E. Newkirk, *J. Am. Chem. Soc.* **82**, 70. (*5*)

J. A. Hren & C. M. Wayman, *Trans. Met. Soc. AIME* **218**, 377. (*23*)

W. J. James & M. E. Straumanis, *J. Electrochem. Soc.* **107**, 69. (*23*)

C. F. Kempter, in Parrish (1960). (*6*)

V. S. Kogan, B. G. Lazarev, & R. F. Bulatova, *Sov. Phys.-JETP* **10**, 485. (*1*)

B. D. Lichter, *Trans. Met. Soc. AIME* **218**, 1015. (*40*)

E. S. Makarov & L. M. Kuznetsov, *J. Struc. Chem.* **1**, 156. (*22*)

L. V. McCarty & D. R. Carpenter, *J. Electrochem. Soc.* **107**, 38. (*5*)

C. J. McHargue & H. L. Yakel, Jr., *Acta Met.* **8**, 637. (*58*)

D. B. McWhan, J. C. Wallmann, B. B. Cunningham, L. B. Asprey, F. H. Ellinger, &
     W. H. Zachariasen, *J. Inorg. Nucl. Chem.* **15**, 185. (*95*)

A. G. Metcalf, *Am. Chem. Soc. Monograph Series* **149**, 157. (*27*)

B. Meyer, Dissertation, Zürich. (*16*)

B. Meyer & E. Schumacher, (a) *Helv. Chim. Acta* **43**, 1333 (*16*); (b) *Nature* **186**, 801. (*16*)

G. Ozolins & A. Ievinš, *Latvijas PSR Zinatnu Akad. Vestis* **1960**, 61. (*14*)

W. Parrish, *Acta Cryst.* **13**, 838. (*6, 14, 74*)

E. Parthé & J. T. Norton, *Z. Krist.* **114**, 480. (*5*)

W. G. Perdok, in Parrish (1960). (*6*)

A. G. Pinkus & J. L. McAtee, Jr., *Chem. Eng. News* **38**, 44. (*16*)

E. G. Ponyatovski, *Sov. Phys.-Cryst.* **5**, 147. (*83*)

F. A. Raal, *Nature* **185**, 523. (*6*)

H. Schäfer & H.-J. Heitland, *Z. anorg. allg. Chem.* **304**, 250. (*77*)

M. Seal, *Nature* **185**, 522. (*6*)

S. A. Semiletov, *Sov. Phys.-Cryst.* **4**, 588. (*34*)

F. H. Spedding, A. H. Daane, G. Wakefield, & D. H. Dennison, *Trans. Met. Soc. AIME*
     218, 608. (*21*)

F. H. Spedding, J. J. McKeown, & A. H. Daane, *J. Phys. Chem.* **64**, 289. (*62*)

E. G. Steward, B. P. Cook, & E. A. Kellett, *Nature* **187**, 1015. (*6*)

M. E. Straumanis & W. James, in Parrish (1960). (*6*)

E. F. Sturcken & B. Post, *Acta Cryst.* **13**, 852. (*92*)

H. E. Swanson, M. I. Cook, T. Isaacs, & E. H. Evans, *Nat. Bur. Stand. Circular 539,* **9**,
     54. (*16*)

C. P. Talley, S. LaPlaca, & B. Post, *Acta Cryst.* **13**, 271. (*5*)

I. Teodorescu & A. Glodeanu, *Phys. Rev. Lett.* **4**, 231. (*28*)

M. Tourarie, in Parrish (1960). (*6*)

M. M. Umanski, D. M. Kheiker, & L. S. Zevin, *Sov. Phys.-Cryst.* 4, 345. (*74*)
M. M. Umanski, V. V. Zubenko, & Z. K. Zolina, *Sov. Phys.-Cryst.* 5, 43. (*74*)
J. Vlach & B. Stehlik, *Coll. Czech. Chem. Comm.* 25, 676. (*47*)
M. Wilkins, in Parrish (1960). (*6*)

1961

S. C. Abrahams, *Acta Cryst.* 14, 311. (*16*)
V. M. Amonenko, V. E. Ivanov, G. F. Tikhinski, V. A. Finkel, & I. V. Spagin, *Phys. Metals Metallogr.* 12, (6), 77. (*4*)
A. Balchan & H. G. Drickamer, *Rev. Sci. Inst.* 32, 308. (*26*)
C. S. Barrett, *Adv. X-Ray Anal.* 5, 33. (*31*)
C. Bonnelle & F. Jacquot, *Compt. rend. Acad. Sci. Paris* 252, 1448. (*28*)
J. P. Burger & M. A. Taylor, *Phys. Rev. Lett.* 6, 185. (*23*)
O. N. Carlson & C. V. Owen, *J. Electrochem. Soc.* 108, 88. (*23*)
A. Caron & J. Donohue, *Acta Cryst.* 14, 548. (*16*)
A. H. Cash, E. W. Hughes, & C. C. Murdock, *Acta Cryst.* 14, 313. (*92*)
N. T. Chebotarev, *Atomnaya Energiya* 10 (1), 43. (*92*)
A. S. Cooper, W. L. Bond, & S. C. Abrahams, *Acta Cryst.* 14, 1008. (*16*)
L. J. Cotta & C. P. Gazzara, *Adv. X-Ray Anal.* 5, 57. (*26*)
H. Curien, A. Rimsky, & A. Defrain, *Bull. Soc. Franc. Mineral. Crist.* 84, 260. (*31*)
G. F. Day, *J. Inst. Met.* 89, 296. (*79*)
V. T. Deshpande & D. B. Sirdeshmukh, *Acta Cryst.* 14, 355. (*50*)
J. Donohue, (a) *Acta Cryst.* 14, 327 (*92*); (b) *Acta Cryst.* 14, 1000. (*7*)
J. Donohue & A. Caron, *J. Polymer Sci.* 50, S17. (*16*)
J. Donohue, A. Caron, & E. Goldish, *J. Am. Chem. Soc.* 83, 3748. (*16*)
A. J. Eatwell & B. L. Smith, *Phil. Mag.* 6, 461. (*54*)
E. O. Eliot & A. C. Larson, *The Metal Plutonium,* A. S. Coffinberry & W. N. Miner, eds., University of Chicago Press, p. 265. (*94*)
D. S. Evans & G. V. Raynor, *J. Nucl. Mater.* 4, 66. (*40, 90*)
J. D. Farr, A. L. Giorgi, M. G. Bowman, & R. K. Money, *J. Inorg. Nucl. Chem.* 18, 42. (*89*)
J. P. Franck, *Phys. Rev. Lett.* 7, 435. (*2*)
H. J. Goldschmidt & J. A. Brand, *J. Less-Common Metals* 3, 44. (*42*)
P. Greenfield, J. A. Lee, P. G. Mardon, & J. A. L. Robertson, *The Metal Plutonium,* A. S. Coffinberry & W. N. Miner, eds., University of Chicago Press, p. 144. (*94*)
R. O. A. Hall, *Acta Cryst.* 14, 1004. (*14*)
E. M. Hörl & L. Marton, *Acta Cryst.* 14, 11. (*7*)
W. J. James & M. E. Straumanis, *Z. physik. Chem.* 29, 134. (*23*)
W. J. James, M. E. Straumanis & P. B. Rao, *J. Inst. Metals* 90, 176. (*24*)
R. H. Johnson & R. W. K. Honeycombe, *J. Nucl. Mater.* 4, 59. (*90*)
G. C. Kennedy & P. N. LaMori, *Progress in Very High Pressure Research,* Wiley, New York. (*81*)
W. King & L. F. Vassamillet, *Adv. X-Ray Anal.* 5, 78. (*47*)
V. S. Kogan, B. G. Lazarev, & R. F. Bulatova, *Sov. Phys.-JETP* 13, 19. (*10*)
V. S. Kogan, B. G. Lazarev, R. P. Ozerov, & G. S. Zhdanov, *Sov. Phys.-JETP* 13, 118. (*1*)
S. T. Konobeevsky & N. T. Chebotarev, *Atomnaya Energiya* 10, 50. (*94*)
D. J. Lam, J. B. Darby, J. W. Downey, & L. J. Norton, *Nature* 192, 744. (*43*)
F. Lihl & H. Ebel, *Arch. Eisenhüttenw.* 32, 489. (*26*)
F. A. Mauer & L. H. Bolz, *J. Res. Nat. Bur. Stand.* 65C, 225. (*10*)
O. D. McMasters & W. L. Larsen, *J. Less-Common Metals* 3, 312. (*73, 90*)

J. A. C. Marples & J. A. Lee, *Nature* **191**, 1391. (*94*)

R. L. Mills & A. F. Schuch, *Phys. Rev. Lett.* **6**, 263. (*2*)

R. P. Ozerov, V. S. Kogan, G. S. Zhdanov, & O. L. Kukhto, *Sov. Phys.-Cryst.* **6**, 507. (*1*)

E. Parthé, *Z. Krist.* **115**, 52. (*15*)

D. T. Peterson & V. G. Fattore, *J. Phys. Chem.* **65**, 2062. (*20*)

D. E. Sands, C. F. Cline, A. Zalkin, & C. L. Hoenig, *Acta Cryst.* **14**, 309. (*5*)

A. F. Schuch & R. L. Mills, (a) *Adv. Cryogenic Engineering* **7**, 311 (*2*); (b) *Phys. Rev. Lett.* **6**, 596. (*2*)

F. H. Spedding, J. J. Hanak, & A. H. Daane, *J. Less-Common Metals* **3**, 110. (*21, 39, 57, 58, 59, 60, 63, 64, 65, 66, 67, 68, 69, 70, 71*)

M. E. Straumanis, P. Gorgeaud, & W. J. James, *J. Appl. Phys.* **32**, 1382. (*14*)

H. Strunz & E. Herda, *Naturwiss.* **48**, 597. (*16*)

J.-J. Trillat, N. Terao, & L. Tertian, *Compt. rend. Acad. Sci. Paris* **253**, 3512. (*28*)

B. M. Vasjutinski, G. N. Kartmazov, & V. A. Finkel, *Phys. Metals Metallogr.* **12**, 141. (*24*)

J. H. Vignos & H. A. Fairbank, *Phys. Rev. Lett.* **6**, 265. (*2*)

R. J. Wasilewski, *Trans. Met. Soc. AIME* **221**, 1081. (*75*)

N. F. Yannoni, *Diss. Abstr.* **22**, 1032. (*5*)

1962

V. M. Amomenko, V. E. Ivanov, G. F. Tikhinski, & V. A. Finkel, *Phys. Metals Metallogr.* **14**, 47. (*4*)

C. S. Barrett, *Adv. X-Ray Anal.* **5**, 33. (*49*)

K. E. Beu, F. J. Musil, & D. R. Whitney, *Acta Cryst.* **15**, 1292. (*14*)

J. Boon, in Tuinstra (1967). (*16*)

L. Bosio, A. Defrain, & I. Epelboin, *J. Phys. Radium* **23**, 877. (*31*)

F. P. Bundy, *Science* **137**, 1055. (*6*)

A. S. Cooper, *Acta Cryst.* **15**, 578. (*13, 16, 32, 92*)

P. Cucka & C. S. Barrett, *Acta Cryst.* **15**, 865. (*83*

V. T. Deshpande & D. B. Sirdeshmukh, *Acta Cryst.* **15**, 294. (*50*)

B. N. Dutta, *Phys. Status solidi* **2**, 984. (*14*)

O. Erämetsä, *Suom. Kemi.* B**35**, 154. (*16*)

S. Ergun & L. E. Alexander, *Nature* **195**, 765. (*6*)

E. J. Freise, *Nature* **193**, 671. (*6*)

H. J. Goldschmidt, *Adv. X-Ray Anal.* **5**, 191. (*26*)

A. H. Graham, R. W. Lindsay, & H. J. Read, *J. Electrochem. Soc.* **109**, 1200. (*28*)

E. R. Grilly & R. L. Mills, *Ann. Phys.* **18**, 250. (*2*)

K. A. Gschneider, R. O. Elliott, & R. R. McDonald, *Phys. Chem. Solids* **23**, 555. (*58*)

E. M. Hörl, *Acta Cryst.* **15**, 845. (*8*)

J. C. Jamieson, *Abstr. Geol. Soc. Am., Houston*, p. 78A. (*14, 32, 50*)

J. C. Jamieson & A. W. Lawson, *J. Appl. Phys.* **33**, 776. (*26*)

V. S. Kogan & A. S. Bulatov, *Sov. Phys.-JETP* **15**, 1041. (*28*)

B. Kolakowski, *Acta Phys. Polon.* **22**, 439. (*5*)

J. van Kranendonk & H. P. Gush, *Phys. Lett.* **1**, 22. (*1*)

J. A. C. Marples & J. A. Lee, *Sov. J. Atom. Energy* **12**, 453. (*94*)

D. B. McWhan, B. B. Cunningham, & J. C. Wallmann, *J. Inorg. Nucl. Chem.* **24**, 1025. (*95*)

S. Minomura & H. G. Drickamer, *J. Phys. Chem. Solids* **23**, 451. (*32*)

M. H. Mueller, R. L. Hitterman, & H. W. Knott, *Acta Cryst.* **15**, 421. (*92*)

T. Niemyski & Z. Olempska, *J. Less-Common Metals* **4**, 235. (*5*)

R. P. Ozerov, G. S. Zhdanov, & V. S. Kogan. Quoted by van Kranendonk & Gush (1962). *(1)*

G. J. Piermarini & C. E. Weir, *J. Res. Nat. Bur. Stand.* **66A**, 325. *(83)*

E. G. Ponyatovski & A. I. Zakharov, *Sov. Phys.-Cryst.* **7**, 367. *(81)*

W. T. Roberts, *J. Less-Common Metals* **4**, 345. *(22)*

E. Rudy, B. Kieffer, & H. Fröhlich, *Z. Metallk.* **53**, 90. *(44, 75)*

E. M. Savicki, M. A. Tylkina, & V. P. Polyakova, *Russ. J. Inorg. Chem.* **7**, 224. *(44, 75)*

J. E. Schirber & C. A. Swenson, *Acta Metall.* **10**, 511. *(80)*

A. F. Schuch & R. L. Mills, *Phys. Rev. Lett.* **8**, 469. *(2)*

D. R. Sears & H. P. Klug, *J. Chem. Phys.* **37**, 3002. *(54)*

B. D. Sharma & J. Donohue, *Z. Krist.* **117**, 293. *(31)*

O. Skjerven, *Z. anorg. allg. Chem.* **314**, 206. *(16)*

A. I. Soklakov, *Ž. Strukt. Khim.* **3**, 559. *(16)*

R. A. Stager, A. S. Balchan, & H. G. Drickamer, *J. Chem. Phys.* **37**, 1154. *(50)*

E. K. Storms & R. J. McNeal, *J. Phys. Chem.* **66**, 1401. *(23)*

W. E. Streib, T. H. Jordan, & W. N. Lipscomb, *J. Chem. Phys.* **37**, 2962. *(7)*

H. Strunz, *Naturwiss.* **49**, 9. *(16)*

A. Taylor, N. J. Doyle, & B. J. Kagle, *J. Less-Common Metals* **4**, 436. *(42, 76)*

B. M. Vasyutinski, G. N. Kartmazov, & V. A. Finkel, *Ukrain. Fiz. Ž.* **7**, 661. *(73)*

R. H. Willens, *Rev. Sci. Instr.* **33**, 1069. *(22)*

R. M. Wood, *Proc. Phys. Soc.* **80**, 783. *(22)*

1963

S. C. Abrahams, *J. Phys. Chem. Solids* **24**, 589. *(46)*

R. A. Alikhanov, *Proc. Reg. Conf. 3rd, Prague, Physics and Techniques of Low Temperatures*, p. 127. *(8)*

V. M. Amonenko, B. M. Vasyutinski, G. N. Kartmazov, N. Smirnov, & V. A. Finkel, *Fiz. Metal. Metalloved* **15**, 444. *(73)*

R. B. Aust & H. G. Drickamer, *Science* **140**, 817. *(6)*

J. D. Barnett, R. B. Bennion, & H. T. Hall, (a) *Science* **141**, 534 *(56)*; (b) *Science* **141**, 1041. *(50)*

C. S. Barrett, P. Cucka, & K. Haefner, *Acta Cryst.* **16**, 451. *(51)*

C. S. Barrett, M. H. Mueller, & R. L. Hitterman, *Phys. Rev.* **129**, 625. *(92)*

L. H. Bolz & F. A. Mauer, *Adv. X-Ray Anal.* **6**, 242. *(10)*

F. P. Bundy, *J. Chem. Phys.* **38**, 618. *(6)*

F. P. Bundy & J. S. Kasper, *Science* **139**, 340. *(32)*

F. J. Darnell, (a) *Phys. Rev.* **130**, 1825 *(64, 66, 67)*; (b) *Phys. Rev.* **132**, 1098. *(65, 68)*

F. J. Darnell & E. P. Moore, *J. Appl. Phys.* **34**, 1337. *(66)*

G. H. Cockett & C. D. Davis, *J. Iron Steel Inst.* **201**, 110. *(47)*

B. W. Delf, *Br. J. Appl. Phys.* **14**, 345. *(74)*

J. Donohue, *J. Am. Chem. Soc.* **85**, 1238. *(22, 40)*

B. N. Dutta & B. Dayal, (a) *Phys. Status Solidi* **3**, 473 *(79)*; (b) *Phys. Status Solidi* **3**, 2253. *(46, 74)*

W. T. Eeles & A. L. Sutton, *Acta Cryst.* **16**, 575. *(92)*

O. Erämetsä, (a) *Suom. Kemi.* **B36**, 6 *(16)*; (b) *Suom. Kemi.* **B36**, 213. *(16)*

L. Gorski, *Phys. Status Solidi* **3**, K316. *(5)*

H. T. Hall, J. D. Barnett, & L. Merrill, *Science* **139**, 111. *(70)*

H. T. Hall & L. Merrill, *Inorg. Chem.* **2**, 618. *(70)*

R. E. Hughes, C. H. L. Kennard, D. B. Sullenger, H. A. Weakliem, D. E. Sands, & J. L. Hoard, *J. Am. Chem. Soc.* **85**, 361. *(5)*

J. C. Jamieson, (a) *Science* **139,** 762 (*14, 32*); (b) *Science* **139,** 1291 (*15*); (c) *Science* **140,** 72. (*22, 40, 72*)

A. Jayaraman, W. Klement, Jr., & G. C. Kennedy, *Phys. Rev.* **132,** 1620. (*20, 38*)

A. Jayaraman, W. Klement, Jr., R. C. Newton, & G. C. Kennedy, *J. Phys. Chem. Solids* **24,** 7. (*31, 49, 81*)

G. C. Kennedy & R. C. Newton, *Solids Under Pressure,* McGraw-Hill, New York, Chap. 7. (*27, 50, 52, 83*)

W. Klement, Jr., A. Jayaraman, & G. C. Kennedy, (a) *Phys. Rev.* **129,** 1971 (*92*); (b) *Phys. Rev.* **131,** 1 (*80*); (c) *Phys. Rev.* **131,** 632 (*83*)

S. T. Konobeevsky & N. T. Chebotarev, *J. Nucl. Energy* **17,** 416. (*94*)

H. L. Luo & P. Duwez, *Can. J. Phys.* **41,** 758. (*27*)

K. J. H. Mackay & N. A. Hill, *J. Nucl. Mater.* **8,** 263. (*4*)

A. R. Marder, *Science* **142,** 664. (*4*)

D. B. McWhan & A. Jayaraman, *Appl. Phys. Lett.* **3,** 129. (*38*)

G. B. Mitra & S. K. Mitra, *Ind. J. Phys.* **37,** 462. (*29*)

J. Niemiec, (a) *Bull. Acad. Polon. Sci.* **11,** 305 (*42*); (b) *Bull. Acad. Polon. Sci.* **11,** 311. (*43, 75*)

H. M. Otte, W. G. Montague, & D. O. Welch, *J. Appl. Phys.* **34,** 3149. (*13*)

J. R. Packard & C. A. Swenson, *J. Phys. Chem. Solids* **24,** 1405. (*54*)

G. J. Piermarini, in Jayaraman, Klement, Newton, & Kennedy (1963). (*81*)

J. A. Prins & F. Tuinstra, (a) *Physica* **29,** 328 (*16*); (b) *Physica* **29,** 884. (*16*)

R. G. Ross & W. Hume-Rothery, *J. Less-Common Metals* **5,** 258. (*24, 42, 44, 45, 72, 74*)

S. Shibata, *J. Phys. Chem.* **67,** 2256. (*17*)

F. A. Smidt & A. H. Daane, *J. Phys. Chem. Solids* **24,** 361. (*68*)

Yu. N. Smirnov & V. A. Finkel, *Phys. Metals Metallogr.* **16,** 139. (*24*)

R. A. Stager & H. G. Drickamer, (a) *Phys. Rev.* **131,** 2524 (*12, 20, 56*); (b) *Phys. Rev.* **132,** 124. (*3, 19, 37*)

V. N. Svechnikov, *Phys. Metals Metallogr.* **15,** 113. (*24*)

N. Terao, *Japan. J. Appl. Phys.* **2,** 156. (*41*)

R. H. Wentorf, Jr., & J. S. Kasper, *Science* **139,** 338. (*14*)

W. C. Wyder & M. Hoch, *Trans. Met. Soc. AIME* **227,** 588. (*24*)

W. H. Zachariasen & F. H. Ellinger, (a) *Acta Cryst.* **16,** 369 (*94*); (b) *Acta Cryst.* **16,** 777. (*94*)

1964

R. A. Alikhanov, (a) *Sov. Phys.-JETP* **18,** 556 (*8*); (b) *J. Phys.* **25,** 449. (*8*)

J. D. Barnett & H. T. Hall, *Rev. Sci. Instr.* **35,** 175. (*50*)

C. S. Barrett & L. Meyer, *J. Chem. Phys.* **41,** 1078. (*18*)

B. J. Beaudry & A. H. Daane, *J. Less-Common Metals* **6,** 322. (*21, 64*)

K. E. Beu, *Acta Cryst.* **17,** 1149. (*74*)

P. Blanconnier, Thèse 3e Cycle, Paris. (*31*)

L. H. Bolz & F. A. Mauer, in G. L. Pollack (1964). (*18, 36, 54*)

E. Bonnier, P. Hicter & S. Aléonard, *Compt. rend. Acad. Sci. Paris* **258,** 166. (*52*)

E. Bonnier, P. Hicter, S. Aléonard, & G. Laugier, *Compt. rend. Acad. Sci. Paris* **258,** 4967. (*34*)

G. Brauer & W.-D. Schnell, *J. Less-Common Metals* **6,** 326. (*23*)

M. Brith & D. White, *Phys. Lett.* **11,** 203. (*1*)

F. P. Bundy, *J. Chem. Phys.* **41,** 3809. (*14, 32*)

R. E. W. Casselton & W. Hume-Rothery, *J. Less-Common Metals* **7,** 212. (*28, 42*)

J. A. Catterall & S. M. Barker, *Plansee Proc.,* 577. (*41, 42, 46*)

R. L. Clendenen & H. G. Drickamer, (a) *J. Phys. Chem. Solids* **25**, 865 (*26, 44*); (b) *Phys. Rev.* **135A**, 1643. (*12*)

B. B. Cunningham & J. C. Wallmann, *J. Inorg. Nucl. Chem.* **26**, 271. (*96*)

A. E. Curzon & A. T. Pawlowicz, (a) *Proc. Phys. Soc.* **83**, 499 (*1*); (b) *Proc. Phys. Soc.* **83**, 888. (*1*)

J. G. Daunt, A. F. Schuch, & R. L. Mills, *Physics Today,* 51. (*2*)

B. L. Davis & L. H. Adams, *J. Phys. Chem. Solids* **25**, 379. (*58*)

V. T. Deshpande & R. R. Pawar, *Curr. Sci.* **33**, 741. (*34*)

J. Donohue & S. H. Goodman, *J. Phys. Chem.* **68**, 2363. (*16*)

J. S. Dugdale & J. P. Franck, *Phil. Trans. Roy. Soc.* **A257**, 1. (*2*)

A. A. Eliseev, E. I. Jarembash, E. S. Vigileva, L. I. Antonova, & A. V. Zachatskaya, *Zh. Neorg. Khim. SSSR* **9**, 1032. (*57*)

B. C. Giessen, R. Koch, & N. J. Grant, *Trans. Met. Soc. AIME* **230**, 1268. (*41*)

M. J. Goringe & U. Valdre, *Phil. Mag.* **9**, 897. (*10*)

K. A. Gschneider & J. T. Waber, *J. Less-Common Metals* **6**, 354. (*58*)

H. T. Hall, L. Merrill, & J. D. Barnett, *Science* **146**, 1297. (*55*)

I. R. Harris & G. V. Raynor, *J. Less-Common Metals* **7**, 11. (*58, 90*)

A. F. Ieviņš, & L. F. Lindinya, *Latv. P. S. R. Zinat. Akad. Vestis, Kim.* 649. (*49*)

R. Jaggi, (a) *Helv. Phys. Acta* **37**, 618 (*83*); (b) *Bull. Soc. Sci. Bret.* **39**, 120. (*83*)

J. C. Jamieson, *Science* **145**, 572. (*66*)

A. Jayaraman & R. C. Sherwood, (a) *Phys. Rev. Lett.* **12**, 22 (*64*); (b) *Phys. Rev.* **134**, A691. (*65, 66, 67, 68*)

T. H. Jordan, H. W. Smith, W. E. Streib, & W. N. Lipscomb, *J. Chem. Phys.* **41**, 756. (*7*)

T. H. Jordan, W. E. Streib, H. W. Smith, & W. N. Lipscomb, *Acta Cryst.* **17**, 777. (*8, 9*)

T. H. Jordan, W. E. Streib, & W. N. Lipscomb, *J. Chem. Phys.* **41**, 760. (*9*)

S. S. Kabalkina & V. P. Mylov, *Sov. Phys.-Dokl.* **8**, 917. (*51*)

S. S. Kabalkina, L. F. Vereshchagin, & B. M. Shulenin, *Sov. Phys. -JETP* **18**, 1422. (*52*)

J. S. Kasper & S. M. Richards, *Acta Cryst.* **17**, 752. (*14, 32*)

V. S. Kogan, A. S. Bulatov, & L. F. Yakimenko, *Sov. Phys.-JETP* **19**, 107. (*1*)

D. B. McWhan & W. L. Bond, *Rev. Sci. Instr.* **35**, 626. (*57*)

B. Meyer, *Chem. Rev.* **42**, 442. (*16*)

L. Meyer, C. S. Barrett, & P. Haasen, *J. Chem. Phys.* **40**, 2744. (*18*)

G. J. Piermarini & C. E. Weir, *Science* **144**, 69. (*57, 58, 59, 60*)

G. L. Pollack, *Rev. Mod. Phys.* **36**, 748. (*36, 54*)

A. K. N. Reddy, *Acta Cryst.* **17**, 443. (*28*)

N. Shaw & Y.-H. Liu, *Acta Phys. Sinica* **20**, 699. (*14, 32*)

J. F. Smith & V. L. Schneider, *J. Less-Common Metals* **7**, 17. (*49*)

C. Susse, R. Epain, & B. Vodar, *Compt. rend. Acad. Sci. Paris* **258**, 4513. (*16*)

T. Takahashi & W. A. Bassett, *Science* **145**, 483. (*26*)

J. Taylor, M. Mack, & W. Parrish, *Acta Cryst.* **17**, 1229. (*14, 74*)

J. R. Thompson, *J. Less-Common Metals* **6**, 94. (*90*)

L. F. Vershchagin & S. S. Kabalkina, *Ž. Eksp. Teor. Fiz., SSSR* **47**, 414. (*51*)

1965

S. C. Abrahams, *Acta Cryst.* **18**, 566. (*16*)

T. Bååk, *Science* **148**, 1220. (*16*)

C. S. Barrett & F. J. Spooner, *Nature* **207**, 1382. (*31*)

C. H. Bates, F. Dachille, & R. Roy, *Science* **147**, 860. (*14, 32*)

P. Blanconnier, L. Bosio, A. Defrain, A. Rimsky, & H. Curien, *Bull. Soc. Fr. Minéral. Crist.* **88**, 145. (*31*)

W. L. Bond, A. S. Cooper, K. Andres, G. W. Hull, T. H. Geballe, & B. T. Matthias, *Phys. Rev. Lett.* **15**, 260. (*75*)

O. Bostanjoglo, *Z. Physik* **187**, 444. (*1*)

A. Brown & S. Rundqvist, *Acta Cryst.* **19**, 684. (*15*)

A. Caron & J. Donohue, *Acta Cryst.* **18**, 562. (*16*)

M. Clouter & H. P. Gush, *Phys. Rev. Lett.* **15**, 200. (*1*)

A. E. Curzon & A. J. Mascall, *J. Appl. Phys.* **16**, 1301. (*1*)

A. E. Curzon & A. T. Pawlowicz, *Proc. Phys. Soc.* **85**, 375. (*8*)

B. C. Deaton & F. A. Blum, *Phys. Rev.* **137**, A1131. (*16*)

V. T. Deshpande & R. R. Pawar, *Physica* **31**, 671. (*52*)

J. Donohue & S. H. Goodman, *Acta Cryst.* **18**, 568. (*17*)

J. Donohue & B. Meyer, *Elemental Sulfur,* B. Meyer, ed., Interscience, New York, Chap. 1. (*16*)

J. S. Dugdale, *Physics of High Pressures and the Condensed Phase,* van Itterbeck, ed., North Holland Pub. Co., Amsterdam, Chap. 9. (*2*)

H. A. Eick, *Inorg. Chem.* **4**, 1237. (*5*)

W. E. Evenson & H. T. Hall, *Science* **150**, 1164. (*24*)

L. Gorski, *Phys. Status Solidi* **9**, K169. (*5*)

R. E. Harris & G. Jura, *Elemental Sulfur,* B. Meyer, ed., Interscience, New York, Chap. 9. (*16*)

P. Hemenger & H. Weik, *Acta Cryst.* **19**, 690. (*28*)

J. C. Jamieson & D. B. McWhan, *J. Chem. Phys.* **43**, 1149. (*52*)

A. Jayaraman, (a) *Phys. Rev.* **137**, 179 (*58*); (b) *Phys. Rev.* **139**, A690. (*57, 59, 60, 62, 64, 65*)

R. W. Lynch & H. G. Drickamer, *J. Phys. Chem. Solids* **26**, 63. (*30, 48*)

J. A. C. Marples, *Acta Cryst.* **18**, 815. (*91*)

D. B. McWhan & A. L. Stevens, *Phys. Rev.* **139**, A682. (*64, 65, 66, 67*)

R. L. Mills & A. F. Schuch, *Phys. Rev. Lett.* **15**, 722. (*1*)

C. E. Montfort, III, & C. A. Swenson, *J. Phys. Chem. Solids* **26**, 623. (*65*)

K. F. Mucker, S. Talhouk, P. M. Harris, D. White, & R. A. Erickson, *Phys. Rev. Lett.* **15**, 586. (*1*)

I. E. Paukov, E. Yu. Tonkov, & D. S. Mirinski, *Dokl. Acad. Nauk SSSR* **164**, 588. (*16*)

M. H. Read & C. Altman, *Appl. Phys. Lett.* **7**, 51. (*73*)

D. E. Sands, *J. Am. Chem. Soc.* **87**, 1395. (*16*)

M. E. Straumanis & S. M. Riad, *Trans. Met. Soc. AIME* **233**, 964. (*47*)

F. N. Tavadze, I. A. Bairamashvili, G. V. Tsagareishvili, K. P. Tsomaya, & N. A. Zoidze, *Sov. Phys.-Cryst.* **9**, 768. (*5*)

J. B. Taylor, S. L. Bennett, & R. D. Heyding, *J. Phys. Chem. Solids* **26**, 69. (*33*)

A. Taylor & N. J. Doyle, *J. Less-Common Metals* **9**, 190. (*42, 74*)

M. Thackray, *Elemental Sulfur,* B. Meyer, ed., Interscience, New York, Chap. 3. (*16*)

L. F. Vereshchagin, S. S. Kabalkina, & Z. V. Troitskaya, *Sov. Phys.-Dokl.* **9**, 894. (*31, 49*)

H. Weik & P. Hemenger, *Bull. Am. Phys. Soc.* **10**, 1140. (*28*)

R. H. Wentorf, Jr., *Science* **147**, 49. (*5*)

J. G. Wright & J. Goddard, *Phil. Mag.* **11**, 485. (*28*)

1966

J. D. Barnett, V. E. Bean, & H. T. Hall, *J. Appl. Phys.* **37**, 875. (*50*)

C. S. Barrett, L. Meyer, & J. Wasserman, *J. Chem. Phys.* **45**, 834. (*1, 10*)

C. Bonnelle & F. Vergand, *Acta Cryst.* **21**, 1001. (*28*)

P. Cherin & P. Unger, *Acta. Cryst.* **21**, A46. (*34*)

M. F. Collins, *Proc. Phys. Soc.* **89**, 415. (*8*)

F. A. Cotton & G. Wilkinson, *Advanced Inorganic Chemistry,* 2nd ed., Interscience, New York, p. 523. (*16*)

P. Denbigh & R. B. Marcus, *J. Appl. Phys.* **37**, 4325. (*73*)

F. H. Ellinger, K. A. Johnson, & V. O. Struebing, *J. Nucl. Mater.* **20**, 83. (*95*)

E. Gebhardt, W. Dürrschnabel, & G. Hörz, *J. Nucl. Mater.* **18**, 119. (*41*)

S. Geller, *Science* **152**, 644. (*16*)

R. N. Jeffery, J. D. Barnett, H. B. Vanfleet, & H. T. Hall, *J. Appl. Phys.* **37**, 3172. (*56*)

A. Kutoglu & E. Hellner, *Ang. Chem. Int. Ed.* **5**, 965. (*16*)

A. G. Leiga, *J. Phys. Chem.* **70**, 3254. (*47*)

C. E. Lundin, A. S. Yamamoto, & J. F. Nachman, *Acta Met.* **13**, 149. (*59, 60*)

H.-K. Mao, W. A. Bassett, & T. Takahashi, *J. Appl. Phys.* **38**, 272. (*26*)

R. L. Mills, A. F. Schuch, & D. A. Depatie, *Phys. Rev. Lett.* **17**, 1131. (*1*)

K. F. Mucker, Dissertation, Ohio State University. (*1*)

K. F. Mucker, S. Talhouk, P. M. Harris, D. White, & R. Erickson, *Phys. Rev. Lett.* **16**, 795. (*1*)

M. Norman, I. R. Harris, & G. V. Raynor, *J. Less-Common Metals* **11**, 395. (*90*)

L. Pauling, *Proc. Nat. Acad. Sci. U. S.* **56**, 1646. (*6*)

O. G. Peterson, D. N. Batchelder, & R. O. Simmons, *Phys. Rev.* **150**, 703. (*18*)

J. C. Raich & H. M. James, *Phys. Rev. Lett.* **16**, 173. (*1*)

M. Schmidt & E. Wilhelm, *Ang. Chem. Int. Ed.* **5**, 964. (*16*)

C. B. Sclar, L. C. Carrison, W. B. Gage, & O. M. Stewart, *J. Phys. Chem. Solids* **27**, 1339. (*16*)

A. F. Schuch & R. L. Mills, *Phys. Rev. Lett.* **16**, 616. (*1*)

H. Steeple & T. Ashworth, *Acta Cryst.* **21**, 995. (*92*)

H. E. Swanson, M. C. Morris, & E. H. Evans, *Nat. Bur. Stand. Monogr. 25,* **4**, 10. (*27*)

H. Thurn & H. Krebs, *Ang. Chem.* **78**, 1101. (*15*)

F. Tuinstra, *Acta Cryst.* **20**, 341. (*16*)

V. V. Vorobiev, Y. N. Smirnov, & v. A. Finkel, *Sov. Phys.-JETP* **22**, 1212. (*64*)

1967

C. S. Barrett, L. Meyer, & J. Wasserman, (a) *Phys. Rev.* **163**, 851 (*8*); (b) *J. Chem. Phys.* **47**, 592. (*8*)

D. N. Batchelder, D. L. Losee, & R. O. Simmons, *Phys. Rev.* **162**, 767. (*10*)

F. van Bolhuis, P. B. Koster, & T. Migchelsen, *Acta Cryst.* **23**, 90. (*53*)

O. Bostanjoglo & R. Kleinschmidt, *J. Chem. Phys.* **46**, 2004. (*1*)

O. Bostanjoglo & B. Lischke, *Z. Naturf.* **22a**, 1620. (*8*)

R. M. Brugger, R. B. Bennion, & T. G. Worlton, *Phys. Lett.* **24A**, 714. (*83*)

F. P. Bundy & J. S. Kasper, *J. Chem. Phys.* **46**, 3437. (*6*)

P. Cherin & P. Unger, (a) *Acta Cryst.* **23**, 670 (*52*); (b) *Inorg. Chem.* **6**, 1589. (*34*)

K. L. Chopra, M. Randlett, & R. H. Duff, *Phil. Mag.* **16**, 261. (*40, 42, 72, 73, 74, 75*)

V. A. Finkel, Y. N. Smirnov, & V. V. Vorobiev, *Sov. Phys.-JETP* **24**, 21. (*65*)

V. A. Finkel & V. V. Vorobiev, *Sov. Phys.-JETP* **24**, 524. (*66*)

C. Frondel & U. B. Marvin, *Nature* **214**, 587. (*6*)

Y. Fukano & K. Kimoto, *J. Phys. Soc. Japan* **23**, 668. (*28, 46, 79*)

C. P. Gazzara, R. M. Middleton, R. J. Weiss, & E. O. Hall, *Acta Cryst.* **22**, 859. (*25*)

R. E. Hanneman, H. M. Strong, & F. P. Bundy, *Science* **155**, 995. (*6*)

E. Hellner, private communication. (*16*)

J. L. Hoard & R. E. Hughes, *The Chemistry of Boron and its Compounds,* E. L. Muetterties, ed., John Wiley, New York, Chap. 2. (*5*)

T. E. Hutchinson & K. H. Olsen, *J. Appl. Phys.* **38**, 4933. (*41*)

M. A. Ilina & E. S. Itskevich, *Sov. Phys.-Solid State* **8**, 1873. (*83*)

W. A. Jesser & J. W. Matthews, *Phil. Mag.* **15**, 1097. (*26*)

K. Kimoto & I. Nishida, *J. Phys. Soc. Japan* **22**, 744. (*24*)

L. D. Kolomiels, *Sov. Phys.-Cryst.* **12**, 132. (*92*)

J. Sosniak, W. J. Polito, & G. A. Rozgonyi, *J. Appl. Phys.* **38**, 3041. (*73*)

A. Taylor & N. J. Doyle, *J. Less-Common Metals* **13**, 511. (*41*)

H. Thurn, Dissertation, Stuttgart. (*15*)

F. Tuinstra, (a) *Structural Aspects of the Allotropy of Sulfur and the Other Divalent Elements,* Uitgeverij Waltman, Delft (*16*); (b) *Physica* **34**, 113. (*16*)

F. Vergand, *J. Chim. Phys.* **64**, 306. (*28*)

K. B. Ward, Jr., & B. C. Deaton, *Phys. Rev.* **153**, 947. (*16*)

D. G. Westlake, *Phil. Mag.* **16**, 905. (*23*)

1968

J. S. Abell & A. G. Crocker, *Scripta Met.* **2**, 419. (*80*)

Z. M. Azarkh & P. I. Gavrilov, *Sov. Phys.-Cryst.* **12**, 972. (*68*)

R. L. Barns, *J. Appl. Phys.* **39**, 4044. (*41*)

L. Bosio, A. Defrain, I. Epelboin, & J. Vidal, *J. Chim. Phys.* **65**, 719. (*31*)

P. R. Doidge & A. R. Eastham, *Phil, Mag.* **18**, 655. (*80*)

A. El Goresy & G. Donnay, *Science* **161**, 363. (*6*)

V. A. Finkel & V. V. Vorobiev, *Sov. Phys.-Cryst.* **13**, 457. (*39*)

D. L. Losee & R. O. Simmons, *Phys. Rev.* **172**, 944. (*36*)

R. B. Marcus & S. Quigley, *Thin Solid Films* **2**, 467. (*73*)

L. Meyer, C. S. Barrett, & S. C. Greer, *J. Chem. Phys.* **49**, 1902. (*9*)

K. F. Mucker, P. M. Harris, D. White, & R. A. Erickson, *J. Chem. Phys.* **49**, 1922. (*1*)

J. N. Pratt, K. M. Myles, J. B. Darby, & M. H. Mueller, *J. Less-Common Metals* **14**, 427. (*46*)

H. F. Schaake, *J. Less-Common Metals* **15**, 103. (*77*)

M. Schmidt, B. Block, H. D. Block, H. Köpf, & E. Wilhelm, *Ang. Chem. Int. Ed.* **7**, 632. (*16*)

H. P. Singh, *Acta Cryst.* **A24**, 469. (*32, 45, 77*)

J. D. Speight, I. R. Harris, & G. V. Raynor, *J. Less-Common Metals* **15**, 317. (*58, 59, 60, 62, 65*)

M. E. Straumanis & R. P. Shodhan, *Trans. Met. Soc. AIME* **242**, 1185. (*42*)

D. Weaire, *Phil. Mag.* **18**, 213. (*80*)

F. Weigel & A. Trinkl, *Radiochim. Acta* **10**, 78. (*88*)

1969

Anonymous, Fundamental Nuclear Energy Research. 1968. A Supplemental Report to the Annual Report to the Congress for 1968 of the United States Atomic Energy Commission, U. S. Government Printing Office, Washington, D. C., p. 241. (*94*)

A. S. Avilov & R. M. Imamov, *Sov. Phys.-Cryst.* **14**, 259. (*34*)

J. Berty, M.-J. David, & L. Lafourcade, *Compt. rend. Acad. Sci. Paris* **269**, 1089. (*31*)

L. Bosio, A. Defrain, H. Curien, & A. Rimsky, *Acta Cryst.* **B25**, 995. (*31*)

M. Brith, A. Ron, & O. Schnepp, *J. Chem. Phys.* **51**, 1318. (*7*)

V. T. Deshpande & R. R. Pawar, *Acta Cryst.* **A25**, 415. (*49*)

J. Donohue, S. H. Goodman, & M. Crisp, *Acta Cryst.* **B25**, 2168. (*16*)

J. Forssell & B. Persson, *J. Phys. Soc. Japan* **27**, 1368. (*24*)

E. Franceschi & E. L. Olcese, *Phys. Rev. Lett.* **22**, 1299. (*58*)

S. Geller & M. D. Lind, *Acta Cryst.* **B25**, 2166. (*16*)

E. M. Hörl, *Acta Cryst.* **B25**, 2515. (*8*)

J. Kumar & O. N. Srivastava, *Acta Cryst.* **B25**, 2654. (*62*)

N. Kunitomi, T. Yamada, Y. Nakai, & Y. Fujii, *J. Appl. Phys.* **40**, 1265. (*25*)

M. D. Lind & S. Geller, *J. Chem. Phys.* **51**, 348. (*16*)

R. M. Middleton & C. P. Gazzara, *Phys. Rev.* **185**, 1230. (*14*)

R. L. Mills & A. F. Schuch, *Phys. Rev. Lett.* **23**, 1154. (7)

S. M. Richards & J. S. Kasper, *Acta Cryst.* **B25**, 237. (*5*)

N. N. Roy & E. G. Steward, *Nature* **224**, 905. (*13*)

D. Schiferl & C. S. Barrett, *J. Appl. Cryst.* **2**, 30. (*33*)

O. Schnepp & A. Ron, *Disc. Faraday Soc.* **48**, 26. (7)

P. K. Smith, W. H. Hale, & M. C. Thompson, *J. Chem. Phys.* **50**, 5066. (*96*)

M. E. Straumanis & L. S. Yu, *Acta Cryst.* **A25**, 676. (*29*)

D. B. Sullenger, K. D. Phipps, P. W. Seabaugh, C. R. Hudgens, D. E. Sands, & J. S. Cantrell, *Science* **163**, 935. (*5*)

H. E. Swanson, H. F. McMurdie, M. C. Morris, & E. H. Evans, *Nat. Bur. Stand. Monogr.* 25, 7, 142. (*25*)

T. Takahashi, H. K. Mao, & W. A. Bassett, *Science* **165**, 1352. (*82*)

H. Thurn & H. Krebs, *Acta Cryst.* **B25**, 125. (*15*)

G. C. Vezzoli, F. Dachille, & R. Roy, (a) *Science* **166**, 218 (*16*); (b) *Inorg. Chem.* **8**, 2658. (*16*)

S. A. Weiner, E. Gürmen, & A. Arrott, *Phys. Rev.* **186**, 705. (*19*)

A. G. Whittaker & P. Kintner, *Science* **165**, 589. (*6*)

1970

J. S. Abell, A. G. Crocker, & H. W. King, *Phil. Mag.* **21**, 207. (*80*)

A. Anderson, T. S. Sun, & M. C. A. Donkersloot, *Can. J. Phys.* **48**, 2265. (*7*)

P. N. Baker, *Thin Solid Films* **6**, R57. (*73*)

C. Boulesteix, P. E. Caro, M. Gasgnier, C. Henry la Blanchetais, B. Pardo, & L. Valiergue, *Acta Cryst.* **B26**, 1043. (*62*)

E. Bucher, P. H. Schmidt, A. Jayaraman, K. Andres, J. P. Marita, & P. D. Dernier, *Phys. Rev.* **B2**, 3911. (*70*)

E. M. Compy, *J. Appl. Phys.* **41**, 2014. (*83*)

V. A. Finkel, V. I. Glamazda, & G. P. Kovtun, *Sov. Phys.-JETP,* **30**, 581. (*23*)

D. Geist, R. Kloss, & H. Follner, *Acta Cryst.* **B26**, 1800. (*5*)

R. J. Gerdes, A. T. Chapman, & G. W. Clark, *Science* **167**, 979. (*74*)

J. L. Hoard, D. B. Sullenger, C. H. L. Kennard, & R. E. Hughes, *J. Solid State Chem.* **1**, 268. (*5*)

R. E. Hughes, Annual Technical Progress Report, Materials Science Center of Cornell University, Ithaca, N. Y., p. 70. (*5*)

I. Kawada & E. Hellner, *Ang. Chem. Int. Ed.* **9**, 379. (*16*)

V. Klechkovskaya, *Sov. Phys.-Cryst.* **15**, 299. (*41*)

W. E. Krull & R. W. Newman, *J. Appl. Cryst.* **3**, 519. (*29*)

P. G. Mardon & C. C. Koch, *Scripta Metallurg.* **4**, 477. (*62*)

J. A. Oberteuffer & J. A. Ibers, *Acta Cryst.* **B26**, 1499. (*25*)

L. Pauling, I. Keaveny, & A. B. Robinson, *J. Solid State Chem.* **2**, 225. (*9*)

M. Schmidt & E. Wilhelm, *Chem. Comm.,* p. 1111. (*16*)

F. Schrey, R. D. Mathis, R. T. Payne, & L. E. Murr, *Thin Solid Films* **5**, 29. (*73*)

A. F. Schuch & R. L. Mills, *J. Chem. Phys.* **52**, 6000. (*7*)

M. I. Sokhor & V. D. Vitol, *Sov. Phys.-Cryst.* **14**, 632. (*6*)
M. E. Straumanis & S. Zyszczynski, *J. Appl. Cryst.* **3**, 1. (*41*)
J. A. Venables, *Phil. Mag.* **21**, 147. (*7*)
W. D. Westwood & F. C. Livermore, *Thin Solid Films* **5**, 407. (*73*)
T. Yamada & Y. Fujii, *J. Phys. Soc. Japan* **28**, 1503. (*25*)

1971

E. Amberger & K. Ploog, *J. Less-Common Metals* **23**, 21. (*5*)
L. B. Asprey, R. D. Fowler, J. D. G. Lindsay, R. W. White, & B. B. Cunningham, *Inorg. Nucl. Chem. Lett.* **7**, 977. (*91*)
M. Y. Bakinov, G. I. Ibaev, & K. A. Agaev, *Sov. Phys.-Cryst.* **15**, 1107. (*34*)
J. Donohue & H. M. Einspahr, *Acta Cryst.* **B27**, 1740. (*92*)
V. A. Finkel & M. I. Palatnik, *Sov. Phys.-JETP* **32**, 828. (*67, 68*)
R. Gupty & T. R. Anantharaman, *J. Less-Common Metals* **25**, 3531. (*64, 65, 66, 67, 68*)
H. H. Hill & F. H. Ellinger, *J. Less-Common Metals* **23**, 92. (*57*)
F. X. Kayser, *Phys. Status Solidi* **8**, 233. (*70*)
E. A. Kellett & B. P. Richards, *J. Appl. Cryst.* **4**, 1. (*6*)
P. G. Pallmer & T. D. Chikalla, *J. Less-Common Metals* **24**, 233. (*61*)
E. Parthé, private communication. (*5*)
J. R. Peterson, J. A. Fahey, & R. D. Baybarz, *J. Inorg. Nucl. Chem.* **33**, 3345. (*97*)
D. E. Sands, private communication. (*16*)
S. Singh, N. C. Khanduri, & T. Tsang, *Scripta Met.* **5**, 167. (*69, 71*)
F. H. Spedding & B. J. Beaudry, *J. Less-Common Metals* **25**, 61. (*21, 39, 64, 65, 66, 67, 68, 69, 71*)
F. Tuinstra, private communication. (*16*)
C. E. Weir, G. J. Piermarini, & S. Block, *J. Chem. Phys.* **54**, 2768. (*31, 55*)
A. G. Whittaker, G. Donnay, & K. Lonsdale, *Ann. Rep. Geophys. Lab. Carneg. Inst. Wash. 1969-70*, p. 311. (*6*)
C. L. Woodard & M. E. Straumanis, *J. Appl. Cryst.* **4**, 201. (*13, 42*)

1972

L. Bosio, H. Curien, M. Dupont, & A. Rimsky, *Acta Cryst.* **B28**, 1974. (*31*)
J. R. Brookeman & T. A. Scott, *Acta Cryst.* **B28**, 983. (*7*)
V. N. Bykov, G. G. Zdorovtseva, V. A. Troyan, & V. S. Khaimovich, *Sov. Phys.-Cryst.* **16**, 699. (*26, 28*)
P. Cherin & P. Unger, *Acta Cryst.* **B28**, 313. (*34*)
G. Das, *Thin Solid Films* **12**, 305. (*73*)
S. LaPlaca & W. C. Hamilton, *Acta Cryst.* **B28**, 984. (*7*)
G. S. Pawley & R. P. Rinaldi, *Acta Cryst.* **B28**, 3605. (*16*)
R. B. Roof, *Aust. J. Phys.* **25**, 335. (*16*)
A. Schaub & M. Roschy, *Thin Solid Films* **12**, 313. (*73*)
P. Schaufelberger, H. Merx, & M. Contré, *High Temp. - High Press.* **4**, 111. (*83*)
D. G. Westlake, S. T. Ockers, M. H. Mueller, & K. D. Anderson, *Met. Trans.* **3**, 1711. (*23*)
A. G. Whittaker & G. M. Wolten, *Science* **178**, 54. (*6*)